SELECTED PAPERS IN BIOCHEMISTRY

SELECTED PAPERS IN BIOCHEMISTRY

Volume 3
CODING

Edited by
Susumu Nishimura
National Cancer Center Research Institute

UNIVERSITY PARK PRESS
Baltimore · London · Tokyo

UTP 3345-67430-5149
Printed in Japan

Originally published in 1971 by
UNIVERSITY OF TOKYO PRESS

UNIVERSITY PARK PRESS
Baltimore · London · Tokyo
ISBN 0-8391-0613-0
Library of Congress Catalog Card Number 78-156228

INTRODUCTORY NOTE

The problem of the genetic code is one of the most essential research subjects in the field of molecular biology. Numerous reports have been made which have played important roles in the development of genetic coding research. It is, therefore, a difficult task to select twenty papers from among them. Moreover, the problem of the genetic code is closely related to many other areas of molecular biology, for example, transfer RNA and protein synthesis, making it difficult to distinguish clearly between these fields. In the face of these difficulties I have selected what I consider the most important papers among those presented before 1968.

It is often said that the problem of the genetic code has been largely solved with the completion of the code table in 1965 and the golden age of molecular biology has passed. However, this cannot be considered a valid judgment. As a matter of fact, research concerning the genetic code is even more active today than in the past, and important discoveries in the mechanisms of chain initiation and chain termination, protein synthesis in animal cells, and determination of the base sequence of RNA in RNA phages have been successively reported. Therefore, readers are advised to acquire the basic concepts of genetic coding and the trends in the development of its research by reading the papers selected here, and at the same time, to peruse the latest papers on the subject.

In the course of the development of the research on the genetic code, there have been many cases where different research groups have independently made similar important discoveries almost simultaneously. In such cases, I have tried to include all the groups concerned.

In addition I have included a bibliography of selected works in this field. The paper by Crick (1) presents an accurate description and expresses his views clearly. Although this paper was published as early as 1963, it is still very much worth reading. If readers want the most up-to-date information, I recommend Jukes's review (5) or Yčas's book (6). The latter, elaborately covering almost all the literature on the genetic code, is regarded as an especially suitable reference book.

August, 1970

Susumu Nishimura

BIBLIOGRAPHY

1. Crick, F. H. C., "The Recent Excitement in the Coding Problem" *in* Progress in Nucleic Acid Research, Davidson, J. N. and Cohn, W. E., Eds., Vol. 1, p. 164, Academic Press, New York and London, 1963.
2. The Genetic Code. *Cold Spring Harbor Symp. Quant. Biol.*, **31**, 1966.
3. Woese, C. R., "The Present Status of the Genetic Code" *in* Progress in Nucleic Acid Research, Davidson, J. N. and Cohn, W. E., Eds., Vol. 7, p. 107, Academic Press, New York and London, 1967.
4. Sadgopal, A., "The Genetic Code After the Excitement" *in* Advances in Genetics, Caspari, E. W., Ed., Vol. 14, p. 326, Academic Press, New York and London, 1968.
5. Jukes, T. H., "Recent Problems in the Genetic Code" *in* Current Topics in Microbiology and Immunology, Vol. 49, p. 178, Springer-Verlag, Berlin, Heidelberg and New York, 1969.
6. Yčas, M., The Biological Code, North-Holland, Amsterdam, 1969.

ま え が き

遺伝情報 (genetic code) の問題は，分子生物学の研究課題のなかでも最も重要なものの一つで，それだけに歴史的にみて，研究発展の重要なポイントとなった報告も数えあげればきりがない．そのなかから 20 ばかりの文献を選び出すのは非常に難しいことである．また遺伝情報の問題は他の色々な研究分野と密接に関連していて (例えばタンパク質生合成とか転移 RNA)，画然と区切ることも難しい．以上のような難点があるにせよ，一応小生の判断で 1968 年までの論文のうちから重要と思われるものを選び出した次第である．

ちまたによく遺伝情報の問題は，1965 年に code 表が完成することによりほぼ終了し，それと平行して分子生物学のはなやかな時代も終ったと言われているが，これは決して正しい判断とは言えない．事実，遺伝情報に関した研究は現在でもますます盛んで，chain initiation, chain termination のメカニズム，動物細胞でのタンパク合成，RNA ファージの塩基配列の決定など，画期的な発見が続々と発表されているのである．従って読者は，ここに選ばれた論文を読むことにより，遺伝情報の基本概念と，その研究の発展の流れを把握するとともに，つとめて最新の文献をも読むことをおすすめする．

歴史的に遺伝情報の発展をふりかえると，いくつかの研究グループが，ほぼ同時に同じような発見をし，それが同様に貢献している場合が多い．従ってそのような主題に関しては，あえて両者とも収録することにした．

なお遺伝情報に関する review としては Bibliography にあげるようなものがある．このうちで 1) の Crick の論文は明確な解説で，しかも彼の考え方が非常にはっきりと表われていて，1963 年の発表とはいえ，現在でも一番読みごたえがあるのではないだろうか．ごく最近までのデータを知りたい人は，5) の Jukes の review か 6) の Yčas の著書がよい．特に Yčas の本は非常に緻密に遺伝情報に関するほとんどの文献を網羅していて参考書として最適である．

1970 年 8 月　　　　西 村　暹

CONTENTS

Volume 3
CODING

Reprinted from ADVANCES IN BIOLOGICAL AND MEDICAL PHYSICS, Vol. 4, 1956

The Problem of Information Transfer from the Nucleic Acids to Proteins[1]

BY GEORGE GAMOW

Department of Physics, The George Washington University, Washington, D.C.

ALEXANDER RICH

Laboratory of Neurochemistry, National Institutes of Health, Bethesda, Maryland

MARTYNAS YČAS

Quartermaster Research & Development Center, U.S. Army, Natick, Massachusetts

I. INTRODUCTION

In this article, we plan to discuss the relationship between the chemical information which can be stored in the nucleic acid molecule and the sequence of amino acids which is found in polypeptides and proteins. In Section II we will discuss some of the known facts which relate the nucleic acids to the proteins; in Section III we list some of the experimental

[1] Most of the considerations presented in this chapter have not previously been published.

information available concerning amino acid sequences; and in Sections IV and V we will describe various coding attempts to interpret amino acid sequences.

The two nucleic acids, deoxyribose nucleic acid (DNA) and ribosenucleic acid (RNA), together with the proteins, comprise the most important group of macromolecules in the cell which are capable of great specificity. The specificity arises from the fact that they are long, polymeric molecules in which the ordering of the specific monomeric elements may be varied. It is generally believed that the chemical activity of the macromolecule is determined in large part by the sequence of amino acids or nucleotides.

If we assume that we have enough evidence to involve the nucleic acids in the synthesis of proteins, the basic problem which we are considering is that of translating chemical information implicit in the structure of a nucleic acid molecule into elements which determine specific sites for the various amino acids. The nucleic acids are polynucleotides composed of four elements; the proteins are polyamino acids which have twenty-odd elements. Our problem is that of translating from a four letter code to a twenty letter code. In its fullest detail, this is the problem of specifying the detailed interactions between amino acids and the nucleic acids which determine site specificity. However, we are able to consider only the more general aspects of the coding problem here.

II. The Relation of the Nucleic Acids to Protein Synthesis

1. Functions of the Nucleic Acids

A great deal has been written about the nucleic acids and their relation to protein synthesis (7, 13, 15), and among the accumulated data there is evidence which both supports and refutes the relation. It is not our purpose critically to review this data, but rather to briefly outline the main trends in current research activities.

DNA is present in the nucleus of all cells. Various facts have pointed towards a genetic role for this substance. The amount of DNA per nucleus is found to be constant in the various tissues of complex organisms except where there are haploid or polyploid chromosomes (40, 62). There, as in germinal tissue and spermatozoa, the amount of DNA per nucleus is one-half that found in somatic tissue.

In its genetic role, DNA is believed to undergo duplication, following which it transmits from cell to cell the chemical information necessary to carry on cellular reproduction.

Support for this belief comes from various sources. During the reproduction of bacteriophage, only the DNA from this virus is injected into the bacterial cell, the protein shell of the virus remaining in the external medium (22). This event is sufficient to bring about reproduction of the entire virus, including both nucleic acid and protein parts. It has been shown that the same virus emerges even if a different bacterial host cell is used (25), which again emphasizes the genetic importance of the viral DNA.

Another demonstration of the genetic importance of DNA is seen in its transforming activity. DNA isolated from one type of bacterial cell can be used to transform a second type of bacterial cell, introducing inheritable characteristics (2). Thus, for example, a DNA extract from a penicillin-resistant strain is capable of conferring permanent penicillin resistance to sensitive strains. Over sixteen such transforming principles have been discovered, and all of them have been highly purified and shown to be DNA (29).

Although these demonstrations show that DNA is capable of carrying out genetic activity, it does not rule out the possible participation of protein. Felix and his co-workers have shown that the protamines in fish sperm are highly heterogeneous, and they believe that these proteins act as genetic components (20).

Although DNA probably has an important genetic role, it is not a necessary component for protein synthesis. Brachet and Chantrenne (8) have shown that protein synthesis continues in the enucleated unicellular alga *Acetabularia mediterranea* and therefore in the absence of DNA. Borsook *et al.* (6) have shown that hemoglobin synthesis continues in the reticulocyte stage of erythrocyte maturation, at which time the cell has no DNA.

These data imply that cytoplasmic protein synthesis can be carried out in the absence of DNA. In some recent experiments, Allfrey (1) has shown that protein synthesis occurs in a system of isolated thymus nuclei. Addition of desoxyribonuclease to the nuclei inhibited protein synthesis, whereas the addition of ribonuclease was without effect. These observations suggest that DNA may have a protein synthetic role in the nucleus although not in the cytoplasm.

A large number of indirect observations serve to link cytoplasmic ribonucleic acid with the synthesis of proteins. Studies by Caspersson (13, 14) and Brachet (7) have correlated the amount of RNA with the amount of protein synthesis in a variety of tissues. Caldwell, Mackor, and Hinshelwood (12) have found that in exponentially growing cells of *Bacterium lactis aerogenes* the ratio of RNA to cellular nitrogen is roughly proportioned to the reciprocal of the mean generation time. This suggests that both RNA and protein follow closely related paths.

Studies by Swenson (58) have shown that the ultraviolet action spectrum for the inhibition of the formation of the adaptive enzyme galactozymase is similar to the absorption spectrum of nucleic acid.

Pardee (41) has also investigated the formation of β-galactosidase as well as glycerol oxidizing enzymes, using various mutant strains of *E. coli* which are dependent for growth upon external sources of either purines, primidines, or phosphate. When one of the essential nucleic acid precursors was omitted, RNA synthesis stopped, and at the same time the synthesis of the induced enzyme was halted even in the presence of the inducing agent. This effect was not due to inhibited amino acid synthesis or to destruction of existing enzyme. When nitrogen mustards which inhibit DNA synthesis are used, induction proceeds nonetheless. He concludes that protein synthesis is coupled with the formation of new RNA. However, he also suggests that the RNA, once formed, is an inert metabolic product.

That RNA is associated with protein synthesis is also shown by cell fractionation. Cytoplasmic particles 50–200Å in diameter may be separated centrifugally and have been shown to contain almost 50% RNA. These particles, called microsomes, have been shown capable of carrying out protein synthesis, as demonstrated by the incorporation of labeled amino acids (30).

Somewhat more direct evidence on the participation of RNA in protein synthetic experiments has been obtained with various bacterial extracts. Gale and Folkes (22) have shown that fragmented cells of *Staphylococcus aureus* are still able to synthesize proteins. Addition of RNA isolated from this bacterium stimulated the production of specific protein enzymes; ribonuclease inhibited their formation. Lester (37) and Beljansky (3) have found that lyzates of *Micrococcus lysodeikticus* will incorporate labelled glycine and alanine, whereas the addition of ribonuclease almost totally suppresses the incorporation. It is interesting to note that when deoxyribonuclease is added to this bacterial system, there results an increase in amino acid incorporation.

There are a large number of such experimental observations which relate the nucleic acids to protein synthesis. However, it should be borne in mind that they are largely of an indirect nature, and, consequently, the direct participation of nucleic acid in protein synthesis remains only an attractive hypothesis.

2. *Structural Considerations*

Recent work on the structure of the nucleic acids and the structure of proteins has made it possible to consider their possible interactions in more detail.

Proteins are large molecules, often globular in form, and are composed of one or more polypeptide chains which are folded in some specific fashion relative to each other. The positions of the amino acids residues undoubtedly determine the stability of such a folded configuration as well as its chemical activity on the macromolecular surface. Recent work by Pauling and Corey (42) has suggested that individual chains themselves exist often in a helical configuration, which is a position of maximum stability. If such a helical chain were unfolded, the polypeptide chain would be extended, and the amino acid residues would lie approximately 3.6Å from each other. A possible scheme for protein synthesis would involve arranging the amino acid residues in such an extended form on another molecule which would serve as a template, so that each amino acid site would be specific for that amino acid. Following the polymerization of such residues, the newly formed chain could then roll up into its stable helical form and subsequently fold together to make a globular molecule. The nucleic acids may participate in the first step of such a mechanism by providing a specific condensing surface.

Both RNA and DNA are composed of nucleotides. The nucleotide itself has three components, a purine or pyrimidine base, a sugar (ribose or deoxyribose), and phosphoric acid. There are four bases in DNA, guanine, adenine, thymine, and cytosine. RNA has the same four bases except that its thymine lacks one methyl group and is called uracil. The other difference between the nucleotides of the two nucleic acids occurs in the sugar molecule on carbon 2′, where DNA lacks a hydroxyl group which is present in RNA. In the nucleotides of both DNA and RNA, the bonds between the base, sugar, and phosphate groups are identical.

Both RNA and DNA are polymers of these nucleotide units, with the principal chain passing through the sugar-phosphate groups while the bases act as side chains. In both molecules, the major internucleotide linkage is through sugar carbon 3′ and 5′ (9, 10). Thus, it is apparent that there are remarkable similarities between the two molecules.

Using x-ray techniques, Wilkins *et al.* have shown that all DNA's have the same structure (64). Rich and Watson have shown that the same is true of RNA (49). Watson and Crick (63) have recently proposed a structure for DNA consisting of a two-stranded helix, each strand being a polymer of nucleotide residues. These two strands are held together by hydrogen bonds through the bases in such a way that a given base is always bonded to another specific base. Thus, adenine links with thymine, and guanine with cytosine. These bases pairs are quite specific and exclude other combinations. This proposed DNA structure has a helical pitch of 34Å with 10 base pairs lying at right angles to the axis in this interval. The planar bases are thus stacked upon each other at a distance of 3.4Å. The phosphate groups on the outside of the helix are 7.2Å apart.

Rich and Watson (50) have shown that the x-ray diffraction pattern of RNA has many similarities to that of DNA. The data suggest that it is also a helical molecule with dimensions similar to that of DNA. However, it is not yet known whether there is a similar pairing of the bases.

Studies have been made of the nucleotide sequence in these molecules (55). The partial hydrolysis evidence shows that all possible nucleotide sequences occur, and hence there are no restrictions on where the four elements may be placed relative to each other in a single chain.

As mentioned above, adjacent amino acid side chains on a polypeptide can be 3.6Å apart or less, depending upon whether the chain is fully extended or coiled. Nucleotide residues on a polynucleotide chain, if extended, are about 7.2Å apart. Again, coiling might shorten this distance. The fact that these separations are of the same order of magnitude suggests that if the nucleic acid acts as a template for an amino acid chain, each amino acid would probably interact with only a small number of nucleotides. Using geometric arguments of this kind, it is possible to postulate a number of such simplified amino acid nucleotide interactions and to examine the coding consequences of such arrangements. In some cases, as shown below, certain configurations can be ruled out, as they are clearly incompatible with the existing polypeptide sequence information. In other cases, however, we do not have enough evidence to rule out a possible coding scheme. At the end of our discussion, we will show how various statistical studies bear upon coding schemes.

The ultimate solution of nucleic acid coding of amino acid sequence (providing it exists) will be a detailed description not only in terms of the configuration of the chains, but in terms of intermolecular forces. Each site might have features such as hydrogen bonding potentialities, exact shape, and charge distributions which would enable only one amino acid to settle there. Unfortunately, we cannot as yet consider these more formidable problems and will confine ourselves to more general aspects.

III. Empirical Evidence

1. Amino Acid Sequences

Amino acid sequences are an important part of the available information which bears on the nucleic acid-protein coding problem. All known sequences are collected in Table I. Most of the information has been obtained very recently, and some of it was not available when the "decoding" work was carried out, as described below. We have excluded from this table sequences which contain "abnormal" amino acids, such as the D forms. These are discussed somewhat by Sanger (52).

TABLE I

Polypeptide Sequences of Various Proteins

Protein	Sequences	References
Actin	His-Ileu-Phe*	Locker, R. H. (1954). *Biochim. et Biophys. Acta* **14**, 533.
Adrenocorticotropin	*Ser-Tyr-Ser-Met-Glu-His-Phe-Arg-Try-Gly-Lys-Pro-Val-Gly-Lys-Lys-Arg-Arg-Pro-Val-Lys-Val-Tyr-Pro-Ala (Gly, Glu, Asp) Asp-Glun-Leu-Ala-Glu-Ala-Phe-Pro-Leu-Glu-Phe*	Bell, P. H. (1954). *J. Am. Chem. Soc.* **76**, 5565.
Carboxypeptidase	*Aspn-Ser Ser-Thr	Thompson, E. O. P. (1953). *Biochim. et Biophys. Acta* **10**, 633.
Clupein †	*Pro-Ser-Arg; Ser-Ala-Arg-Arg*; Arg-Arg-Arg-Arg	Waldschmidt-Leitz, E. (1953). *Angew. Chem.* **65**, 349. Felix, K., Hirohata, R., and Dirr, K. (1933). *Z. physik. Chem.* **218**, 269.
Cytochrome C	Lys-Cys-Ala-Glun-Cys-His-Thr-Val-Glu-Lys	Tuppy, H., and Bodo, G. (1954). *Monatsh.* **85**, 1024, 1182.
Gelatin †	Arg-Gly-Gly; Lys-Pro-Gly; Ala-Ala-Gly; Gly-Pro-Ala; Ala-Gly-Ala; Ala-Arg; Ala-Asx; Ala-Lys; Asx-Arg; Glx-Ala; Glx-Gly; Gly-Ala; Gly-Asx; Gly-Glx; Pro-Ser; Pro-Thr; Ser-Arg; Thr-Ala; Thr-Gly; Val-Glx; Val-Gly; Gly-Gly; Gly-Pro; Leu-Ala	Schroeder, W. A., Kay, L. M., LeGette, J., Honnen, L., and Green, F. C. (1954). *J. Am. Chem. Soc.* **76**, 3556.
Globulin	*Ala-Leu-Val-Asx	Porter, R. R. (1950). *Biochem. J.* **46**, 473.
Hemoglobin (horse)	*Val-Glun-Leu; *Val-Leu	Porter, R. R., and Sanger, F. (1949). "Haemoglobin" (F. J. W. Houghton and J. C. Kendrew, eds.). Interscience, New York.

TABLE I.—(*Continued*)

Protein	Sequences	References
Insulin	"A" Chain Thr-Ser-Ileu (pig) *Gly-Ileu-Val-Glu-Glun-Cys-Cys-Ala-Ser-Val-Cys-Ser-Leu- Ala-Gly-Val (sheep) Tyr-Glun-Leu-Glu-Aspn-Tyr-Cys-Aspn* (bovine) "B" Chain *Phe-Val-Aspn-Glun-His-Leu-Cys-Gly-Ser-His-Leu-Val-Glu-Ala-Leu-Tyr-Leu-Val-Cys-Gly-Glu-Arg-Gly-Phe-Phe-Tyr-Thr-Pro-Lys-Ala*	Sanger, F., and Tuppy, H. (1951). *Biochem. J.* **49,** 481. Sanger, F., and Thompson, E.O.P. (1952). *Biochem. J.* **52,** iii. Harfenist, E. J. (1953). *J. Am. Chem. Soc.* **75,** 5528. We are indebted to Dr. Sanger for information on the species differences and positions of amide groups in insulin prior to publication.
Lysozyme	*Lys-Val-Phe-Gly-Arg; Ser-Phe-Asx; Ser-Asp-Gly-Met-Asx; Thr-Asx-Val-Glun-Ala; Ileu-Glun-Leu-Ala-Leu; Thr-Glun-Ala; Asx-Glun-Ala; Leu-Thr-Ala; Ala-Met-Lys-Cys-Arg; Gly-Phe-Glun-Asx-Ileu; Arg-Cys-Glun-Ala; Thr-Pro-Gly; Arg-His-Lys; Ala-Ala; Ala-Lys; Asx-Ala; Asx-Arg; Asx-Leu; Arg-Asx; Arg-Leu; Cys-Ala; Cys-Asp; Cys-Lys; Gly-Leu; Gly-Lys; Ileu-Asx; Ileu-Arg; Ileu-Val; Leu-Leu; Phe-Asx; Ser-Ala; Ser-Arg; Ser-Leu; Ser-Val; Thr-Gly	Acher, R., Laurilla, U. R., Thaureaux, J., and Fromageot, C. (1954). *Biochim. et Biophys. Acta* **14,** 151. Ohno, K. (1954). *J. Biochem.* (*Japan*) **41,** 345. Thompson, A. R. (1954). *Biochim. et Biophys. Acta* **14,** 581.
Ovalbumin	Asx-SerP-Glx-Ileu-Ala; Ala-Gly-Val-Asx-Ala-Ala; Cys-Ala; Cys-Val; Cys-Phe; Cys-Glx; Cys-Gly; Thr-Cys; Val-Cys; Phe-Cys; Glx-Cys; Asx-Cys; Ser-Cys	Flavin, M., and Anfinsen, C. B. (1954). *J. Biol. Chem.* **211,** 375. Flavin, M. (1954). *J. Biol. Chem.* **210,** 771; Ottessen, M., and Wollenberg, A. (1953). *Compt. rend. Trav. Lab. Carlsberg Ser. chim.* **28,** 463.
Oxytocin (bovine)	*Cys-Tyr-Ileu-Glun-Aspn-Cys-Pro-Leu-Gly-NH_2*	DuVigneaud, V., Ressler, C., and Trippett, S. (1953). *J. Biol. Chem.* **205,** 949.

TABLE I.—(*Continued*)

Protein	Sequences	References
Papain	*Ileu-Pro-Glx	Thompson, E. O. P. (1954). *J. Biol. Chem.* **207,** 563.
Pepsin	Thr-SerP-Glx; *Leu-Gly-Asx-Asx	Flavin, M. (1954). *J. Biol. Chem.* **210,** 771; Passmann, J. M., and Williamson, M. B. (Sept. 1954). Abstracts of Papers, American Chemical Society, p. 91C.
Ribonuclease	*Lys-Glu-Thr-Ala-Ala-Ala-Lys; Ser-Arg-Aspn-Leu-Thr-Lys-Asp-Arg	Hirs, W. H., Moore, S., and Stein, W. (1954). Abstracts of Papers, American Chemical Society, p. 89C. Anfinsen, C. B., Redfield, R. R., Choate, W. C., Page, J., and Carroll, W. R. (1954). *J. Biol. Chem.* **207,** 201. Redfield, R. R. Personal communication.
Salmine†	*Pro-Arg-Arg; Arg-Pro-Val-Arg-Arg; Pro-Ileu-Arg; Arg-Val-Ser-Arg; Arg-Ileu-Arg; Arg-Ala-Ser-Arg; Arg-Gly-Gly-Arg; Arg-Ser-Ser-Arg; Val-Gly	Monier, R., and Jutisz, M. (1954). *Biochim. et Biophys. Acta* **15,** 62.
Serum Albumins	*Asx-Ala(human); *Asx-Thr(bovine)	Thompson, E. O. P. (1954). *J. Biol. Chem.* **208,** 565.
Silk Fibroin	Gly-Ser-Pro-Tyr-Pro; Tyr-Pro-Ser-Tyr; Ser-Gly-Ala-Gly; Ala-Gly-Ala-Gly; Thr-Gly; Ala-Ala; Gly-Val-Gly; Gly-Tyr; Tyr-Gly; Phe-Gly	Abderhalden, E., and Bahn, A. (1932). *Z. physiol. Chem.* **210,** 246; (1933). **219,** 72. Kay, L. M., and Schroeder, W. A. (1954). *J. Am. Chem. Soc.* **76,** 3564; also Schroeder, W. A. Personal communication.

TABLE I.—(*Continued*)

Protein	Sequences	References
Tropomyosin	Ala-Ileu-Met-Thr-Ser-Ileu*	Locker, R. H. (1954). *Biochim. et Biophys. Acta* **14**, 533.
Trypsinogen †	*Val-Asp-Asp-Asp-Asp-Lys	Davie, E. W., and Neurath, H. (1953). *Biochim. et Biophys. Acta* **10**, 442. Davie, E. W. (1954). Ph.D. Thesis, Univ. of Washington.
Vasopressin	*Cys-Tyr-Phe-Glun-Aspn-Cys-Pro-Arg-Gly-NH_2* (bovine); *Cys-Tyr-Phe-Glun-Aspn-Cys-Pro-Lys-Gly-NH_2* (pig)	Popenoe, E. A., and DuVigneaud, V. (1954). *J. Biol. Chem.* **207**, 563.
Wool†	Glx-Asx; Glx-Glx; Glx-Ala; Glx-Tyr; Glx-Leu; Glx-Cys; Asx-Glx; Ser-Glx; Gly-Glx; Ala-Glx; Tyr-Glx; Val-Glx; Leu-Glx; Cys-Glx; Asx-Val; Asx-Leu; Ser-Asx; Leu-Asx; Asx-Cys; Gly-Cys; Ser-Cys; Thr-Cys; Ala-Cys; Leu-Cys; Phe-Cys; Cys-Gly; Cys-Ala; Cys-Val; Cys-Leu	Consden, R., Gordon, A. H., and Martin, A. J. P. (1949). *Biochem. J.* **44**, 548. Consden, R., and Gordon, A. H. (1950). *Biochem. J.* **46**, 8.

* Denotes a terminal residue, the N terminus at the left and the C terminus at the right. † Denotes material not used in Section V 3. Glutamine = Glun Asparagine = Aspn, Glutamic acid = Glu, Aspartic acid = Asp. Glx and Asx denotes that no information is available on whether the residues are in the acid or amide form. Phosphoserine = SerP. Other abbreviations are those used by Brand and Edsall (Brand, E., and Edsall. J. T., *Ann. Rev. Biochem.* **16**, 224, 1947).

12

The experimental techniques for determining sequences are relatively new. Various difficulties encountered in these studies are discussed by Sanger (52). It is possible that there are a small number of errors in Table I. This should be borne in mind, as they might have a serious effect on the decoding procedures described later.

The data in Table I can be used to determine which amino acid sequences, if any, are forbidden. For this purpose, the sequences in Table I have been broken down into dipeptides. Thus, a sequence such as Ala-Leu-Val-Ser is listed as three dipeptide units, Ala-Leu, Leu-Val, and Val-Ser. The results are summarized in Table II.

Examination of this table shows that the maximum *known* number of different amino acids following any one amino acid is sixteen, as seen after serine. This is associated with the fact that this compilation is biased, since serine residues can be readily analyzed. Ten amino acids form dipeptides with themselves, and two (arginine and aspartic, Table I) form iterative sequences of more than three residues.

A question of importance is whether any of the blank spaces in Table II represent forbidden combinations. Inspection of the table shows that there are several blank spaces among what should be common dipeptides. Some of these absences may be the result of the lability of certain bonds (e.g., Glu-Ser). Apart from this, however, blank spaces can be expected because of the inevitably biased manner in which the data have been collected. The data in Table II are not drawn from a single homogeneous protein but from several proteins and parts of proteins, some having clearly nonramdom structures. Thus, a large part of the arginine in our "sample population" is segregated in salmine and clupein, proteins which have no glutamic acid. This probably accounts for the absence of the dipeptide Arg-Glu. The inclusion of "structural" proteins such as salmine, gelatin, and silk, which contain abnormal amounts of a few amino acids, tends to raise the number of occurrences of certain combinations. On the other hand, quantitative data (16) indicate that sequences such as Glu-Glu, for example, occur with a high frequency in wool, but they are counted only once in Table I. In view of this bias, this tabulation can be misleading. There can be no question, however, that many of the blank spaces in Table II are a result of inadequate sampling, and we feel that if forbidden combinations exist they must be few in number. Fortunately, the rapid accumulation of data may be expected to answer this question in the near future. We shall discuss this statistically in Section V 3.

It should be noted that the existence of such dipeptides as A-B and B-C does not *necessarily* indicate that the tripeptide A-B-C is an allowed combination. A tabulation of alternate neighbors would be necessary to yield this information. This will be discussed further in Section V 3.

TABLE II

Neighbor Distribution Plot for All Sequences

First Amino Acid	Second Amino Acid																		Marginal Totals
	Gly	Ala	Glu	Arg	Asp	Cys	Ser	Leu	Val	Pro	Thr	Lys	Ileu	Phe	Tyr	His	Met	Try	
Gly	3	3	2	2	2	1	2	1	3	2		3	1	2	1		1		54
Ala	5	6	3	2	1	1	3	3				3	1	1		1	1		56
Glu	2	8	2	2	4	4		5			1	1	1	1	1	2			49
Arg	5	1		6	2	1	1	1	1	2			1					1	44
Asp	1	3	4	3	5	3	2	3	1		1	1	1		1				44
Cys	4	5	3	1	2	1	1	1	2	1	1	1		1	1	1			40
Ser	1	2	3	7	2	2	1	2	2	1	1		2	1	2	1	1		32
Leu	2	3	3		1	2		1	3		2				2				30
Val	4		7	1	4	3	1	1				1		1	1				29
Pro	2	2	1	2			3	2	3		1	2	1		1				22
Thr	3	3	1		1	2	3		1	2		1							19
Lys	1	1	2	1	1	2			2	2		1							20
Ileu		1	2	3	1	1			2	1				1			1		15
Phe	2		2	1	2	2			1	1				1	1				16
Tyr	1		2			1	1	1		3	1		1	1					15
His								2			1	1	1	1					6
Met			1		1						1	1							4
Try	1																		1

Neighbor distribution for all the different dipeptides obtained from data in Table I. Numbers in the grid give the total number of occurrences (thus, Glu-Arg occurs twice, etc.). Marginal totals at right indicate number of times the amino acid appears in the data. Very similar sequences, such as insulins of different species, oxytocin, and vasopressin are counted only once, except for the dipeptides by which they differ. Glutamic acid and glutamine, and aspartic acid and asparagine are counted as one amino acid, and the two occurrences of phosphoserine in Table I are counted here as serine.

2. Amino Acid Replacements

Besides the evidence furnished by amino acid sequences, there is another type of evidence which may throw light on the mechanism of protein synthesis. It has been known for some time that homologous proteins of different species of organisms differ somewhat in their properties. Table III shows the differences in sequence of homologous proteins and

TABLE III

Sequence Information on Homologous Proteins

Protein	Species
Insulin*	
. . . Thr-Ser-Ileu . . .	Pig
. . . Ala-Ser-Val . . .	Cattle
. . . Ala-Gly-Val . . .	Sheep
Serum Albumin (59)	
Asp-Ala . . .	Man
Asp-Thr . . .	Cattle
Virus Protein (C-Terminal Residue)*	
. . . Thr	Tobacco Mosaic Virus
. . . Ala	Cucumber Virus
Protamines (19)†	
Gly_2 Ser_3 Ala_2 Pro_5 Val_4 $Ileu_1$	*Salmo irideus*
Gly_2 Ser_3 Ala_2 Pro_5 Val_5 $Ileu_0$	*Salmo trutta*
Hemoglobin (N-Terminal Residues) (45)	
. . . Val	Horse, Donkey
. . . Val . . . Met	Sheep, Cattle, Goat
Vasopressin (44)	
. . . Pro-Arg-Gly	Cattle
. . . Pro-Lys-Gly	Pig

* We are indebted to Drs. F. Sanger and C. A. Knight for this information and for permission to use it prior to publication.

† Composition, not sequence.

peptides. Insulin and vasopressin demonstrate that species differences do not necessarily involve changes in the chain length or a general scrambling of the amino acid sequence, but rather one or more replacements of one residue by another.

The question naturally arises whether replacements are random or show some discernible regularity. From Table III the known replacements can be summarized as follows:

Gly ↔ Ser
Ala ↔ Thr (three occurrences)
Val ↔ Ileu (two occurrences)
Val ↔ Met
Lys ↔ Arg

Given the small number of cases known, it is rather striking that two of these replacements occur more than once, suggesting that the replacement process is not a random one. Before discussing the possible nature of any regularity, it is worthwhile to consider whether more evidence of replacements may not be available from a comparison of proteins that are not strictly homologous.

There exist in the same organism certain proteins and peptides which, though different, are sufficiently similar in structure and function to suggest some kind of relationship between them. For example, the amino acid sequences of oxytocin and vasopressin of cattle are as follows:

Cys-Tyr-Ileu-Glun-Aspn-Cys-Pro-Leu-Gly-NH_2
Cys-Tyr-Phe-Glun-Aspn-Cys-Pro-Arg-Gly-NH_2

The sequences differ only in two residues. These hormones are both produced in the posterior pituitary gland, and one possibility is that they have differentiated from a common phylogenetic precursor by a process of replacement analogous to the differentiation of homologous proteins of related species. Whatever may be the correct explanation in this particular case, the occurrence of some such process is almost a necessity if evolution involves an increase in the number of different kinds of proteins in organisms.

A consideration of the structure and distribution of hemoglobin seems to show that such a process may in fact exist. Hemoglobin (36) occurs in three different forms. Noncirculating, intracellular hemoglobin (myoglobin) has a molecular weight of 17,600 and one heme group per molecule. The hemoglobin of vertebrates, present in the erythrocyte, has four such units as far as heme groups and molecular weight is concerned. Hemoglobin circulating in solution in the serum, as in annelids, has molecular weights of the order of several hundred thousand or more, but the analytical values for iron indicate, nevertheless, that units of 17,600 molecular weight are still present. The taxonomic distribution of noncirculating hemoglobin is very wide,[2] and, taken in conjunction with the varied distribution of the circulating form, there is a strong suggestion that the myoglobin molecule of molecular weight 17,600 is similar to the phylogenetically primitive form. Since hemoglobins are made up of subunits of this molecular weight, it suggests that they arose by a process of polymerization of a proto-myoglobin. Myoglobin differs in amino acid composition from the circulating hemoglobin of the same species (60). Furthermore, the N-terminal sequences of the subunits of horse hemoglobin are Val-Leu . . . and Val-Glun-Leu, indicating that at least two types of subunits occur differing in amino acid composition (46). These differences may be accounted for by an evolutionary replacement of amino acid residues which differentiates a molecule of myoglobin from one of hemoglobin, etc.

Since a number of proteins are known to be composed of subunits which are straight chain peptides, it is of interest to examine insulin from this point of view. If one writes the sequences of the two insulin chains in juxtaposition, the following is obtained:

Gly-Ileu-Val-Glu-Glun-Cys-Cys-Ala-Ser-Val-Cys-Ser-Leu-Tyr-Glun-Leu
Phe-Val-Aspn-Glun-His-Leu-Cys-Gly-Ser-His-Leu-Val-Glu-Ala-Leu-Tyr
1 2 3 4 5 6 7 8 9 10 11 13 13 14 15 16
-Glu-Aspn-Tyr-Cys-Aspn
-Leu-Val-Cys-Gly-Glu . . .
17 18 19 20 21

It will be noted that Cys and Ser occur in both chains at 7 and 9, respectively, and Glu or Glun at 4. Furthermore, there are three cases of "pair association" occurring more than once (Glu ↔ Leu at 13, 15, 17; Asp ↔ Val at 3, 18; and Cys ↔ Leu at 6, 11). Also to be noted is the occurrence of the sequences Leu-Tyr-Leu and Glun-Leu-Glu in the same position (15–17). At position 2 is the pair Ileu ↔ Val, which occurs as a replacement twice in homologous proteins. While individually none of these coincidences is convincing, taken together the evidence suggests that the two insulin chains may have had a phylogenetically common ancestor from which they have diverged by replacement. It would be of interest to have data on the insulins from lower classes of vertebrates and to examine them from this point of view.

In the above we have discussed two types of replacements, those from homologous proteins and those from similar proteins within the same species. Taking these together, certain structural relationships can be seen to exist between amino acids that replace each other. Thus, if one

[2] Hemoglobin has recently been found to be present in yeast, *Neurospora crassa*, and several protozoa (31–33).

writes the replacement pairs (Fig. 1), it will be seen that in both cases neighboring members of a homologous series (Gly, Ala and Val, Leu) are replaced by neighboring members of another homologous series (Ser, Thr and Asp, Glu). Another such series is not to be expected since none of the other amino acids are members of homologous series. The replacement Val ↔ Ileu, however, also involves amino acids differing by one CH_2 group in the side chain.

The structural chemical basis for these replacement regularities is not known. If we assume that they depend on some property of the protein-forming template, it may indicate that the template of each amino acid is a compound structure, the members of replacement pairs having templates which are identical except for one element.[3] We have used these

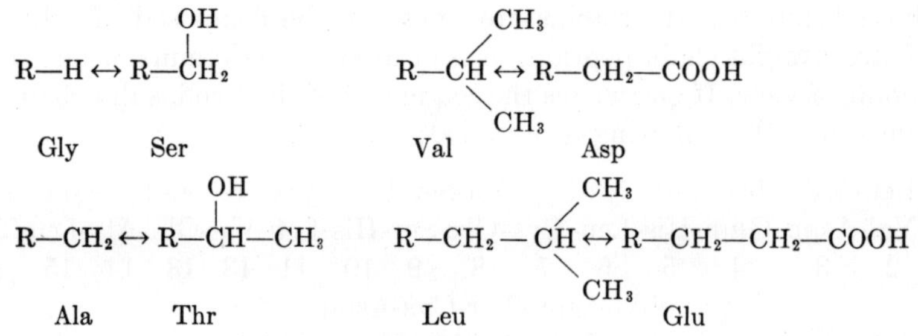

FIG. 1. Structural formulas of replacement pairs.

ideas in discussing the major-minor code in Section IV 4 below. Whatever the correct explanation may prove to be, evidence of replacements—preferably much expanded by future work—should provide an additional criterion by which the correctness of any proposed nucleic acid protein correlation can be evaluated.

We know that many enzymes and hormones are able to function when several groups on their side chains are blocked or altered, and from this we conclude that a considerable portion of the protein molecule is irrelevant to the particular function which we are measuring. In the same way, it is probable that these amino acid replacement sites in the various homologous proteins represent positions which are not enzymatically active. It is likely that a replacement which would not, for example, permit insulin to function would constitute a lethal mutation and would be difficult to detect if it did occur.

[3] That the templates of amino acids have points of similarity is suggested by the fact that in the two cases in which the side chains of amino acids have an optically active carbon atom (threonine and isoleucine), the orientation in space of the CH_3 and OH groups around it is the same (48, 56, 61).

IV. Nucleic Acid-Protein Coding Schemes

1. Introduction

Although a great deal of indirect evidence has related protein synthesis to the nucleic acids, there have been in the past very few detailed considerations of the "coding" requirements necessary in their interactions.

Caldwell and Hinshelwood (11) have considered the problem of determining over 20 amino acid sites from the 2 purine, 2 pyrimidine, and 1 ribose phosphate residues found in RNA. They have pointed out that if 2 of these elements determined 1 amino acid site, this would give rise to 25 sites, an adequate number for the amino acids. This system, of course, permits all amino acid sequences to occur.

Dounce (18) has put forward a detailed method for synthesizing protein using a nucleic acid template. In his construction, a given amino acid residue is determined both by the nucleotide to which it is attached and by the 2 adjoining neighbors. Although he points out that this would give rise to at least 40 possible specific sites, he does not consider the restrictions which this scheme places on the adjoining amino acid neighbors. In our consideration of the major-minor code below, we will show how such an arrangement can be analyzed and compared with the existing amino acid sequence data.

Haurowitz (27) has postulated an entirely different role for the nucleic acids in protein synthesis. He believes that the long chain polymeric nucleic acid molecule simply acts as a linear rod, serving to hold a polypeptide chain in its extended form. The extended polypeptide chain then serves as a template for the production of an identical chain, where Van der Waals forces between similar amino acid residues hold them in place. This role for the nucleic acids divorces them from any chemical specificity.

However, we believe that the specificity inherent in the nucleic acid molecule may somehow determine the specificity of proteins so that, to each arrangement of bases in the polynucleotide chain, there corresponds a unique arrangement of amino acids in the polypeptide chain synthesized by it.

One can imagine that the various amino acids, or perhaps some suitable derivatives, possess some kind of specific affinity for particular loci of the polynucleotide chain and, being attracted to these particular spots on the surface of the nucleic acid molecule, are held in that position long enough to permit the formation of peptide bonds between them. Since the distances between neighboring amino acids in a polypeptide chain is comparable to the distance between neighboring bases in a polynucleotide chain, one may assume, for example, that the nucleic acid template and the protein molecule manufactured by it possess an equal number of constituent units; thus, the polypeptide chain of insulin A consisting of twenty-one amino acids may be manufactured by a nucleic acid molecule formed by a sequence of about the same number of bases.

If we distinguish between 4 different bases by means of numerals 1, 2, 3, and 4, and use the reduced English alphabet (*a*, *b*, *c*, *d*, *e*, *f*, *g*, *h*, *i*, *j*, *k*, *l*, *m*, *n*, *o*, *p*, *r*, *s*, *t*, *u*, *v*) for 20 different amino acids, we may represent the correlation between the structure of the template and the structure of the corresponding protein by a long work written under a long number formed by only 4 different digits. Thus, in the case of insulin *A* the correlation could be:

3 1 1 2 4 3 4 1 1 2 3 1 4 2 2 2 1 3 4 2 1
l m a k s i i e k a i k n t s n k f t i f

where the letters in the second line are derived from the digits in the first one by some kind of coding rule. The number of all possible arrangements of 4 digits into a sequence of 21 digits is $4^{21} = 4.4 \cdot 10^{12}$, while the number of all possible arrangements of 20 letters into a sequence of the same length is $20^{21} = 2.1 \cdot 10^{27}$. Thus, for any given code, only 1 out of $5 \cdot 10^{14}$ arbitrary arrangements of letters can be "decoded," i.e., represented as the result of coding of a certain number written in the 4 digital system. Thus, if protein molecules are actually manufactured by a nucleic acid template having the same number of bases as there are residues in the protein, the order of amino acids in the known polypeptide sequences must be far from arbitrary, and there must exist a strong correlation between the neighboring amino acids. The situation is similar to that existing in any language where sensible sentences like:

"Amino acids form proteins."
or:
"Read Alice in Wonderland."

each containing 21 letters, represent only a negligible fraction of all possible sequences of the same length formed by the letters of the alphabet. Since, as was indicated in the previous section, the proteins and polypeptide fragments in which the order of amino acids is known add up to a sequence a few hundred amino acids long, we here face difficulties similar to those encountered by an Armed Force's Intelligence Office trying to break an enemy code on the basis of a single message less than two printed lines long. Of course, the number of such known amino acid sequences is rapidly growing, but this can hardly improve the situation very much in the near future, since, in order to make the decoding easier, the length of known sequences must increase not by a factor of two or three, but rather by two or three orders of magnitude.

It would be helpful, of course, if one knew some sequences of bases in polynucleotide chains, especially those connected with proteins which have already been analyzed, but at present this knowledge is completely lacking. Thus, difficult as it is, the only way open at present is to attack the problem on the basis of very meager material concerning amino acid sequences, using in addition all other available evidence on the structure and interrelations between polypeptide and polynucleotide chains. In the following we present various attempts which have been made in this direction.

2. Diamond Code

The first attempt to devise a workable code on the basis of the DNA model of Watson and Crick was made by Gamow (23). As discussed in the previous section, biological evidence suggests that DNA is probably not directly involved in the process of protein synthesis, but perhaps acts through the intermediary of RNA, the exact structure of which is still unknown. However, one may presume that RNA molecules are topologically similar (in the sense of amino acid sequences) to the DNA molecules from which they are perhaps derived. Also, as discussed above, there is no conclusive evidence (except in the case of plant viruses) against ascribing to RNA a two stranded structure similar to that of DNA (50).

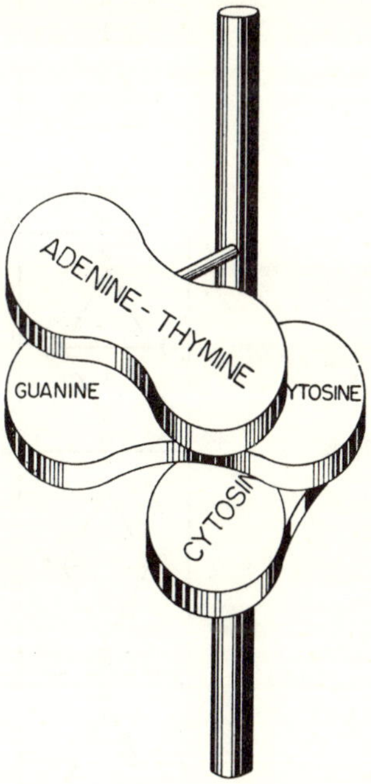

DNA.

FIG. 2. Schematic diagram of DNA.

The idea of Gamow is shown schematically in Fig. 2, in which 3 successive pairs of bases are represented by 3 biscuit-shaped plates. It is seen that the combination of a pair of bases forming the middle plate and the 2 protruding ends of the upper and lower plates form a diamond shaped structure which can be symbolized as: $\begin{pmatrix} 2 \\ 1\text{—}2 \\ 4 \end{pmatrix}$. Since the upper and the lower protruding extremities can correspond to any of the 4 bases, while the 2 middle bases must be bound by relations, 1—2 and 3—4, the number of all differently shaped combinations equals the number of different trip-

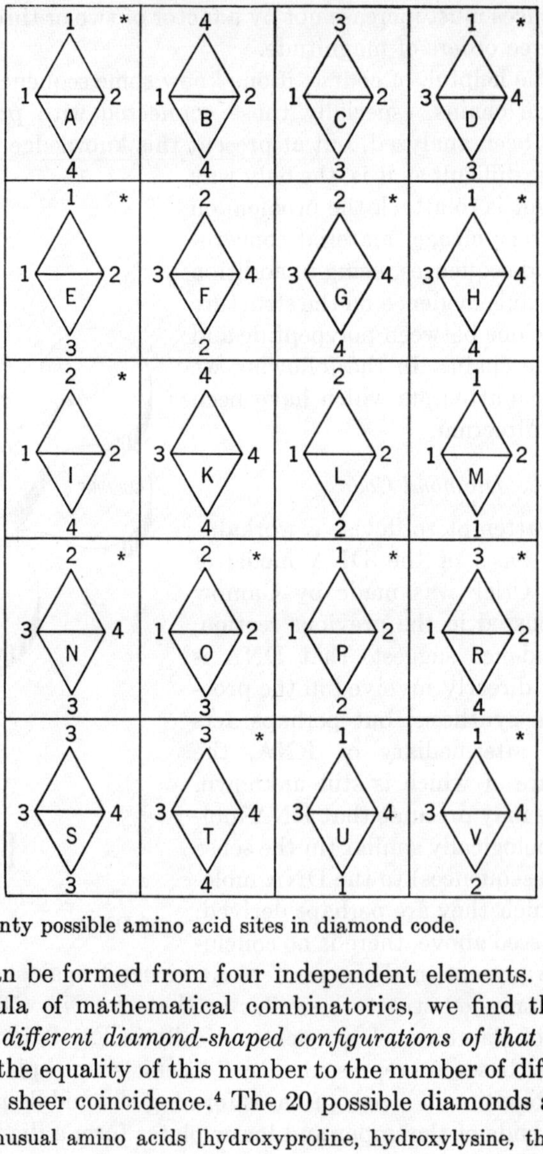

FIG. 3. Twenty possible amino acid sites in diamond code.

lets which can be formed from four independent elements. Using a well known formula of mathematical combinatorics, we find that *there are altogether 20 different diamond-shaped configurations of that kind*, and it may be that the equality of this number to the number of different amino acids is *not* a sheer coincidence.[4] The 20 possible diamonds are shown in

[4] Certain unusual amino acids [hydroxyproline, hydroxylysine, thyroxine, phosphoserine (43), tyrosine-O-sulfate (4), alpha amino adipic acid (65)] occur in a restricted number of proteins only. F. H. C. Crick (personal communication) suggests that they are formed from the common amino acids after incorporation into the protein chain. The structure of the rare amino acids is favorable to such an interpretation, which is experimentally supported in the case of thyroxine (5, 26, 38). Work with labeled compounds indicates that this may also be true of hydroxyproline (57) and phosphoserine (21), although in these cases the evidence is not conclusive.

Fig. 3, and it will be noticed that, while 8 diamonds possess bilateral symmetry, the remaining 12 (marked by an asterisk) are antisymmetrical. If one considers the 2 bilaterally antisymmetrical forms such as, for example $(3 \cdot \frac{1}{4} \cdot 4)$ and $(4 \cdot \frac{1}{4} \cdot 3)$ as 2 different entities, the number of different diamonds will be raised to 32. However, inspection of the structural formulae of amino acids indicates that their side-chains possess a bilateral symmetry[5] (in other words, the side chains are not optically active), so that the same amino acid residue can be attached to either of the 2 antisymmetrical diamonds simply by turning it around the α-C bond by 180°. The situation is illustrated schematically in Fig. 4.

Since 2 adjacent diamonds in the sequence possess a common side, there exists a strong intersymbol correlation between the letters in the long words formed by them. Thus, whereas the name "GAMOW" can

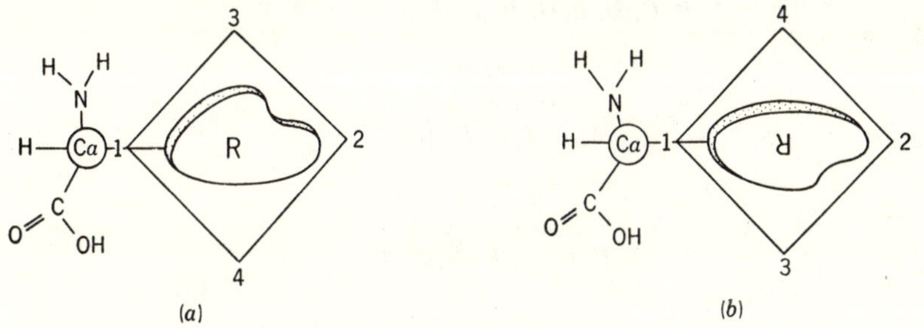

FIG. 4. Bilateral symmetry of amino acid side chains.

be consistently composed (by prearrangement, of course) in this particular diamond code, the chances that any other name of equal length can be written in that code are 1 out of 3125. (And, indeed, nobody has yet succeeded in doing it.) The list of all possible diamond-pairs is given in Table IV, and we see that the letters fall into a group of 8 which can have 14 different neighbors, and the remaining 12 which can have only 7 neighbors each. This table also shows that letters *B*, *C*, *R*, *F*, *U*, *V*, cannot be double (i.e., do not combine with themselves), and that the other *12* can occur in pairs, whereas *P* and *T* can repeat consecutively an indefinite number of times.

The problem of decoding protein sequences in terms of the diamond-code consists of finding a one-to-one correlation between 20 possible diamonds, and 20 different amino acids which would permit one to write down these sequences in terms of diamonds in accordance with the combination rules given in Table IV. If the code is correct, there must be only one possible solution. If the code is incorrect, the decoding attempt must lead to a contradiction. The number of possible assignments is immensely

[5] Except threonine, isoleucine, and proline.

large, 20! = 2.3×10^{17}, and if one could spend only one second to check each assignment, one would have to work continuously for about five billion years, which is the present age of our Universe! However, as often happens in coded messages, one may find some particular subsequences of code symbols which make the solution of the problem much easier. One

TABLE IV

Nearest Neighbor Combinations in Diamond Code

A, E, I, O	associates with	*A, D, E, F, G, H, I, L, M, N, O, P, U, V*
D, G, H, N	associates with	*A, B, C, D, E, G, H, I, K, N, O, R, S, T*
L, M, P	associates with	*A, E, I, O, L, M, P*
K, S, T	associates with	*K, G, H, K, N, S, T*
B, C, R	associates with	*D, F, G, H, N, U, V*
F, U, V	associates with	*A, B, C, E, I, O, R*

such peculiar subsequence of amino acids can be found in the insulin A chain and runs as follows:

-Val-Glu-Glu-Cys-Cys-Ala-Ser-Val-Cys-

According to the diamond code rules, there are only four pairs of letters which can stand side by side in a doubled form: *A-A-N-N-; L-L-M-M-; K-K-T-T;* and *T-T-S-S*. In each of these four cases we also know the letter assignment of the fourth amino acid to the right from Cys-Cys, which is Cys. This restricts the choice of letters for Val, since it must precede both Glu and Cys. Proceeding this way, it is possible to decode this particular sequence in a unique way as shown in Fig. 5. However, the attempt to continue this decoding for the rest of the molecule runs into an immediate difficulty since the next amino acid on the right is Ser, which was already assigned the letter *C*, while on the other hand, according to Table IV, the letter *C* does not combine with the letter *I*, which represents Cys. The assignment shown in Fig. 5 also brings us to contradictions when we try to use it for the insulin B sequence.

It should be noted, however, that the contradiction reached here is a consequence of our using the same symbol for glutamic acid and glutamine, whereas, in fact, the two neighboring Glus in insulin A belong to these two different groups. If, however, we assume that glutamic acid and

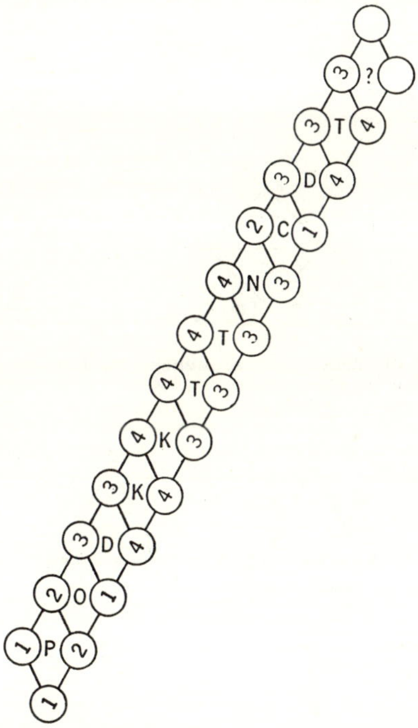

FIG. 5. Decoding of insulin segment.

glutamine correspond to two different diamond-letters, we lose the advantage of two neighboring pairs in the sequence, and any attempt at decoding becomes tremendously more difficult.

Another check of the correctness of the diamond code can be made by considering the relationship between three triplets from the insulin molecule: (a) -Leu-Tyr-Leu (ins. B) (b) -Tyr-Glu-Leu- (ins. A) (c) -Phe-Phe-Tyr- (ins. B). Let us write *a* and *a*′, *b* and *b*′, etc., for any pair of bases related by the 1—2 and 3—4 rule. In this case, Tyr can be written in the most general way as $\left(a \frac{b}{c} a'\right)$ where *a*, *b*, and *c* may or may not be the same. But since Tyr is flanked on both sides by two Leu's possessing the same diagonal pair (Fig. 6 left), we must have either $b = c$ and $b' = c'$ or $b = c'$ and $c = b'$. These two possibilities are shown on the right of Fig. 6. Since, however, the two flanking diamonds must represent the same amino acid (Leu), we must assign to the circled vertices letters a and a', as indicated in Fig. 6. This means that in Leu, the 1—2 and 3—4 relationship also holds for the vertical diagonal, so that it can only be one of the following four shapes: α, β, γ, or δ (Fig. 7). It can be shown, however, that each of the four assumptions leads to a contradiction.

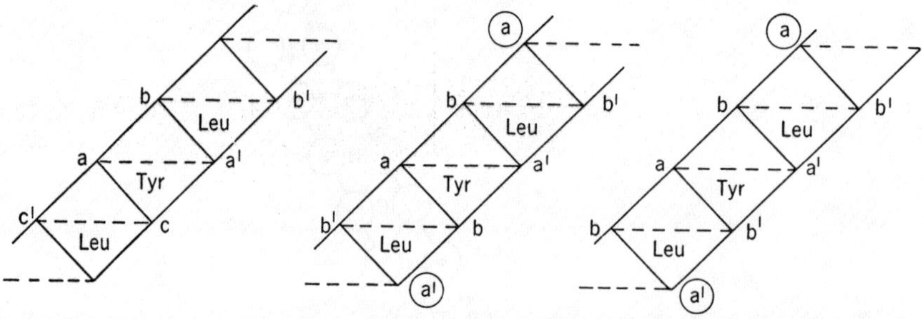

FIG. 6. Assignments of bases to the sequence Leu-Tyr-Leu assuming the diamond code.

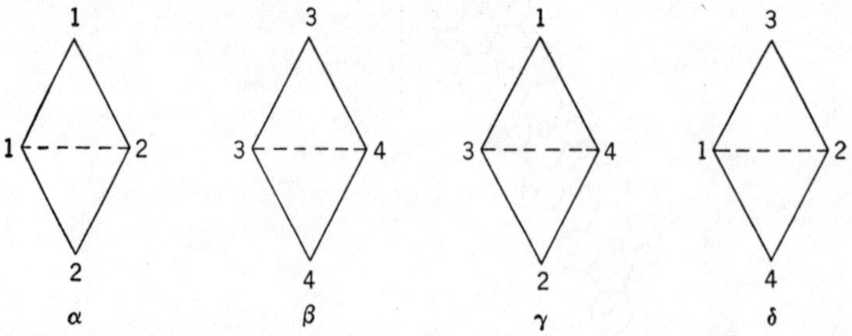

FIG. 7. Possible assignments of bases determining leucine in the sequence shown in Fig. 6.

1. If Leu corresponds to α-diamond, the sequence a shows that Tyr must be represented by one of the two diamonds ϵ or ζ (Fig. 8), in which case, the sequence b leads to the conclusion that Glu must correspond to ζ-diamond if Tyr corresponds to ϵ and vice versa. The contradiction occurs in considering sequence c, since inspection shows that, using the above limitations, one cannot possibly interpret that sequence in such a manner that the diamond assigned to Phe is not identical with either Glu or Leu.

2. If one assumes that Leu corresponds to β-diamond, the argument is identical with bases 1 and 2 substituted by 3 and 4.

3. If Leu corresponds to γ-diamond, sequence (a) leads to three possibilities for Tyr η, θ, and ι (Fig. 9), but then the construction of sequence (b) becomes impossible.

4. The assumption that Leu corresponds to δ-diamond is disposed of similarly, as in 3.

In connection with the above discussion one must mention, however, that one of the triplets (Leu-Tyr-Leu) used in the argument has never actually been found as a tripeptide in Sanger's analysis of insulin chains.[6] In fact, this is the only missing link in the entire analysis, making this particular part of the sequence somewhat less certain than the rest. On the other hand, the possibility that this particular part of the chain can

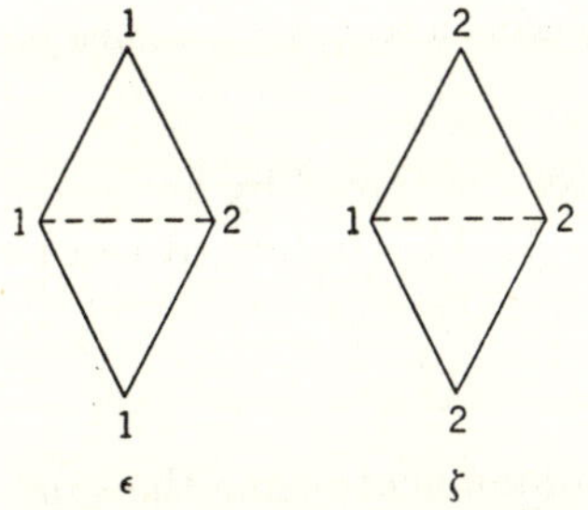

FIG. 8. Possible assignments of bases determining tyrosine in the sequence shown in Fig. 6.

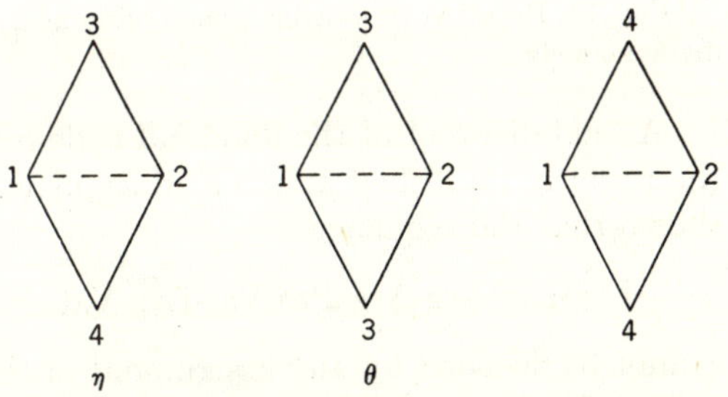

FIG. 9. Possible assignments of bases determining tyrosine in the sequence shown in Fig. 6.

be something like -Leu-Tyr-Tyr-Leu-, or Leu-Tyr-X-Tyr-Leu- seems to be precluded by amino acid analyses of insulin.

A third argument against the correctness of the diamond code is based on the fact that there are known cases in which two related proteins differ in a single amino acid with both its neighbors being the same (compare the oxytocin and vasopressin chains given in Table I). In fact, since each diamond in a chain has two of its vertices common with the neighbor on the right, and two others with the neighbor on the left, there should be only one diamond which fits between two given neighbors. This argument, which is undoubtedly correct if such replacements take place near the middle of a long molecule, may not hold, however, for amino acids standing not too far from one of the terminals. Figure 10 shows, for example, how, by rotating the diamonds, one can change one single amino acid at the third position from the free amino end (as in the case of the Phe—Ileu replacement between oxytocin and vasopressin) without changing the rest of the sequence.

[6] The sequence is obtained by putting end-to-end the observed fragments of insulin B without an overlapping peptide.

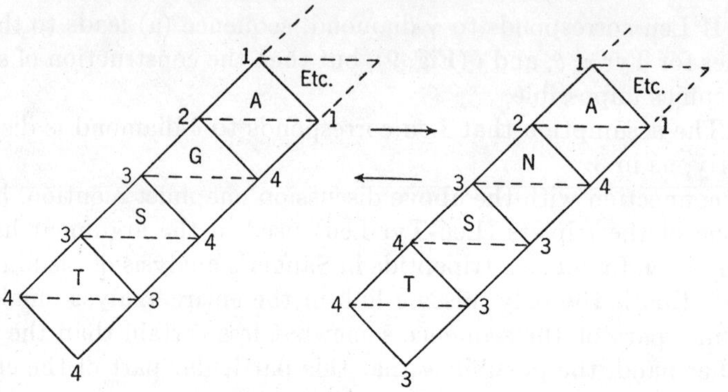

FIG. 10. Rotation producing a one amino acid change in a sequence assuming the diamond code.

A final disproof of the diamond code has been obtained by analyzing the recently published amino acid sequence in ACTH. In fact, it can be shown that the sequence:

-Lys-Lys-Arg-Arg-Pro-Val-Lys-Val-

cannot be decoded by any assignment of the amino acids using the symbols of the diamond code.

We have discussed the diamond code here in some detail mainly to indicate logical methods to be used in the problem of protein decoding, since several other more complicated codes which have been proposed since are in a considerably less developed stage.

3. *Triangular Code* (*Compact and Loose*)

Whereas the diamond code is essentially based on the assumption of a two stranded template identical or similar to the model of the DNA

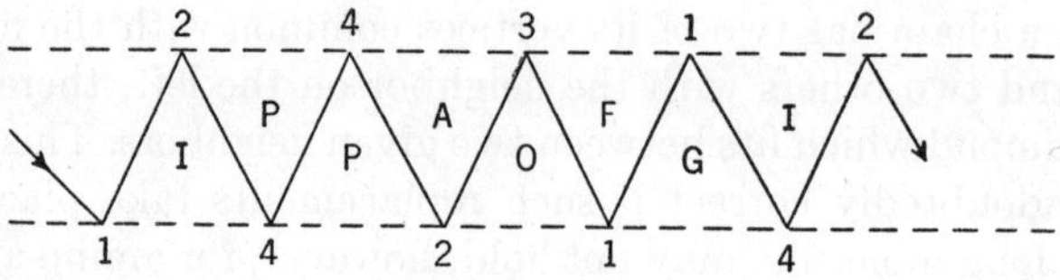

FIG. 11. Schematic diagram of triangular code.

molecule, the triangular code pertains to single sequences of bases and may indeed be applicable if protein synthesis is accomplished by a one stranded molecule of RNA. The principle of this code is shown schematically in Fig. 11, where the sequence of numbers 1, 2, 4, 4, 2, 3, 1, 1, 4 . . . corresponds to a sequence of bases in the single stranded molecule of nucleic acid. Just as in the case of the diamond code, there are 20 different triangles which can be formed by 3 consecutive bases; these triangles are shown in Fig. 12 with the corresponding letter assignment. Again we have assumed that the template is invariant with respect to rotation or reflex-

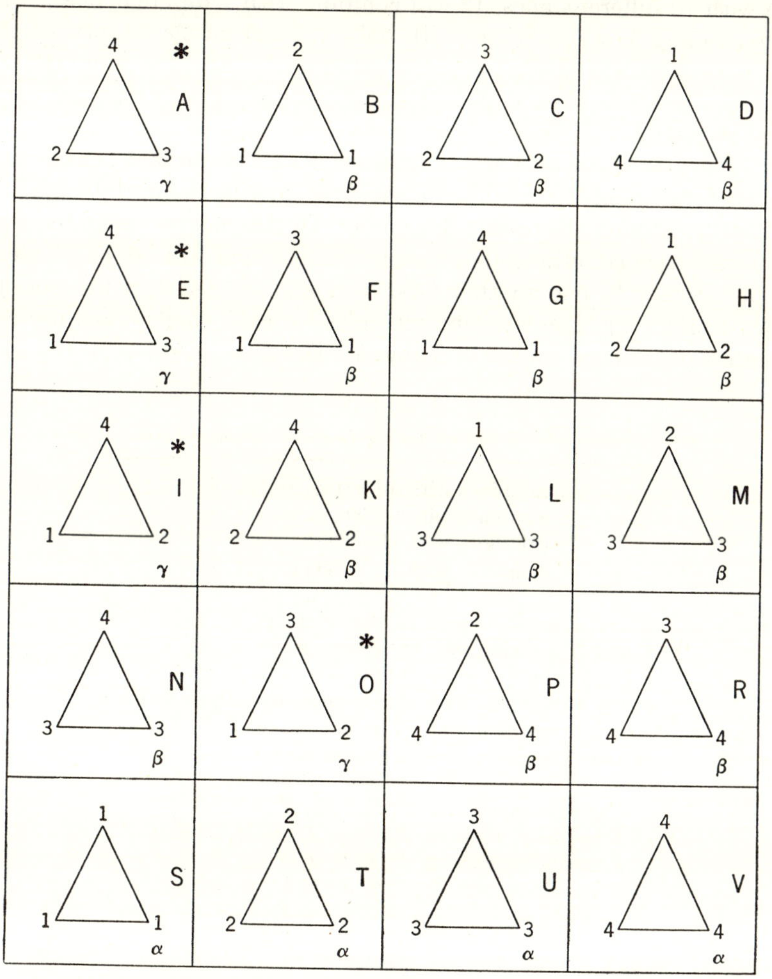

FIG. 12. Twenty possible triads of triangular code.

ion. The rules of combination between triangles are different, however, from those in the case of diamonds and depend on the assumptions we make as to how the amino acids arranged along this template unite to form the polypeptide chains. Two different assumptions can be made:

1. The *compact triangular code* is obtained if we assume that amino

acids unite as they are laid out along the template so that the template shown in Fig. 11 will produce the protein: *I-P-P-A-O-F-G-I*.

The intersymbol combination rules in this case are similar to those between the diamonds, though more restrictive. Four triangles can combine with 10 different ones, 12 can combine with 7 different ones, while the remaining 4 can combine with only 4. Since in the known protein sequences there are cases in which a given amino acid has more than 10 different neighbors, the compact triangular code evidently does not correspond to reality.

2. The *loose triangular code* is obtained if we assume that peptide linkages are formed only between amino acids occupying the upright triangles or those occupying the upside down ones. In this case, a single template may synthesize simultaneously two proteins which will be, so to speak, complementary to one another (*-I-P-O-G* and *-P-A-F-I* in the example shown in Fig. 11). The combination rules for each single chain are considerably relaxed, as shown in Table V. Thus, we see that in the loose

TABLE V

Intersymbol Relations for the Loose Triangular Code

A combines with all letters except: *S*
E combines with all letters except: *T*
I combines with all letters except: *U*
O combines with all letters except: *D*
B and *H* combine with all letters except: *N*, *R*, *U*, *V*
C and *D* combine with all letters except: *C*, *M*, *T*, *U*
F and *L* combine with all letters except: *K*, *P*, *T*, *V*
G and *M* combine with all letters except: *D*, *G*, *S*, *V*
K and *P* combine with all letters except: *F*, *L*, *S*, *V*
N and *R* combine with all letters except: *B*, *H*, *S*, *T*
S combines with: *B*, *D*, *E*, *F*, *G*, *H*, *J*, *L*, *O*, *S*
T combines with: *A*, *B*, *C*, *H*, *J*, *K*, *M*, *O*, *P*, *T*
V combines with: *A*, *D*, *E*, *G*, *J*, *K*, *N*, *P*, *R*, *V*
U combines with: *A*, *C*, *E*, *F*, *L*, *M*, *N*, *O*, *R*, *U*

triangular code, 4 letters (vowels) combine with 19 others, 12 letters combine with 16 others, and the remaining 4 combine with only 10 others.

It should be noted also that in this code (in contrast to diamonds and compact triangles) there are twice as many bases in the template as there are amino acids in a single polypeptide chain produced by it.

For example, to produce insulin A, we will need not 21 but rather 42 consecutive bases in the RNA template. Thus, while the number of possible arrangements of amino acids into the chain of 21 remains the same

(2.1×10^{27}) as given earlier in this section, the number of possible templates formed by 42 bases becomes 4^{42} or 1.5×10^{25}. Thus, in the loose triangular code about 1 per cent of all possible sequences of amino acids should be "decodable" in terms of the template (in contrast to 10^{-12} per cent in the case of the diamond and compact triangular codes).

Because of the much smaller restrictiveness of the loose diamond code, the work of actually decoding known polypeptide sequences becomes more difficult, and no whole-hearted attempt in this direction has been made yet.

It may be worth pointing out that the possible existence of "complementary" proteins, which follows from the nature of the loose triangular code, may have some relation to the empirical findings concerning replacements in related proteins as discussed in Section III 2. Selecting, as an example, a random sequence of 42 bases (4, 3, 2, 4, 4, 3, 4, 2, 4, 3, 4, 1, 1, 4, 1, 2, 4, 3, 3, 1, 2, 1, 3, 4, 1, 2, 2, 2, 3, 2, 1, 4, 1, 1, 2, 4, 1, 3, 4, 1, 1, 3) and constructing, according to Fig. 11, 2 complementary proteins, one gets the following 2 "polypeptides";

```
A-P-R-K-A-E-G-I-A-L-B-E-I -T-C-I-G-I-E-E-F
↕   ↕ ↕ ↕ ↕ ↕ ↕       ↕             ↕ ↕ ↕
A-R-A-K-R-G-G-I-N-O-O-E-H-C-O-G-B-I-E-G-E-
```

In these 2 "complementary sequences" there are 7 cases of identical residues occupying the same position, and 2 pairs of "pair association" (both indicated by arrows) which are not unlike the regularities observed between the A and B chains of insulin as previously discussed. "Complementary chains" are therefore an alternative explanation for the replacement regularities observed in related protein chains. On the other hand, it should be pointed out that one, not two, proteins are synthesized during growth of plant viruses, and that the absence of a complementary protein here where it should be comparatively easy to recognize speaks rather strongly against the whole idea of complementary chains and indirectly against codes producing them.

4. Major-Minor Code

Another type of code, based on the assumption of a single-stranded nature of RNA, was proposed by L. Orgel and studied by him, in some detail, in collaboration with R. Feynman and the authors of the present article. It was essentially based on the fact that single amino acids in certain polypeptide chains are often interchanged with other amino acids, leaving the adjoining neighbors unchanged. Thus, for example, oxytocin has the sequence:

-Cys-Tyr-Ileu-Glu-Asp-

and the closely related vasopressin:

-Cys-Tyr-Phe-Glu-Asp-

where Phe has replaced Ileu. The simplest assumption is that this represents the replacement of one nucleotide in the template by another.

In this code, each amino acid has a particular nucleotide which acts as a major determining factor in the positioning of that amino acid. The interaction between the nucleotide and the amino acid may involve hydrogen bonding, electrostatic interactions, or others, but this need not be specified in a formal development. Of course, several different amino acids must have the same major determinant.

However, in addition to the major determinant, the adjoining nucleotides may also influence the positioning of the amino acids. Thus, for example, certain adjoining nucleotides may sterically block one amino acid but permit another to gain a position which is favorable for polymerization. In this way, each amino acid may be represented formally by a triplet of numbers, where the numbers ranging from 1 to 4 represent the various nucleotides. For example, Ileu may be translated into 3 1 4 where 1 is the *major determinant* and 3, 4 represent *minor determinants.* In this way, the oxytocin sequence may be represented by:

-Cys-Tyr-Ileu-Glu-Asp-
-1 - 3 - 1 - 4 - 2-

Each amino acid has one major determinant but several minor determinants. Thus, Tyr in the above is represented by 1 3 1 but could also be represented by 1 3 2, where the second minor determinant has been changed. Similarly, Glu is represented by 1 4 2 or, for example, 2 4 2. In this way, if Phe had a major determinant of 2, the similar sequence in vasopressin could be represented by:

-Cys-Tyr-Phe-Glu-Asp-
-1 - 3 - 2 - 4 - 2-

Hence, this code permits a one-nucleotide change to appear as a single amino acid replacement.

The essential problem in solving this code consists in filling out the following grid (Table VI) with the various amino acids in such a way that no contradictions develop in the sequence data. As an example, we have made assignments in Table VI for the amino acids in the vasopressin-oxytocin sequences cited above.

From this code it follows that a sequence of two amino acids, A, B, in the triplet $A\ B\ X$, would permit only four other amino acids to fill position X. Similarly for sequences $X\ A\ B$ or $A\ X\ B$.

There are 64 blanks in the grid and 20 amino acids to fill in the spaces. The assignment of amino acids can best be carried out by arranging all the known amino acid sequences in triplets. In doing this, we find sequences such as Leu-Tyr-Glu and Leu-Val-Glu. This means that the major determinant of Val is not the same as that of Tyr, since they are both flanked by the same amino acids, and one writes: (Maj. Val *is not* Maj. Tyr).

TABLE VI

Major-Minor Grid for Amino Acids

	Major Determinants			
Minor Determinants	1	2	3	4
1 1			Tyr	
1 2			Tyr	Glu
1 3				
1 4				
2 1				
2 2				Glu
2 3				
2 4				
3 1				
3 2				
3 3				
3 4	Ileu	Phe		
4 1				
4 2				
4 3				
4 4				

Using a selected set of more-or-less well established tripeptides, a number of relationships of the type (Maj. *X is not* Maj. *Y*) were obtained, and it was found that these relationships are inconsistent with the basic principles governing the code. The possibility, however, is not excluded that this was due to a mistake in chemical analysis of one of the tripeptides used in these studies, since, indeed, a single error in a coded message would inevitably lead to a complete failure of decoding even though one makes a correct guess concerning the coding principle.

5. *Sequential Code*

In an attempt to overcome the difficulties caused by the existence of single replacements, another code was proposed by E. Teller. The idea of this code is that an amino acid is defined by two bases and the previous amino acid. This implies that there are certain preferred sequences and suggests that certain inter-symbol relationships are more important than others. This code shows some similarity to the major-minor code previously discussed, and since no new principles are involved, it will be treated only briefly.

One formulation for this coding involves an equal number of amino acids and nucleotide bases, in which each amino acid is situated between 2 adjoining bases. If the preceding amino acid did not aid in determining the following one, then a given amino acid could only be followed by 4 others. However, if it is the preceding amino acid together with the 2 adjoining bases which determine the next amino acid, a much larger number of possibilities are allowed. In the most general form of this code, all sequential possibilities can occur, although one might expect that certain sequences would be favored. Thus, for example, if amino acids with large, bulky side chains prevented the seating of a second amino acid with a similarly large side chain, then these combinations would occur less frequently than large side chain-small side chain sequences.

The decoding procedure for this scheme is best carried out by studying the experimentally determined sequences to see if there are any systematics present. When this is done with a larger collection of data, it may be possible to infer which influences are most important.

Information bearing on this point will be presented in the statistical discussions in Section V 3.

6. The Use of Symbolic Logic

Returning to the problem of straightforward decoding of the known protein sequences, we have again to face the fact that the number of all possible assignments between amino acids and various code symbols is so tremendously large that the direct testing of all possibilities represents a prohibitive task not only for conventional methods but also for the modern electronic computing machines. In fact, a computor put to work in the days of the Roman Empire and testing one million assignments per second, 24 hours a day, all year round, would not yet be finished with the job.

It was, however, recently indicated by Ledley (35), that the task can be considerably shortened by expressing the problem in terms of the equations of symbolic logic and by solving these equations under the conditions of constraint supplied by the known amino acid sequences. As a simple example, he considered only three letters: $O = (1,2,3)$, $I = (1,2,4)$, and $A = (2,3,4)$ of the compact triangular code and assumed the existence of the following four sequences: *Y-X-X-X*; *Y-Y-Y-X*; *Z-Z-Z-Y*; formed by three different amino acids: *X*, *Y*, and *Z*. A comparatively short hand calculation following Ledley's method leads to a unique solution: $X = A$, $Y = O$, $Z = I$. Of course, this simple problem could be also quickly solved by a direct test of 3! (= 6) possible assignments between A, B, C, and X, Y, Z. The point is, however, that with the increasing number of different amino acids the time necessary for direct testing of all possibilities becomes prohibitively long; however, the solution of the corresponding system of logical equations under the conditions of constraint can still be performed within a reasonable time limit. It was estimated, in fact, that in the case of 20 amino acids, using a code such as one of the above, and all the known protein sequences, the solution would require only a few hundred hours of electronic computation time. It is hoped that some attempts in this direction will be made in the near future.

V. Various Statistical Considerations

1. *Correlation between Nucleic Acid and Protein Composition*

If nucleic acids function as protein templates, a correlation should exist between their composition and that of proteins. The most promising biological material for the study of such correlations is presented by the viruses since, in this case, there is reason to believe that the protein forming the shell of the particle is synthesized according to the information carried by the nucleic acid constituting its interior. Two viruses for which the composition of the nucleic acid as well as that of the protein have been studied are those of Tobacco Mosaic and Turnip Yellow. Tables VII and VIII give the base—and amino acid—composition of these two viruses.

TABLE VII

Base Composition of Tobacco Mosaic and Turnip Yellow Viruses (17, 39)

Virus	Percentage of Bases			
	Adenine	Guanine	Cytasine	Uracil
TMV	0.30	0.25	0.18	0.26
TYV	0.23	0.17	0.38	0.22

Molar content normalized to unity.

It should be noted here that experiments carried out on the RNA of the Tobacco Mosaic virus have shown that it is a single molecular species rather than a mixture of molecules (53).

Table VII shows that RNA composition of TYV differs from that of TMV by a large increase of *Cy*, a moderate decrease of *Ad* and *Gu*, and a small decrease of *Ur*. If the sequence of amino acids in the protein is determined by the sequence of bases in the nucleic acid, we should expect that TYV will have an increased amount of those amino acids which are determined by base configuration containing one or more molecules of Cy. On the other hand, amino acids determined exclusively by the other three bases must be decreased in their amount. Inspection of Table VIII would thus lead to a tentative conclusion that amino acids such as *His*, *Meth*, *Lys*, *Thr* and *Cys* are determined by base configuration containing one or more molecules of Cy, whereas *Arg*, *Asp*, *Pro*, and *Glu* are determined mostly by the other three bases.

TABLE VIII

Protein Composition of Tobacco Mosaic and Turnip Yellow Viruses (34, 51)
(Molar content normalized to unity)

Amino Acid	TMV	TYV	Amino Acid	TMV	TYV
Ala	0.070	0.070	Leu	0.087	0.086
Arg	0.069	0.015	Lys	0.012	0.080
Asp	0.120	0.041	Meth	0	0.020
Aspn			Phe	0.002	0.025
Cys	0.010	0.023	Pro	0.061	0.010
Glu	0.091	0.071	Ser	0.083	0.084
Glun			Thr	0.102	0.138
Gly	0.031	0.042	Try	0.012	0.006
His	0	0.015	Tyr	0.026	0.015
Ileu	0.061	0.089	Val	0.097	0.078

In order to subject this conclusion to a more quantitative test, the following procedure was followed. Assuming that each of the amino acids is determined by the combination of three bases corresponding to one of the *20* triangles given in Fig. 12, and, furthermore, that the order of bases in the RNA template does not strongly deviate from random, we would expect that relative abundances of various amino acids would be given by the products of relative abundances of three bases in the corresponding triplet. Using Table VII, the expected abundances of various base triplets were calculated (with statistical weight factors 1, 3, and 6 for triplets of aaa, aab, and abc types) both for TMV and TYV, and are plotted in decreasing order together with the observed abundances of amino acids taken from Table VIII. The resulting curves (Figs. 13 and 14) run very closely parallel to one another, suggesting that there may exist a correlation between amino acids and base triplets. We notice, however, that the data for TMV and TYV lead to different and contradictory assignments between the amino acids and the base triplets. One can attempt to remove this discrepancy by introducing some intermediary assignment-set which will satisfy *both* TMV and TYV data with somewhat less precision than is shown in Figs. 13 and 14. In changing the assignments of amino acids to base triplets, one should give priority to those with the greatest abun-

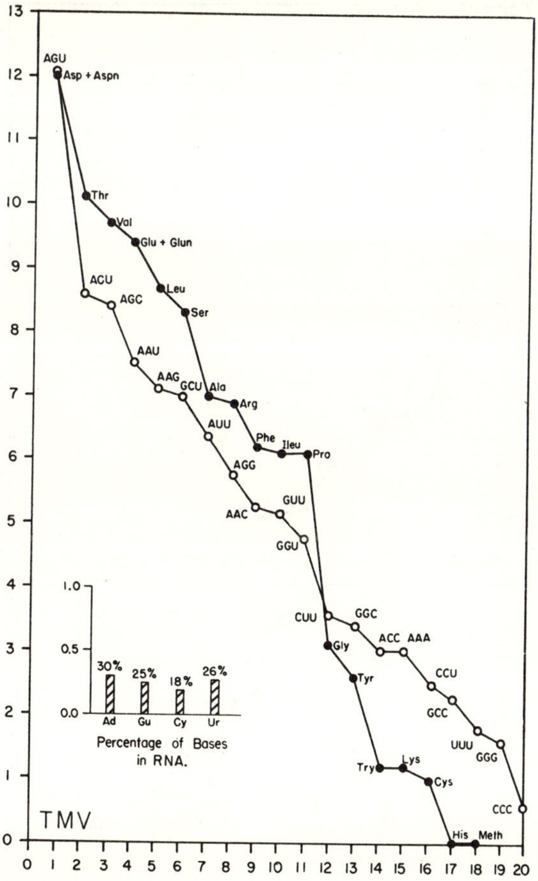

Fig. 13. Distribution of amino acids and base triads in Tobacco Mosaic Virus.

dances, since they are expected to have smaller relative errors. Following this principle, we arrive at an assignment set:

Ala-AAC	Gly-CUU	Pro-CCU
Asp-AGU	Ileu-ACC	Ser-GCU
Arg-AGG	Leu-AGC	Thr-ACU
Glu-AUU	Lys-GCC	Val-AAU
Glun-GGC	Phe-GUU	

where A, G, C, and U stand for Ad, Gu, Cy, and Ur. Using this set of assignments for *both* TMV and TYV, we find that the average deviation from the observed amino acids percentages is about 20 per cent, which is satisfactory, considering possible experimental errors in analyzing nucleic

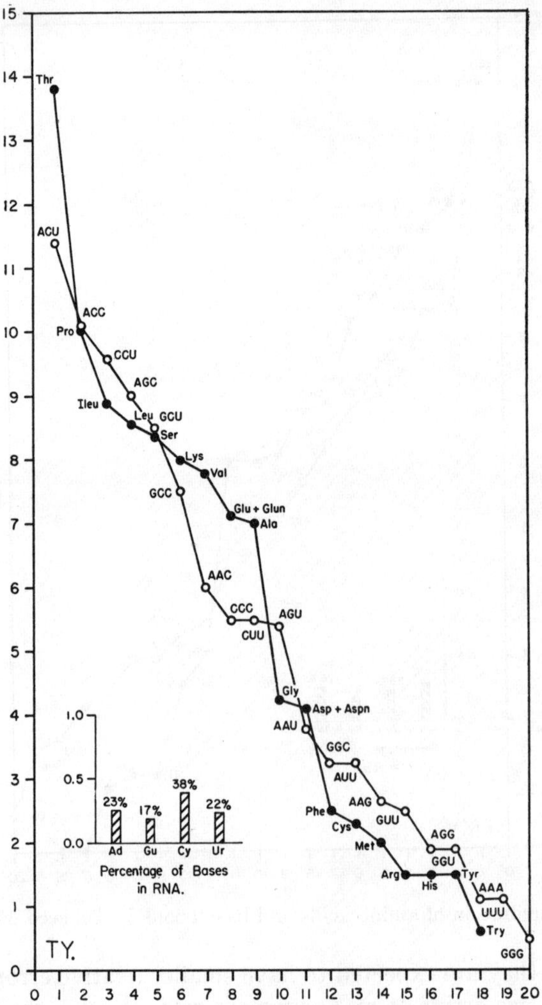

FIG. 14. Distribution of amino acids and base triads in Turnip Yellow Virus.

acid and protein constitution. But, of course, the correctness of the above given assignments, as well as the validity of the entire point of view on which it is based, can be ascertained only when protein and nucleic acid analyses are made of many other plant viruses.

2. Neighbor Saturation Curves

One of the possible approaches to the problem of decoding protein structure consists in a statistical analysis of the known sequences and of the results obtained by using various coding methods so far proposed. In fact, any intersymbol correlation between amino acids in polypeptide chains must be expected to influence, in some degree, the number of different neighbors which can be possessed by any given amino acid in the sequence; the stronger the restrictions governing the associations between the amino acids in the sequence, the smaller must be the number of different neighbors as compared with a random distribution.

Studies in this direction were undertaken by Gamow and Metropolis (24), the computations being carried out on the electronic computor "Maniac" in the Los Alamos Scientific Laboratory of the Atomic Energy Commission. The protein sequences used in this analysis were selected from the most reliable data available at that date, and included: insulin A and B (beef, sheep, and pig), oxytocin (beef), vasopressin (beef and pig), ACTH, and eleven known fragments of lysozyme. This group was selected because in it Glu and Glun had been differentiated chemically. Asp and Aspn were not differentiated, but since they occur with such small frequency it is not significant. For each of the twenty amino acids, the occurrence[7] in the selected sequences and the number of different neighbors on both sides were counted. One finds, for example, that Gly occurs 11.5 times and possesses 14 different neighbors, whereas Arg occurs only 6.5 times and has 8 different neighbors. The results of this analysis are shown in Fig. 15 by heavy circles labeled with the names of amino acids. (The limiting lines indicate that the number of different neighbors cannot be larger than twice the occurrence and also cannot exceed the total number of different amino acids.)

In order to find out whether or not the empirically determined numbers of different neighbors are smaller than would be expected from a purely random distribution of amino acids in polypeptide chains, the following statistical test was made. Assuming that all 20 amino acids occur in proteins with equal probability, 47 random sequences of corresponding length were formed, and in each sequence the occurrences and number of different neighbors were counted. The results are also shown in Fig. 15 by thin circles, the numbers standing within each circle indicating how many times (out of 47 runs) this particular number of neighbors was obtained. Inspection of the relative plots of theoretical and empirical

[7] In cases when an amino acid occurred at the end of a sequence or of a fragment, the occurrence was counted as being one half, since only one of the two possible neighbors could be observed.

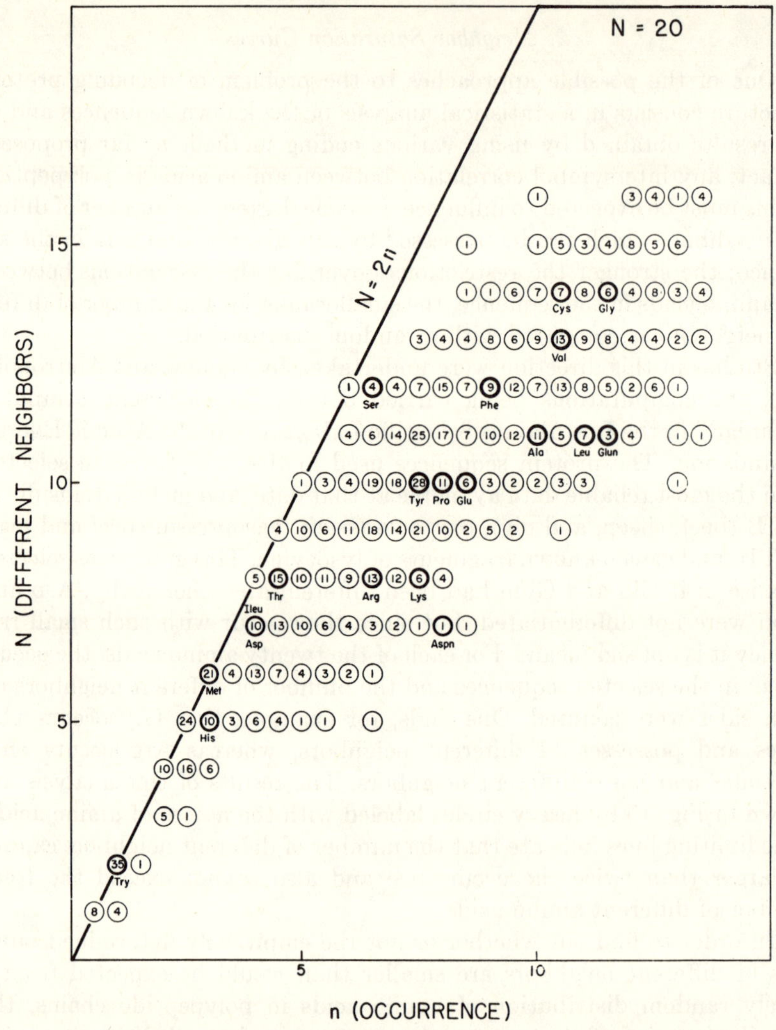

FIG. 15. Neighbor saturation plot. Heavy circles = experimental data. Light circles = data on random polypeptide sequences.

data seems to suggest that the distribution of amino acids in the known protein sequences is purely random.

To check this conclusion, a similar diagram was constructed (Fig. 16) for artificial amino acid sequences derived by means of the loose triangular code from a random sequence of four bases. (The data used in the construction of this diagram does not include ACTH.) In spite of rather strong intersymbol restrictions in the code, no noticeable difference between calculated and observed results can be seen. This seems to indicate that this method of counting the number of different neighbors is not sensitive enough to distinguish between random distributions and those subject to certain intersymbol restrictions. In the next section we will discuss a more sensitive method.

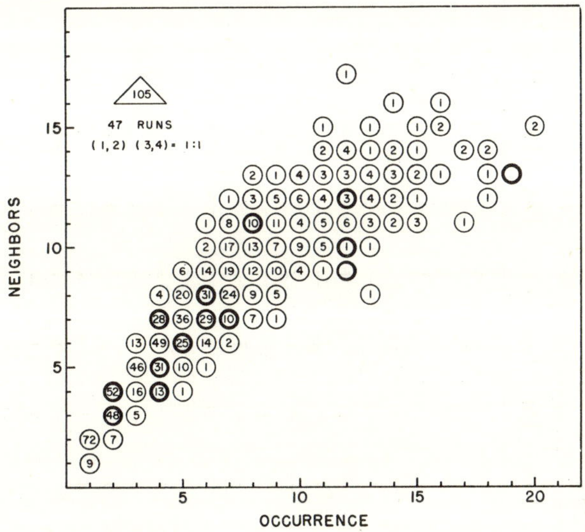

FIG. 16. Neighbor saturation plot. Heavy circles = experimental data. Light circles = data for triangular code.

3. *Neighbor Distribution Plots*

Another method for studying the neighbor distribution in polypeptide chains is suggested by Table II, which lists all the amino acid sequences of Table I. However, two kinds of proteins are listed in Table I, both structural and nonstructural proteins. The former, such as silk, gelatin (a collagen derivative), salmine, and wool, all have highly specialized amino acid contents. Thus, for example, gelatin has proline and hydroxyproline for one third of its residues, salmine has a very high arginine content, etc.

To overcome the bias which these specialized proteins introduced into Table II, in this section we will not consider them and will study only nonstructural proteins which have a more general composition. The excluded proteins are listed with the symbol † in Table I. Many dipeptides are included in Table I because one residue has been easily detected, as, for example, serine in the ovalbumin sequences. Again to obtain a more representative distribution, we will study only the nonstructural amino acid sequences which are tripeptides or longer.

TABLE IX

Neighbor Distribution Plot for Nonstructural Proteins

First Amino Acid	Second Amino Acid																			
	Gly	Ala	Glu	Glun	Arg	Asp	Cys	Ser	Leu	Val	Pro	Thr	Lys	Ileu	Phe	Tyr	His	Met	Try	Aspn
Gly	—	—	1	—	1	1	—	1	—	2	—	—	2	1	2	—	—	1	—	—
Ala	2	1	1	—	—	—	—	1	3	—	—	—	—	1	1	—	1	1	—	—
Glu	—	2	—	1	1	—	—	—	—	—	—	—	—	1	1	—	1	—	—	1
Glun	—	4	—	—	—	1	1	—	4	—	—	1	—	—	—	—	1	—	—	1
Arg	2	—	—	—	1	—	1	—	—	—	1	—	—	—	—	—	—	—	1	—
Asp	1	1	—	2	—	1	—	1	—	1	—	—	—	1	—	—	—	—	—	—
Cys	2	2	—	1	1	—	1	1	—	—	1	1	—	—	—	1	—	—	—	1
Ser	—	—	2	—	—	1	—	—	1	1	—	—	—	2	1	1	1	1	—	—
Leu	2	2	2	—	—	—	1	—	—	3	—	1	—	—	—	2	—	—	—	—
Val	2	—	2	2	—	2	3	—	—	—	—	—	1	—	1	1	—	—	—	1
Pro	1	1	1	—	1	—	—	—	2	2	—	—	2	—	—	—	—	—	—	—
Thr	—	2	—	1	—	1	—	3	—	—	2	—	—	—	—	—	—	—	—	—
Lys	1	1	—	1	1	—	1	—	—	2	1	—	1	—	—	—	—	—	—	—
Ileu	—	1	—	2	—	—	1	—	1	1	1	—	—	—	1	—	—	1	—	—
Phe	1	—	—	2	1	1	—	—	—	1	1	—	—	—	1	1	—	—	—	—
Tyr	—	—	—	1	—	—	1	1	1	—	1	1	—	1	1	—	—	—	—	—
His	—	—	—	—	—	—	—	1	1	—	—	—	1	1	1	—	—	—	—	—
Met	—	—	1	—	—	1	—	—	—	—	—	1	1	—	—	—	—	—	—	—
Try	1	—	—	—	—	—	—	—	—	—	—	—	—	—	—	—	—	—	—	—
Aspn	—	—	—	1	—	—	1	—	—	—	—	—	—	—	—	1	—	—	—	—

In Table IX, we have decomposed all the amino acid sequences into dipeptide units. A sequence such as:

Ala-Leu-Val-Try

would give rise to three units, Ala-Leu; Leu-Val, and Val-Try. In this way, the total data has given rise to 177 dipeptide nearest neighbors.[8] Table IX has 20 amino acid sites, for glutamine and asparagine have been listed as separate amino acids. (This may introduce a small error in some cases.) The entire grid has $20 \times 20 = 400$ spaces, and the average density of points is $\bar{n} = 177/400 = 0.442$.

The distribution of points in Table IX has been found to follow the Poisson distribution.[9] The Poisson distribution is random (as would be expected, for example, for the numbers of hailstones falling on the individual squares of a fine grid). It would arise if there were no strong preferences found between the amino acids so that, for example, the number of Ala-Glu points on the grid is not greater than would be expected to arise by chance from the sample studied.[10]

If $\bar{n}$ is the average density of points, and N is the total number of spaces available, the Poisson distribution is:

$$P(r) = N \frac{e^{-n}(\bar{n})^r}{r!}$$

where $P(r)$ is the expected number of spaces which contain r units. If we compare the observed density distribution with that predicted, we find the results as listed in Table X, columns A and E. The agreement between the observed and expected frequencies is measured quantitatively by using the χ^2 test.[11] The data in Table X, column A, show that $\chi^2 = 7.64$, which is less than that necessary for the 5% significance limit for 3 degrees of freedom.

We have constructed an additional table to look for preferred amino acid neighbors located one away. A sequence Ala-Leu-Val-Gly-Ser would

[8] The initial compilation did not include the cytochrome C and some of the lysozyme sequences which were unavailable when the list was compiled. As in Table II, only the different sequences of similar proteins (e.g. beef, pig, and sheep insulin) were listed.

[9] We are indebted to S. Brenner for pointing out the usefulness of the Poisson distribution in this context.

[10] We have recently obtained additional sequences for cytochrome C, lysozyme, and ribonuclease. When this information is added to Table IX, the results still follow the Poisson distribution.

[11] χ^2 is defined as $\sum \frac{(o_i - e_i)^2}{e_i}$ where i = number of observations, $o_i = i^{\text{th}}$ observation, $e_i = i^{\text{th}}$ expected value.

give rise to three units: Ala-Val, Leu-Gly, and Val-Ser. Using the same data as the above, we have obtained 143 units, so that $\bar{n} = 143/400 = 0.357$. This is compared with a Poisson distribution in Table XI, and again, it fits this distribution very closely.

TABLE X

Comparison of Observed and Predicted Distributions. Average Density $\bar{n}$ = 0.442

No. of units per grid box	*A* Nearest neighbor, observed distribution	*B* Loose triangle code, observed distribution	*C* Diamond code, observed distribution	*D* English language, observed distribution	*E* Poisson distribution calculated
0	264	276	301	305	264.2
1	103	86	54	55	116
2	27	26	28	23	25.6
3	4	9	11	7	3.77
4	2	3	3	3	0.42
5	0	0	1	3	0.037
6	0	0	0	2	0.0027
7	0	0	2	0	1.7×10^{-4}
8	0	0	0	2	9.5×10^{-6}
χ^2 (obs.)	7.64	22.7	2.4×10^4	4.2×10^5	
χ^2 (calc.) for 5% significance	7.82	7.82	11.1×10^4	12.6×10^5	

TABLE XI

Comparison of Observed and Predicted Distribution. Average Density $\bar{n}$ = 0.357

No. of units per grid box	Alternate neighbor, observed distribution	Poisson distribution, calculated
0	283	280
1	96	100
2	16	17.9
3	5	2.13

χ^2 (obs.) = 4.25, χ^2 (calc.) for 5% significance level = 5.99

If we were to make such a neighbor distribution table using 20 letters and choose the sequence of letters at random, we would find that they follow a Poisson distribution. In a "language" which introduces certain preferred neighbors, the resulting restrictions bring about deviations from the Poisson distribution. As an example of this, we determined the neighbor frequency in the English language, using the beginning of Milton's *Paradise Lost.* The letter grid was filled up to $\bar{n} = 0.442$, and the results are listed in Table X, column *D*. Even though repeated words were deleted from the tabulation, the distribution nonetheless deviates widely from Poisson. Sequences in English such as *th, be,* and *an* occur more frequently than in a random compilation.

Neighbor frequency distributions have been listed for several languages and are employed extensively in various cryptographic tests (47). In no written language, however, does the letter neighbor distribution follow that of Poisson.

These neighbor frequency plots can be used in studying any prospective nucleic acid-protein decoding system. If we choose a sequence of the four nucleic acid bases at random, they have maximal disorder. From this sequence, we can translate into an amino acid sequence, using a particular coding procedure, and obtain a synthetic polypeptide which has in it sequence relations which are brought in due to the translating method. We have constructed such synthetic polypeptide sequences using both the open triangle and diamond code schemes. Sequences have been taken which have 177 bits of sequential information, and these were subjected to the same neighbor frequency analysis as the experimentally observed sequences of Table I. The results are listed in Table X, columns *B* and *C*, together with the χ^2 figures. It can readily be seen that both of these decoding methods lead to distributions which deviate noticeably from the Poisson and consequently deviate from our experimental sample. Using this indirect statistical approach, we can therefore show some decoding incompatabilities which were not evident in the curves shown in Figs. 15 and 16.

Two questions must be considered here. First, is our sample of experimental data a reliable representation of amino acid sequences? Second, what is the significance of a Poisson distribution in amino acid sequential data?

It is, of course, apparent that the observed experimental data is largely accidental, influenced by such things as availability, molecular weight, stability, etc. The long sequences, ACTH, lysozyme, and insulin, account for about two-thirds of the data used. Whether these are representative of the large class of nonstructural proteins is unknown. We can only await the accumulation of further data to ascertain the representative character of our present information.

The occurrence of a Poisson distribution is disheartening relative to the possibility of discovering a decoding scheme. Most codes express themselves by restrictions and neighbor preferences, and, in the absence of the latter, the work is made much more difficult. If all the 400 blanks on the grid are eventually filled, it will mean that any amino acid can reside next to any other one. If there are no preferred sequences, it will imply that the preceding amino acid probably does not aid in determining its neighbor, and, furthermore, that a given nucleotide does not aid in determining the position of two adjoining amino acids. One can still have a decoding scheme in which two or three nucleic acid elements determine a given amino acid site, but in which these sites are in a sense independent of each other.

However, it may be that the real significance of the above lies in a consideration of how much "information" a given polypeptide sequence can convey. We use the term "information" in the sense of molecular specificity, that which distinguishes one protein from another. In their *Mathematical Theory of Communication* (54), Shannon and Weaver developed quantitatively the concept of the entropy of a given informational source or language. In systems in which certain sequences are preferred, such as *th*, or perhaps Ala-Leu, there is a loss of potential entropy. Thus, they have estimated that the relative entropy of the English language is 50%. To convey a given message, more symbols must be used when a language is employed in which the entropy is low, that is, in which certain sequential patterns are used very often. Those languages which are the most flexible and economical are those which have very few restrictions. In this case, it may be that Nature uses a polypeptide "language" in which the entropy is high, and the concomitant potential specificity for a given polypeptide chain is very great. This would, of course, not apply to the class of structural proteins, in which their restricted composition has of necessity reduced their informational entropy.

VI. Conclusions

This chapter summarizes recent efforts to find the formal mechanism by which hereditary chemical information presumably stored in long polynucleotide chains is transferred to corresponding polypeptide chains which form the cellular proteins. Although none of the proposed "information-transfer models" (codes) seem to be adequate to reproduce the observed amino acid sequences, a number of methods have been developed which will permit one to check the validity of any transfer mechanism proposed in the future.

In particular, the statistical studies of the last section suggest that the correct transfer-mechanism must be much less restrictive than the codes which have been proposed previously. In fact, the distribution of amino acids in the known sequences seems to be indistinguishable from a random one. This appears to imply that neighboring amino acids share no bases between them, and that the ratio of bases in the template to residues in the resulting protein is at least three.

These results stress the necessity for further experimental studies of the relative nucleic acid and protein compositions in simple organisms such as the viruses.

References

1. Allfrey, V. G. (1954). *Proc. Natl. Acad. Sci. (U.S.)* **40,** 881.
2. Avery, O. T., MacLeod, C. N., and McCarty, M. (1944). *J. Exptl. Med.* **79,** 137.
3. Beljansky, M. (1954). *Biochim. et Biophys. Acta* **15,** 425.
4. Bettelheim, F. R. (1954). *J. Am. Chem. Soc.* **76,** 2838.
5. Block, P. (1940). *J. Biol. Chem.* **135,** 51.
6. Borsook, H., Deasy, C. L., Haagen-Smit, A. J., Keighley, G., and Lowy, P. H. (1952). *J. Biol. Chem.* **196,** 669.
7. Brachet, J. (1947). "Embryologie Chimique." Masson, Paris.
8. Brachet, J., and Chantrenne, H. (1951). *Nature* **168,** 950.
9. Brown, D. M., and Todd, A. R. (1952). *J. Chem. Soc.* p. 52.
10. Brown, D. M., Fassman, G. D., Magrath, D. I., Todd, A. R., Cochran, W., and Woolfson, M. M. (1953). *Nature* **172,** 1184.
11. Caldwell, P. C., and Hinshelwood, C. N. (1950). *J. Chem. Soc.* p. 3156.
12. Caldwell, P. C., Mackor, E. L., and Hinshelwood, C. N. (1950). *J. Chem. Soc.* p. 3151.
13. Caspersson, T. (1950). "Cell Growth and Cell Function." Norton, New York.
14. Caspersson, T. (1947). *Symposia Soc. Exptl. Biol.* **1,** 127.
15. Chantrenne, H. (1952). Symposium sur la biogenese des proteins, *2nd Congr. Intern. Biochim.* Sedes, Paris.
16. Consden, R., Gordon, A. H., and Martin, A. J. P. (1949). *Biochem. J.* **44,** 548.
17. Dorner, R. W., and Knight, C. A. (1953). *J. Biol. Chem.* **205,** 959.
18. Dounce, A. L. (1952). *Enzymologia* **15,** 251.
19. Felix, K. (1952). *Experientia* **8,** 312.
20. Felix, K., Fisher, H., Hartlieb, J., Krekels, A., Ramen, H. M., and Stamm, W. (1952). *2nd Intern. Congr. Biochem.* **190,** 199.
21. Flavin, M. (1954). *J. Biol. Chem.* **210,** 771.
22. Gale, E. F., and Folkes, J. P. (1954). *Nature* **173,** 1223.
23. Gamow, G. (1954). *Nature* **173,** 318; (1954). *Dan. Biol. Medd.* **22,** 3.
24. Gamow, G., and Metropolis, N. (1954). *Science* **120,** 779.
25. Gaw, H. Z., and Stanley, W. M. (1947). *J. Biol. Chem.* **167,** 765.
26. Harrington, C. R., and Rivers, R. V. P. (1939). *Nature* **144,** 205.
27. Haurowitz, F. (1950). "Chemistry and Biology of Proteins," p. 344. Academic Press, New York.
28. Hershey, A. D., and Chase, M. (1952). *J. Gen. Physiol.* **36,** 39.
29. Hotchkiss, R. D. (1952). *Exptl. Cell. Research Suppl.* **2,** 383.
30. Hultin, T. (1950). *Exptl. Cell Research* **1,** 376.
31. Keilin, D. (1953). *Nature* **172,** 390.
32. Keilin, D., and Tissières, A. (1953). *Nature* **172,** 393.
33. Keilin, D., andRyley, J. F. (1953). *Nature* **172,** 451.
34. Knight, C. A. (1947). *J. Biol. Chem.* **171,** 297.
35. Ledley, R. S. (1955). *Proc. Nat. Acad. Sci. (U.S.).* **41,** 498.

36. For a discussion of the structure and function of Hemoglobin, see: Lemberg, R., and Legge, J. W. (1949). "Hematin Compounds and Bile Pigments." Interscience, New York; Florkin, M. (1949). "Biochemical Evolution." Academic Press, New York.
37. Lester, R. L. (1953). *J. Am. Chem. Soc.* **75**, 5448.
38. Ludwig, W., and von Mutzenbecker, P. (1936). *Z. physiol. Chem.* **244**, 4; (1939) **258**, 195.
39. Markham, R., and Smith, J. D. (1951). *Biochem. J.* **49**, 401.
40. Mirsky, A. E., and Ris, H. (1949). *Nature* **163**, 666.
41. Pardee, A. B. (1954). *Proc. Natl. Acad. Sci.* (*U.S.*) **40**, 263.
42. Pauling, L., and Corey, R. B. (1953). *Proc. Roy. Soc.* **141**, 21.
43. Perlman, R. R. (1952). *In* "Phosphorus Metabolism" (W. D. McElroy and B. Glass, eds.), Vol. 2. Johns Hopkins Press, Baltimore.
44. Popenoe, E. A., and Du Vigneaud, V. (1954). *J. Biol. Chem.* **206**, 353.
45. Porter, R. R., and Sanger, F. (1948). *Biochem. J.* **42**, 287.
46. Porter, R. R., and Sanger, F. (1949). *In* "Haemoglobin" (F. J. W. Roughton and J. C. Kendrew, eds.). Interscience, New York.
47. Pratt, F. (1939). "Secret and Urgent." Blue Ribbon Books, New York.
48. Rabinovitz, M. P., Olson, M. E., and Greenberg, D. M. (Sept. 1954), Abstracts of papers, American Chemical Society, p. 3C.
49. Rich, A., and Watson, J. D. (1954). *Nature* **173**, 995.
50. Rich, A., and Watson, J. D. (1954). *Proc. Natl. Acad. Sci.* (*U.S.*) **40**, 759.
51. Roberts, E., and Ramasarma, G. B. (1952). *Proc. Soc. Exptl. Biol. Med.* **80**, 101.
52. Sanger, F. (1952). *Advances in Protein Chem.* **7**, 1.
53. Schramm, G., and Kerekjarto, V. (1952). *Z. Naturforsch.* **76**, 589.
54. Shannon, C. E., and Weaver, W. (1949). "Mathematical Theory of Communication." Univ. of Illinois Press, Urbana.
55. Sinsheimer, R. L. (1954). *J. Biol. Chem.* **208**, 445.
56. Stalberg-Stenhagen, S., and Stenhagen, E. (1947). *Arkiv. Kemi Mineral. Geol.* **24B**, 1.
57. Stetten, M. R. (1949). *J. Biol. Chem.* **181**, 31.
58. Swenson, P. A. (1950). *Proc. Natl. Acad. Sci.* (*U.S.*) **36**, 699.
59. Thompson, E. O. P. (1954). *J. Biol. Chem.* **208**, 565.
60. Tristram, G. R. (1949). *In* "Haemoglobin" (F. J. W. Roughton and J. C. Kendrew, eds.). Interscience, New York.
61. Trommel, J., and Bijvoet, J. M. (1954). *Acta Cryst.* **7**, 703.
62. Vendreley, R., and Vendreley, C. (1949). *Experientia* **5**, 327.
63. Watson, J. D., and Crick, F. H. C. (1953). *Nature* **171**, 737.
64. Wilkins, M. H. F., Seeds, W. E., Stokes, A. R., and Wilson, H. R. (1953). *Nature* **172**, 759.
65. Windsor, E. (1951). *J. Biol. Chem.* **192**, 595.

GENERAL NATURE OF THE GENETIC CODE FOR PROTEINS

By Dr. F. H. C. CRICK, F.R.S., LESLIE BARNETT, Dr. S. BRENNER
and Dr. R. J. WATTS-TOBIN

Medical Research Council Unit for Molecular Biology,
Cavendish Laboratory, Cambridge

THERE is now a mass of indirect evidence which suggests that the amino-acid sequence along the polypeptide chain of a protein is determined by the sequence of the bases along some particular part of the nucleic acid of the genetic material. Since there are twenty common amino-acids found throughout Nature, but only four common bases, it has often been surmised that the sequence of the four bases is in some way a code for the sequence of the amino-acids. In this article we report genetic experiments which, together with the work of others, suggest that the genetic code is of the following general type:

(*a*) A group of three bases (or, less likely, a multiple of three bases) codes one amino-acid.

(*b*) The code is not of the overlapping type (see Fig. 1).

(*c*) The sequence of the bases is read from a fixed starting point. This determines how the long sequences of bases are to be correctly read off as triplets. There are no special 'commas' to show how to select the right triplets. If the starting point is displaced by one base, then the reading into triplets is displaced, and thus becomes incorrect.

(*d*) The code is probably 'degenerate'; that is, in general, one particular amino-acid can be coded by one of several triplets of bases.

The Reading of the Code

The evidence that the genetic code is not overlapping (see Fig. 1) does not come from our work, but from that of Wittmann[1] and of Tsugita and Fraenkel-Conrat[2] on the mutants of tobacco mosaic virus produced by nitrous acid. In an overlapping triplet code, an alteration to one base will in general change three adjacent amino-acids in the polypeptide chain. Their work on the alterations produced in the protein of the virus show that usually only one amino-acid at a time is changed as a result of treating the ribonucleic acid (RNA) of the virus with nitrous acid. In the rarer cases where two amino-acids are altered (owing presumably to two separate deaminations by the nitrous acid on one piece of RNA), the altered amino-acids are not in adjacent positions in the polypeptide chain.

Brenner[3] had previously shown that, if the code were universal (that is, the same throughout Nature), then all overlapping triplet codes were impossible. Moreover, all the abnormal human hæmoglobins studied in detail[4] show only single amino-acid changes. The newer experimental results essentially rule out all simple codes of the overlapping type.

If the code is not overlapping, then there must be some arrangement to show how to select the correct triplets (or quadruplets, or whatever it may be) along the continuous sequence of bases. One obvious suggestion is that, say, every fourth base is a 'comma'. Another idea is that certain triplets make 'sense', whereas others make 'nonsense', as in the comma-free codes of Crick, Griffith and Orgel[5]. Alternatively, the correct choice may be made by starting at a fixed point and working along the sequence of bases three (or four, or whatever) at a time. It is this possibility which we now favour.

Experimental Results

Our genetic experiments have been carried out on the *B* cistron of the r_{II} region of the bacteriophage *T*4, which attacks strains of *Escherichia coli*. This is the system so brilliantly exploited by Benzer[6,7]. The r_{II} region consists of two adjacent genes, or 'cistrons', called cistron *A* and cistron *B*. The wild-type phage will grow on both *E. coli B* (here called *B*) and on *E. coli K*12 (λ) (here called *K*), but a phage which has lost the function of either gene will not grow on *K*. Such a phage produces an *r* plaque on *B*. Many point mutations of the genes are known which behave in this way. Deletions of part of the region are also found. Other mutations, known as 'leaky', show partial function; that is, they will grow on *K* but their plaque-type on *B* is not truly wild. We report here our work on the mutant *P* 13 (now re-named *FC* 0) in the *B*1 segment of the *B* cistron. This mutant was originally produced by the action of proflavin[8].

We[9] have previously argued that acridines such as proflavin act as mutagens because they add or delete a base or bases. The most striking evidence in favour of this is that mutants produced by acridines are seldom 'leaky'; they are almost always completely lacking in the function of the gene. Since our note was published, experimental data from two sources have been added to our previous evidence: (1) we have examined a set of 126 r_{II} mutants made with acridine yellow; of these only 6 are leaky (typically about half the mutants made with base analogues are leaky); (2) Streisinger[10] has found that whereas mutants of the lysozyme of phage *T*4 produced by base-analogues are usually leaky, all lysozyme mutants produced by proflavin are negative, that is, the function is completely lacking.

If an acridine mutant is produced by, say, adding a base, it should revert to 'wild-type' by deleting a base. Our work on revertants of *FC* 0 shows that it usually

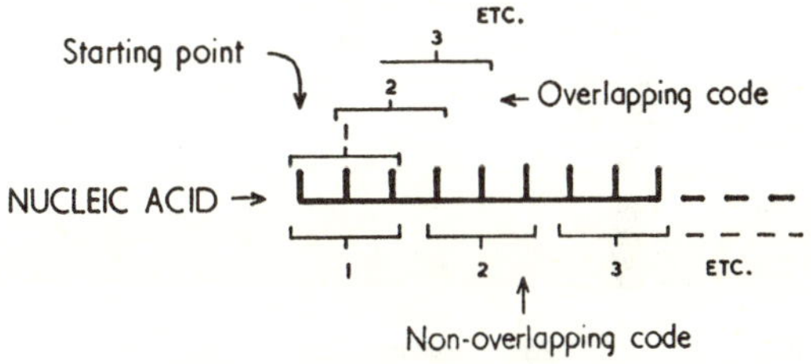

Fig. 1. To show the difference between an overlapping code and a non-overlapping code. The short vertical lines represent the bases of the nucleic acid. The case illustrated is for a triplet code

reverts not by reversing the original mutation but by producing a second mutation at a nearby point on the genetic map. That is, by a 'suppressor' in the same gene. In one case (or possibly two cases) it may have reverted back to true wild, but in at least 18 other cases the 'wild type' produced was really a double mutant with a 'wild' phenotype. Other workers[11] have found a similar phenomenon with r_{II} mutants, and Jinks[12] has made a detailed analysis of suppressors in the h_{III} gene.

The genetic map of these 18 suppressors of *FC* 0 is shown in Fig. 2, line *a*. It will be seen that they all fall in the *B*1 segment of the gene, though not all of them are very close to *FC* 0. They scatter over a region about, say, one-tenth the size of the *B* cistron. Not all are at different sites. We have found eight sites in all, but most of them fall into or near two close clusters of sites.

In all cases the suppressor was a non-leaky *r*. That is, it gave an *r* plaque on *B* and would not grow on *K*. This is the phenotype shown by a complete deletion of the gene, and shows that the function is lacking. The only possible exception was one case where the suppressor appeared to back-mutate so fast that we could not study it.

Each suppressor, as we have said, fails to grow on *K*. Reversion of each can therefore be studied by the same procedure used for *FC* 0. In a few cases these mutants apparently revert to the original wild-type, but usually they revert by forming a double mutant. Fig. 2, lines *b*–*g*, shows the mutants produced as suppressors of these suppressors. Again all these new suppressors are non-leaky *r* mutants, and all map within the *B*1 segment for one site in the *B*2 segment.

Once again we have repeated the process on two of the new suppressors, with the same general results, as shown in Fig. 2, lines *i* and *j*.

All these mutants, except the original *FC* 0, occurred spontaneously. We have, however, produced one set (as suppressors of *FC* 7) using acridine yellow as a mutagen. The spectrum of suppressors we get (see Fig. 2, line *h*) is crudely similar to the spontaneous spectrum, and all the mutants are non-leaky *r*'s. We have also tested a (small) selection of all our mutants and shown that their reversion-rates are increased by acridine yellow.

Thus in all we have about eighty independent *r* mutants, all suppressors of *FC* 0, or suppressors of suppressors, or suppressors of suppressors of suppressors. They all fall within a limited region of the gene and they are all non-leaky *r* mutants.

The double mutants (which contain a mutation plus its suppressor) which plate on *K* have a variety of plaque types on *B*. Some are indistinguishable from wild, some can be distinguished from wild with difficulty, while others are easily distinguishable and produce plaques rather like *r*.

We have checked in a few cases that the phenomenon is quite distinct from 'complementation', since the two mutants which separately are phenotypically *r*, and together are wild or pseudo-wild,

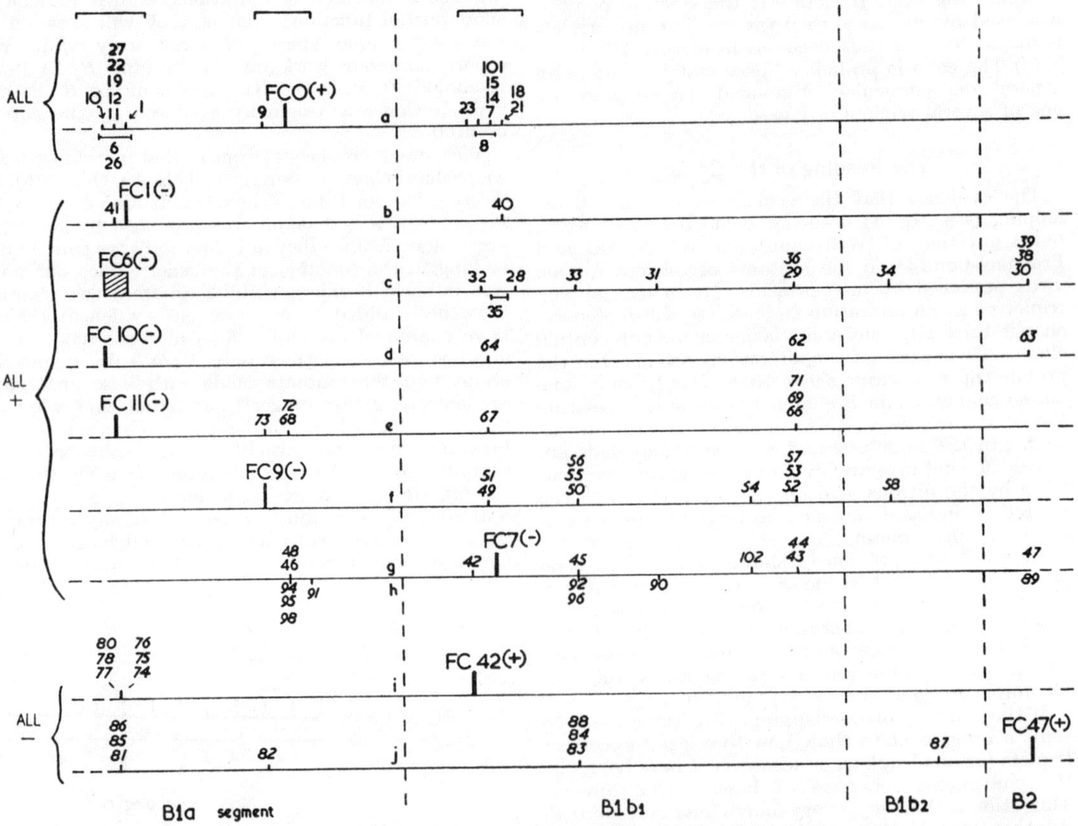

Fig. 2. A tentative map—only very roughly to scale—of the left-hand end of the *B* cistron, showing the position of the *FC* family of mutants. The order of sites within the regions covered by brackets (at the top of the figure) is not known. Mutants in italics have only been located approximately. Each line represents the suppressors picked up from one mutant, namely, that marked on the line in bold figures

must be put together in the same piece of genetic material. A simultaneous infection of K by the two mutants in separate viruses will not do.

The Explanation in Outline

Our explanation of all these facts is based on the theory set out at the beginning of this article. Although we have no direct evidence that the B cistron produces a polypeptide chain (probably through an RNA intermediate), in what follows we shall assume this to be so. To fix ideas, we imagine that the string of nucleotide bases is read, triplet by triplet, from a starting point on the left of the B cistron. We now suppose that, for example, the mutant *FC* 0 was produced by the insertion of an additional base in the wild-type sequence. Then this addition of a base at the *FC* 0 site will mean that the reading of all the triplets to the right of *FC* 0 will be shifted along one base, and will therefore be incorrect. Thus the amino-acid sequence of the protein which the B cistron is presumed to produce will be completely altered from that point onwards. This explains why the function of the gene is lacking. To simplify the explanation, we now postulate that a suppressor of *FC* 0 (for example, *FC* 1) is formed by deleting a base. Thus when the *FC* 1 mutation is present by itself, all triplets to the right of *FC* 1 will be read incorrectly and thus the function will be absent. However, when both mutations are present in the same piece of DNA, as in the pseudo-wild double mutant *FC* (0 + 1), then although the reading of triplets between *FC* 0 and *FC* 1 will be altered, the original reading will be restored to the rest of the gene. This could explain why such double mutants do not always have a true wild phenotype but are often pseudo-wild, since on our theory a small length of their amino-acid sequence is different from that of the wild-type.

For convenience we have designated our original mutant *FC* 0 by the symbol + (this choice is a pure convention at this stage) which we have so far considered as the addition of a single base. The suppressors of *FC* 0 have therefore been designated −. The suppressors of these suppressors have in the same way been labelled as +, and the suppressors of these last sets have again been labelled − (see Fig. 2).

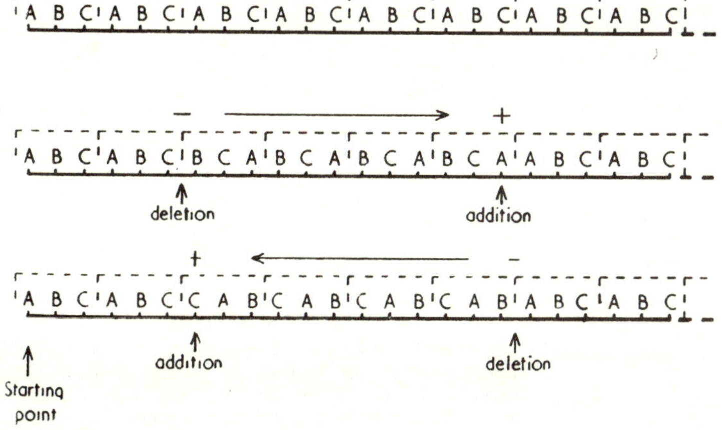

Fig. 3. To show that our convention for arrows is consistent. The letters A, B and C each represent a different base of the nucleic acid. For simplicity a repeating sequence of bases, ABC, is shown. (This would code for a polypeptide for which every amino-acid was the same.) A triplet code is assumed. The dotted lines represent the imaginary 'reading frame' implying that the sequence is read in sets of three starting on the left

Double Mutants

We can now ask: What is the character of any double mutant we like to form by putting together in the same gene any pair of mutants from our set of about eighty? Obviously, in some cases we already know the answer, since some combinations of a + with a − were formed in order to isolate the mutants. But, by definition, no pair consisting of one + with another + has been obtained in this way, and there are many combinations of + with − not so far tested.

Now our theory clearly predicts that all combinations of the type + with + (or − with −) should give an r phenotype and not plate on K. We have put together 14 such pairs of mutants in the cases listed in Table 1 and found this prediction confirmed.

Table 1. Double Mutants having the *r* Phenotype

− With −	+ With +	
FC (1 + 21)	*FC* (0 + 58)	*FC* (40 + 57)
FC (23 + 21)	*FC* (0 + 38)	*FC* (40 + 58)
FC (1 + 23)	*FC* (0 + 40)	*FC* (40 + 55)
FC (1 + 9)	*FC* (0 + 55)	*FC* (40 + 54)
	FC (0 + 54)	*FC* (40 + 38)

At first sight one would expect that all combinations of the type (+ with −) would be wild or pseudo-wild, but the situation is a little more intricate than that, and must be considered more closely. This springs from the obvious fact that if the code is made of triplets, any long sequence of bases can be read correctly in one way, but incorrectly (by starting at the wrong point) in two different ways, depending whether the 'reading frame' is shifted one place to the right or one place to the left.

If we symbolize a shift, by one place, of the reading frame in one direction by → and in the opposite direction by ←, then we can establish the convention that our + is always at the head of the arrow, and our − at the tail. This is illustrated in Fig. 3.

We must now ask: Why do our suppressors not extend over the whole of the gene? The simplest postulate to make is that the shift of the reading frame produces some triplets the reading of which is 'unacceptable'; for example, they may be 'nonsense', or stand for 'end the chain', or be unacceptable in some other way due to the complications of protein structure. This means that a suppressor of, say, *FC* 0 must be within a region such that no 'unacceptable' triplet is produced by the shift in the reading frame between *FC* 0 and its suppressor. But, clearly, since for any sequence there are *two* possible misreadings, we might expect that the 'unacceptable' triplets produced by a → shift would occur in different places on the map from those produced by a ← shift.

Examination of the spectra of suppressors (in each case putting in the arrows → or ←) suggests that while the → shift is acceptable anywhere within our region (though not outside it) the shift ←, starting from points near *FC* 0, is acceptable over only a more limited stretch. This is shown in Fig. 4. Somewhere in the left part of our region, between *FC* 0 or *FC* 9 and the *FC* 1 group, there must be one or more unacceptable triplets when a ← shift is made; similarly for the region to the right of the *FC* 21 cluster. Thus we predict that a combination of a + with a − will be wild or pseudo-wild if it involves a → shift, but that such pairs involving a ← shift will be phenotypically *r* if the arrow crosses one or more of the forbidden places, since then an unacceptable triplet will be produced.

Table 2. Double Mutants of the Type (+ with −)

− \ +	*FC* 41	*FC* 0	*FC* 40	*FC* 42	*FC* 58*	*FC* 63	*FC* 38
FC 1	*W*	*W*	*W*		W		W
FC 86		W	W	W	W	W	
FC 9	*r*	*W*	W	W	*W*		W
FC 82	*r*		W	W	W	W	
FC 21	*r*	*W*			W		W
FC 88	*r*	*r*			W	W	
FC 87	*r*	*r*	*r*	*r*			W

W, wild or pseudo-wild phenotype; *W*, wild or pseudo-wild combination used to isolate the suppressor; *r*, *r* phenotype.
* Double mutants formed with *FC* 58 (or with *FC* 34) give sharp plaques on *K*.

We have tested this prediction in the 28 cases shown in Table 2. We expected 19 of these to be wild, or pseudo-wild, and 9 of them to have the *r* phenotype. In all cases our prediction was correct. We regard this as a striking confirmation of our theory. It may be of interest that the theory was constructed before these particular experimental results were obtained.

Rigorous Statement of the Theory

So far we have spoken as if the evidence supported a triplet code, but this was simply for illustration. Exactly the same results would be obtained if the code operated with groups of, say, 5 bases. Moreover, our symbols + and − must not be taken to mean literally the addition or subtraction of a single base.

It is easy to see that our symbolism is more exactly as follows:

+ represents $+m$, modulo n
− represents $-m$, modulo n

where n (a positive integer) is the coding ratio (that is, the number of bases which code one amino-acid) and m is any integral number of bases, positive or negative.

It can also be seen that our choice of reading direction is arbitrary, and that the same results (to a first approximation) would be obtained in whichever direction the genetic material was read, that is, whether the starting point is on the right or the left of the gene, as conventionally drawn.

Triple Mutants and the Coding Ratio

The somewhat abstract description given above is necessary for generality, but fortunately we have convincing evidence that the coding ratio is in fact 3 or a multiple of 3.

This we have obtained by constructing triple mutants of the form (+ with + with +) or (− with − with −). One must be careful not to make shifts

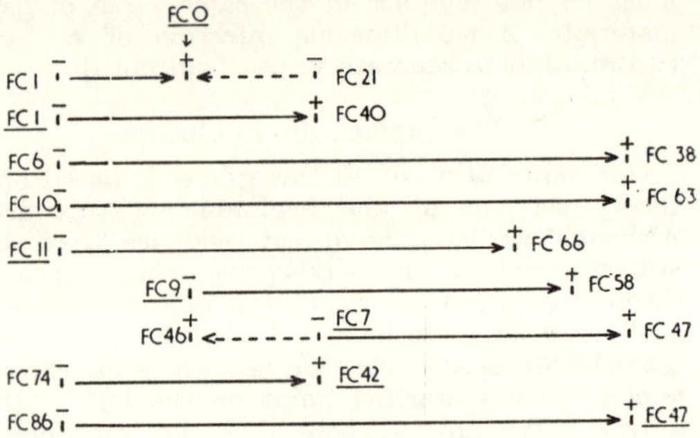

Fig. 4. A simplified version of the genetic map of Fig. 2. Each line corresponds to the suppressor from one mutant, here underlined. The arrows show the range over which suppressors have so far been found, the extreme mutants being named on the map. Arrows to the right are shown solid, arrows to the left dotted

across the 'unacceptable' regions for the ← shifts, but these we can avoid by a proper choice of mutants.

We have so far examined the six cases listed in Table 3 and in all cases the triples are wild or pseudo-wild.

Table 3. Triple Mutants having a Wild or Pseudo-wild Phenotype

FC (0 + 40 + 38)
FC (0 + 40 + 58)
FC (0 + 40 + 57)
FC (0 + 40 + 54)
FC (0 + 40 + 55)
FC (1 + 21 + 23)

The rather striking nature of this result can be seen by considering one of them, for example, the triple (*FC* 0 with *FC* 40 with *FC* 38). These three mutants are, by themselves, all of like type (+). We can say this not merely from the way in which they were obtained, but because each of them, when combined with our mutant *FC* 9 (—), gives the wild, or pseudo-wild phenotype. However, either singly or together in pairs they have an *r* phenotype, and will not grow on *K*. That is, the function of the gene is absent. Nevertheless, the combination of all three in the same gene partly restores the function and produces a pseudo-wild phage which grows on *K*.

This is exactly what one would expect, in favourable cases, if the coding ratio were 3 or a multiple of 3.

Our ability to find the coding ratio thus depends on the fact that, in at least one of our composite mutants which are 'wild', at least one amino-acid must have been added to or deleted from the polypeptide chain without disturbing the function of the gene-product too greatly.

This is a very fortunate situation. The fact that we can make these changes and can study so large a region probably comes about because this part of the protein is not essential for its function. That this is so has already been suggested by Champe and Benzer[13] in their work on complementation in the r_{II} region. By a special test (combined infection on *K*, followed by plating on *B*) it is possible to examine the function of the *A* cistron and the *B* cistron separately. A particular deletion, 1589 (see Fig. 5) covers the right-hand end of the *A* cistron and part of the left-hand end of the *B* cistron. Although 1589 abolishes the *A* function, they showed that it allows the *B* function to be expressed to a considerable extent. The region of the *B* cistron deleted by 1589 is that into which all our *FC* mutants fall.

Joining two Genes Together

We have used this deletion to re-inforce our idea that the sequence is read in groups from a fixed starting point. Normally, an alteration confined to the *A* cistron (be it a deletion, an acridine mutant, or any other mutant) does not prevent the expression of the *B* cistron. Conversely, no alteration within the *B* cistron prevents the function of the *A* cistron. This implies that there may be a region between the two cistrons which separates them and allows their functions to be expressed individually.

We argued that the deletion 1589 will have lost this separating region and that therefore the two (partly damaged) cistrons should have been joined together. Experiments show this to be the case, for now an alteration to the left-hand end of the *A* cistron, if combined with deletion 1589, can prevent the *B* function from appearing. This is shown in Fig. 5. Either the mutant *P*43 or *X*142 (both of which revert strongly with acridines) will prevent the *B* function when the two cistrons are joined, although both of these mutants are in the *A* cistron. This is also true of *X*142 *S*1, a suppressor of *X*142 (Fig. 5, case *b*). However, the double mutant (*X*142 with *X*142 *S*1), of the type (+ with −), which by itself is pseudo-wild, still has the *B* function when combined with 1589 (Fig. 5, case *c*). We have also tested in this way the 10 deletions listed by Benzer[7], which fall wholely to the left of 1589. Of these, three (386, 168 and 221) prevent the *B* function (Fig. 5, case *f*), whereas the other seven show it (Fig. 5, case *e*). We surmise that each of these seven has lost a number of bases which is a multiple of 3. There are theoretical reasons for expecting that deletions may not be random in length, but will more often have lost a number of bases equal to an integral multiple of the coding ratio.

It would not surprise us if it were eventually shown that deletion 1589 produces a protein which consists of part of the protein from the *A* cistron and part of that from the *B* cistron, joined together in the same polypeptide chain, and having to some extent the function of the undamaged *B* protein.

Is the Coding Ratio 3 or 6 ?

It remains to show that the coding ratio is probably 3, rather than a multiple of 3. Previous rather rough extimates[10,14] of the coding ratio (which are admittedly very unreliable) might suggest that the coding ratio is not far from 6. This would imply, on our theory, that the alteration in *FC* 0 was not to one base, but to two bases (or, more correctly, to an even number of bases).

We have some additional evidence which suggests that this is unlikely. First, in our set of 126 mutants produced by acridine yellow (referred to earlier) we have four independent mutants which fall at or

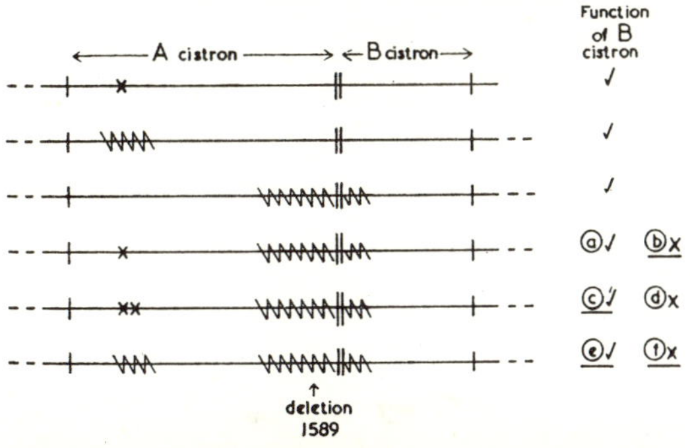

Fig. 5. Summary of the results with deletion 1589. The first two lines show that without 1589 a mutation or a deletion in the *A* cistron does not prevent the *B* cistron from functioning. Deletion 1589 (line 3) also allows the *B* cistron to function. The other cases, in some of which an alteration in the *A* cistron prevents the function of the *B* cistron (when 1589 is also present), are discussed in the text. They have been labelled (*a*), (*b*), etc., for convenience of reference, although cases (*a*) and (*d*) are not discussed in this paper. ✓ implies function; × implies no function

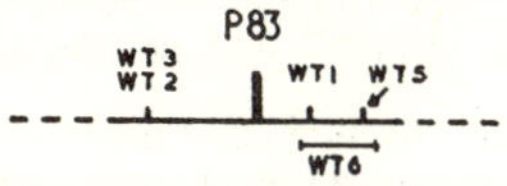

Fig. 6. Genetic map of *P* 83 and its suppressors, *WT* 1, etc. The region falls within segment *B*9*a* near the right-hand end of the *B* cistron. It is not yet known which way round the map is in relation to the other figures

close to the *FC* 9 site. By a suitable choice of partners, we have been able to show that two are + and two are −. Secondly, we have two mutants (*X*146 and *X*225), produced by hydrazine[15], which fall on or near the site *FC* 30. These we have been able to show are both of type −.

Thus unless both acridines and hydrazine usually delete (or add) an even number of bases, this evidence supports a coding ratio of 3. However, as the action of these mutagens is not understood in detail, we cannot be certain that the coding ratio is not 6, although 3 seems more likely.

We have preliminary results which show that other acridine mutants often revert by means of close suppressors, but it is too sketchy to report here. A tentative map of some suppressors of *P* 83, a mutant at the other end of the *B* cistron, in segment *B* 9*a*, is shown in Fig. 6. They occur within a shorter region than the suppressors of *FC* 0, covering a distance of about one-twentieth of the *B* cistron. The double mutant *WT* (2 + 5) has the *r* phenotype, as expected.

Is the Code Degenerate?

If the code is a triplet code, there are 64 (4 × 4 × 4) possible triplets. Our results suggest that it is unlikely that only 20 of these represent the 20 amino-acids and that the remaining 44 are nonsense. If this were the case, the region over which suppressors of the *FC* 0 family occur (perhaps a quarter of the *B* cistron) should be very much smaller than we observe, since a shift of frame should then, by chance, produce a nonsense reading at a much closer distance. This argument depends on the size of the protein which we have assumed the *B* cistron to produce. We do not know this, but the length of the cistron suggests that the protein may contain about 200 amino-acids. Thus the code is probably 'degenerate', that is, in general more than one triplet codes for each amino-acid. It is well known that if this were so, one could also account for the major dilemma of the coding problem, namely, that while the base composition of the DNA can be very different in different micro-organisms, the amino-acid composition of their proteins only changes by a moderate amount[16]. However, exactly how many triplets code amino-acids and how many have other functions we are unable to say.

Future Developments

Our theory leads to one very clear prediction. Suppose one could examine the amino-acid sequence of the 'pseudo-wild' protein produced by one of our double mutants of the (+ with −) type. Conventional theory suggests that since the gene is only altered in two places, only two amino-acids would be changed. Our theory, on the other hand, predicts that a string of amino-acids would be altered, covering the region of the polypeptide chain corresponding to the region on the gene between the two mutants. A good protein on which to test this hypothesis is the lysozyme of the phage, at present being studied chemically by Dreyer[17] and genetically by Streisinger[10].

At the recent Biochemical Congress at Moscow, the audience of Symposium I was startled by the announcement of Nirenberg that he and Matthaei[18] had produced polyphenylalanine (that is, a polypeptide all the residues of which are phenylalanine) by adding polyuridylic acid (that is, an RNA the bases of which are all uracil) to a cell-free system which can synthesize protein. This implies that a sequence of uracils codes for phenylalanine, and our work suggests that it is probably a triplet of uracils.

It is possible by various devices, either chemical or enzymatic, to synthesize polyribonucleotides with defined or partly defined sequences. If these, too, will produce specific polypeptides, the coding problem is wide open for experimental attack, and in fact many laboratories, including our own, are already working on the problem. If the coding ratio is indeed 3, as our results suggest, and if the code is the same throughout Nature, then the genetic code may well be solved within a year.

We thank Dr. Alice Orgel for certain mutants and for the use of data from her thesis, Dr. Leslie Orgel for many useful discussions, and Dr. Seymour Benzer for supplying us with certain deletions. We are particularly grateful to Prof. C. F. A. Pantin for allowing us to use a room in the Zoological Museum, Cambridge, in which the bulk of this work was done.

1 Wittman, H. G., Symp. 1, Fifth Intern. Cong. Biochem., 1961, for refs. (in the press).

2 Tsugita, A., and Fraenkel-Conrat, H., *Proc. U.S. Nat. Acad. Sci.*, **46**, 636 (1960); *J. Mol. Biol.* (in the press).

3 Brenner, S., *Proc. U.S. Nat. Acad. Sci.*, **43**, 687 (1957).

4 For refs. see Watson, H. C., and Kendrew, J. C., *Nature*, **190**, 670 (1961).

5 Crick, F. H. C., Griffith, J. S., and Orgel, L. E., *Proc. U.S. Nat. Acad. Sci.*, **43**, 416 (1957).

6 Benzer, S., *Proc. U.S. Nat. Acad. Sci.*, **45**, 1607 (1959), for refs. to earlier papers.

7 Benzer, S., *Proc. U.S. Nat. Acad. Sci.*, **47**, 403 (1961); see his Fig. 3.

8 Brenner, S., Benzer, S., and Barnett, L., *Nature*, **182**, 983 (1958).

9 Brenner, S., Barnett, L., Crick, F. H. C., and Orgel, A., *J. Mol. Biol.*, **3**, 121 (1961).

10 Streisinger, G. (personal communication and in the press).

11 Feynman, R. P.; Benzer, S.; Freese, E. (all personal communications).

12 Jinks, J. L., *Heredity*, **16**, 153, 241 (1961).

13 Champe, S., and Benzer, S. (personal communication and in preparation).

14 Jacob, F., and Wollman, E. L., *Sexuality and the Genetics of Bacteria* (Academic Press, New York, 1961). Levinthal, C. (personal communication).

15 Orgel, A., and Brenner, S. (in preparation).

16 Sueoka, N. *Cold Spring Harb. Symp. Quant. Biol.* (in the press).

17 Dreyer, W. J., Symp. 1, Fifth Intern. Cong. Biochem., 1961 (in the press).

18 Nirenberg, M. W., and Matthaei, J. H., *Proc. U.S. Nat. Acad. Sci.*, **47**, 1588 (1961).

Reprinted from NATURE, Vol. 180, 1957

GENE MUTATIONS IN HUMAN HÆMOGLOBIN: THE CHEMICAL DIFFERENCE BETWEEN NORMAL AND SICKLE CELL HÆMOGLOBIN

By Dr. V. M. INGRAM

Medical Research Council Unit for the Study of the Molecular Structure of Biological Systems, Cavendish Laboratory, University of Cambridge

I REPORTED recently[1] that the globins of normal and sickle cell anæmia human hæmoglobins differed only in a small portion of their polypeptide chains. I have now found that out of nearly 300 amino-acids in the two proteins, only one is different; one of the glutamic acid residues of normal hæmoglobin is replaced by a valine residue in sickle cell anæmia hæmoglobin. The latter is an abnormal protein which is inherited in a strictly Mendelian manner; it is now possible to show, for the first time, the effect of a single gene mutation as a change in one amino-acid of the hæmoglobin polypeptide chain for the manufacture of which that gene is responsible.

In previous experiments[1], tryptic digests of the two proteins had been prepared; the resulting mixtures of small peptides were separated on a sheet of paper, using electrophoresis in one direction and partition chromatography in the other. These paper chromatograms derived from the two proteins, which I had called 'finger-prints', showed all peptides to have identical electrophoretic and chromatographic properties, except for one spot, peptide No. 4. This occupied different positions in the 'finger-prints' of normal (Hb *A*) and sickle cell anæmia (Hb *S*) hæmoglobins, indicating that the difference between the two proteins was located there. I have now determined the chemical constitution of these No. 4 peptides derived from both hæmoglobin *A* and *S*.

The hæmoglobin *A* No. 4 peptide was prepared by elution from the neutral fraction of many 'finger-prints', followed by cooled paper electrophoresis in pyridine/acetic acid/water (*p*H 3·6 [2]) on Whatman No. 3 *MM* paper at 30 V./cm. for 75 min. The peptide was obtained as a well-separated band and eluted. The hæmoglobin *S* No. 4 peptide could be produced in a fairly pure state by eluting the slowest positively charged band from an extended one-dimensional paper electrophoresis of the peptide mixture in the *p*H 6·4 buffer[1]. In both cases qualitative amino-acid analysis by paper chromatography showed the presence of histidine, valine, leucine, threonine, proline, glutamic acid and lysine. There was more glutamic acid in the hæmoglobin *A* peptide, but more valine in the hæmoglobin *S* peptide. In view of the known specificity of trypsin[3], it was to be expected that these peptides, obtained by tryptic hydrolysis, had lysine at the C-terminal end. This agreed with all the results from the partial acid hydrolysis studies, as reported below.

Partial hydrolysis in 12 *N* hydrochloric acid at 37° C. for two or three days, followed by 'finger-printing'[1], gave the peptides indicated in Fig. 1, and

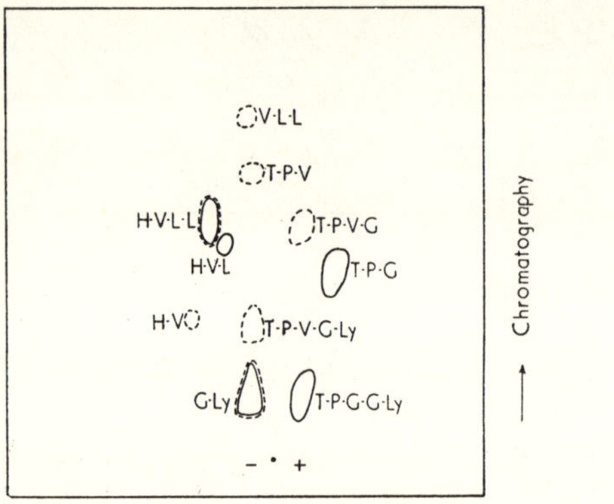

Fig. 1. Acid degradation and structure of the No. 4 peptides from hæmoglobins *A* and *S*. Hæmoglobin *A* (full lines): His^{++}—Val—Leu—Leu—Thr—Pro—*Glu*$^{-}$—Glu^{-}—Lys^{+-}. Hæmoglobin *S* (broken lines): His^{+-}—Val—Leu—Leu—Thr—Pro—*Val*—Glu^{-}—Lys^{+-}

also free amino-acids, which are omitted from the figure. The N-terminal amino-acids of most of these peptides were determined by the fluoro-2,4-dinitrobenzene method[4]. Together with the amino-acid compositions, these fragments indicated the sequences of the No. 4 peptides of hæmoglobins *A* and *S* shown in Fig. 1. The only ambiguity was the amino-acid following threonine. Here the relevant products from the hydrochloric acid splitting—threonyl–prolyl–glutamyl–glutamyl–lysine and threonyl–prolyl–valyl–glutamyl–lysine—were subjected, on paper strips, to a stepwise Edman degradation for two cycles[4,5]. The results indicated the sequence threonyl–prolyl– in both cases. The charge distribution of the two No. 4 peptides shown in Fig. 1 was deduced from the electrophoretic behaviour of the two peptides, especially in relation to the behaviour of the smaller split peptides. The only difference found between the two No. 4 peptides is that the first glutamic acid residue of the hæmoglobin *A* peptide is replaced by valine in the hæmoglobin *S* peptide.

It is known from X-ray crystallographic[6] and from chemical[7] studies that the human hæmoglobin molecule of molecular weight 66,700 is composed of two identical half-molecules, each approximately 33,000. It is believed that this substitution, which occurs in each of the two identical half-molecules, constitutes the only chemical difference between normal and sickle cell anæmia hæmoglobins. Certainly the hæm groups of the two proteins are the same[8]. The fact that in each half-molecule a glutamic acid is replaced by the neutral amino-acid valine agrees with previous findings that the whole hæmoglobin *S* molecule has two to three carboxyl groups fewer than the normal protein[9,10]. All the other peptides of the tryptic digest occupy identical, and characteristic, positions in the two 'finger-prints'. Qualitative amino-acid analyses of these peptides have now been carried out, but have failed to reveal any differences between them. It would seem probable, therefore, that they have identical structures, leaving the two No. 4 peptides as the only ones that differ.

About 30 per cent of the hæmoglobin molecule is not susceptible to attack by trypsin and does not appear on the 'finger-print'. To eliminate the possibility that an additional difference resides in these large hæmoglobin *A* and *S* fragments, they were digested with chymotrypsin, which attacks them readily. Again two peptide mixtures were obtained, which were examined both by 'finger-printing' and by careful paper chromatography of the neutral peptides. No differences between them could be detected.

We owe to Pauling and his collaborators[9] the realization that sickle cell anæmia is an example of an inherited 'molecular disease' and that it is due to an alteration in the structure of a large protein molecule, an alteration leading to a protein which is by all criteria still a hæmoglobin. It is now clear that, per half-molecule of hæmoglobin, this change consists in a replacement of only one of nearly 300 amino-acids, namely, glutamic acid, by another, valine—a very small change indeed.

Differences between closely related proteins, involving only a very small number of amino-acids, are known; the clearest examples are the differences between horse, whale, sheep, pig and cattle insulins, which show changes in only one sequence of three amino-acids[11]. However, since these are inter-species differences, the genetic mechanism underlying them is by no means clear and cytoplasmic inheritance has not yet been ruled out. The abnormal human hæmoglobins, on the other hand, are a group of very closely related proteins within the same species. It is certain that the inheritance of these proteins is Mendelian in character and occurs through the chromosomal genes. Neel[12] has shown that a single mutational step of such a gene, the one responsible for making hæmoglobin, produces the new abnormal sickle cell anæmia hæmoglobin. Previous investigations on the normal and the sickle cell anæmia protein could not decide whether the difference between them is due to a difference in folding of identical polypeptide chains or to a difference in the amino-acid sequences of the two chains. While there may also be changes in folding, it has now been definitely established that the amino-acid sequences of the two proteins differ, and differ at only one point. Thus it can be seen that an alteration in a Mendelian gene causes an alteration in the amino-acid sequence of the corresponding polypeptide chain. In the case of sickle cell anæmia hæmoglobin, this is the smallest alteration possible—only one amino-acid is affected—reflecting, presumably, a change in a very small portion of the hæmoglobin gene. It is not known, but it may well be that this involves a replacement of no more than a single base-pair in the chain of the deoxyribonucleic acid of the gene.

It is well known that mutations lead to very small chemical differences between, for example, flower pigments[13]. It seems likely that these mutations produce first a change in a protein, in this case probably an enzyme, which in turn causes the production of a changed flower pigment. These enzymes, which have not yet been investigated for differences, stand in closer relationship to the gene than do the flower pigments themselves. The protein hæmoglobin is just as close. It has therefore been called the first gene product, and is probably the first protein to be made by the gene.

The divisibility of genes in a virus was shown previously in bacteriophage by Benzer[14] and Streisinger[15], who studied the effects of many different mutations of a gene on the behaviour of the virus. Such sub-units in genes have also been shown in *Aspergillus*[16] and *Neurospora*[17]. The results presented in this communication are certainly what one would expect on the basis of the widely accepted hypothesis of gene action ; the sequence of base-pairs along the chain of nucleic acid provides the information which determines the sequence of amino-acids in the polypeptide chain for which the particular gene, or length of nucleic acid, is responsible. A substitution in the nucleic acids leads to a substitution in the polypeptide.

The abnormally low solubility of reduced hæmoglobin *S*, which causes the sickling of the erythrocytes in the anæmia, is presumably a function of the charge distribution on the surface of the molecule. The replacement of two charged glutamic acid residues for two uncharged valines is presumably enough to alter this distribution towards one favouring abnormally easy crystallization.

It is hoped that similar studies of other abnormal human hæmoglobins[18] will provide further insight into the effects of gene mutations.

Full details of this work will be published shortly elsewhere. I am indebted to Drs. S. Brenner and G. Seaman, Cambridge, for supplying blood from patients with homozygous sickle cell anæmia. It is a pleasure to acknowledge the constant interest and encouragement shown by Drs. M. F. Perutz and F. H. C. Crick. Some of the enzymatic digestion and 'finger-prints' were done by Mr. J. A. Hunt ; Miss Rita Prior rendered invaluable assistance.

1 Ingram, V. M., *Nature*, **178**, 792 (1956).
2 Michl, H., *Monatsh. Chem.*, **82**, 489 (1951).
3 Sanger, F., and Tuppy, H., *Biochem. J.*, **49**, 463, 481 (1951). Sanger, F., and Thompson, E. O. P., *Biochem. J.*, **53**, 353, 356 (1953).
4 Fraenkel-Conrat, H., Harris, J. I., and Levy, A. L., "Methods of Biochemical Analysis" **2**, 359 (1955).
5 Edman, P., *Acta Chem. Scand.*, **4**, 277, 283 (1950).
6 Perutz, M. F., Liquori, A. M., and Eirich, F., *Nature*, **167**, 929 (1951).
7 Schroeder, W. A., Rhinesmith, H. S., and Pauling, L. (in the press).
8 Havinga E., and Itano, H. A., *Proc. U.S. Nat. Acad. Sci.*, **39**, 65 (1953).
9 Pauling, L., Itano, H. A., Singer, S. J., and Wells, I. C., *Science*, **110**, 543 (1949).
10 Scheinberg, I. H., Harris, R. S., and Spitzer, J. L., *Proc. U.S. Nat. Acad. Sci.*, **40**, 777 (1954).
11 Harris, J. I., Sanger, F., and Naughton, M. A., *Arch. Biochem. Biophys.*, **65**, 427 (1956).
12 Neel, J. V., *Science*, **110**, 64 (1949).
13 Haldane, J. B. S., "Biochemistry of Genetics" (Allen and Unwin, London, 1954).
14 Benzer, S., in "The Chemical Basis of Heredity", edit. by McElroy, W. D., and Glass, B., 70 (Johns Hopkins Press, Baltimore, 1957).
15 Streisinger, G., and Franklin, N. C., *Cold Spring Harbor Symp. Quant. Biol.*, **21**, 103 (1956).
16 Pritchard, R. H., *Heredity*, **9**, 343 (1955).
17 Giles, N. H., Partridge, C. W. H., and Nelson, N. J., *Proc. U.S. Nat. Acad. Sci.*, **43**, 305 (1957).
18 Itano, H. A., *Ann. Rev. Biochem.*, **25**, 331 (1956).

Reprinted from PROCEEDINGS OF THE NATIONAL ACADEMY OF SCIENCES, Vol. 51, 1964

ON THE COLINEARITY OF GENE STRUCTURE AND PROTEIN STRUCTURE*

BY C. YANOFSKY, B. C. CARLTON,† J. R. GUEST,‡ D. R. HELINSKI,† AND U. HENNING†

DEPARTMENT OF BIOLOGICAL SCIENCES, STANFORD UNIVERSITY

Communicated by Victor Twitty, December 18, 1963

The pioneering studies of Beadle and Tatum with *Neurospora crassa*[1] led to the concept that there is a 1:1 relationship between gene and enzyme. Subsequent studies on the structure of proteins[2] and genetic material[3] permitted a restatement of this relationship in molecular terms;[4] the linear sequence of nucleotides in a gene specifies the linear sequence of amino acids in a protein.

Several years ago studies were initiated with the A gene–A protein system of the tryptophan synthetase of *Escherichia coli* with the intention of examining this concept of a colinear relationship between gene structure and protein structure. A large number of mutant strains which produced altered A proteins were isolated, and genetic and protein primary structure studies were performed with these strains to locate the positions of the alterations within the A gene and the A protein.[5–10] It was hoped that with information of this type it would be possible to determine whether a genetic map and the primary structure of the corresponding protein were colinear. Recently, Kaiser[11] has demonstrated the correspondence of the genetic map with the sequence of blocks of nucleotides in DNA. Thus if a colinear relationship could be established between a genetic map and the primary structure of a protein, it would be reasonable to conclude that this relationship extends to the nucleotide sequence corresponding to the genetic map.

In previous reports on studies with the tryptophan synthetase A protein, conclusive evidence was presented for the colinearity of a segment of the A gene and a segment of the A protein.[12, 13] The present communication deals with more extensive data with 16 mutants with mutational alterations in one segment of the A gene and the A protein.

Materials and Methods.—Mutant strains: Of the A-protein mutants examined in detail in this paper, strains A23, A27, A28, A36, A46, A58, A78, A90, A94, A95, A169, A178, and A187 were isolated following ultraviolet irradiation of the K-12 wild-type strain of *E. coli*, and strain A223 was isolated following treatment of the wild-type strain with ethylmethanesulfonate. Mutant A446 (previously designated PR8)[14] and mutant A487 were initially isolated as spontaneous second-site reversions and subsequently were separated from the original A mutants with which they had been associated. Strains $anth_1^-$ and $anth_2^-$ are blocked prior to anthranilic acid in the tryptophan pathway and respond to anthranilic acid, indole, or tryptophan. V_1^R and V_1^R *tryp*⁻ deletion mutants were isolated by treatment of T1-sensitive populations of the various mutants with phage T1h⁺. All of the V_1^R mutations mentioned are very closely linked to the A gene, and the V_1^R *tryp*⁻ deletions include the V_1^R locus and some segment of the A gene.[15] A stock of unrestricted T1 phage (uT1)[16] was kindly supplied by J. R. Christensen.

Protein studies: The altered A proteins were isolated and examined for primary structure changes as described previously.[6, 7, 9] The ordering of the tryptic peptides mentioned in the paper will be described in detail elsewhere.[17, 18]

Genetic studies: Recombination experiments were performed with the temperate-transducing phage P1kc.[19] Recombination distances between A mutants were obtained by determining the frequency of appearance of $tryp^+$ transductants. Transduction from $his^- \rightarrow his^+$ was scored in each experiment for internal reference, and the ratio of $tryp^+/his^+$ transductants calculated.[20]

Each value was halved to correct for the difference in relative frequency of transduction in the *his* and *tryp* regions.[20] In transduction experiments with leaky mutants (A169, A223, A446, and A487) the plating medium was supplemented with 0.1 μg/ml DL-5-methyltryptophan to suppress growth of the leaky mutants. This supplement has little or no effect on the growth of wild-type recombinants and does not appear to affect the recombination values obtained with nonleaky mutants. In spite of the presence of 5-methyltryptophan, it was frequently difficult to score recombination in experiments with leaky mutants. Anthranilic acid requirement was scored either by picking and streaking or by replication to appropriate test media. Resistance to phage T1 (V_1^R) was scored by picking, streaking, and spot testing with phage uT1. As shown by Drexler and Christensen,[16] P1 lysogeny does not prevent the multiplication of uT1.

Results and Discussion.—Relative order of mutational alterations in the A gene: The genetic map based on recombination frequencies, deletion mapping, and three-point crosses is shown in Figure 1.

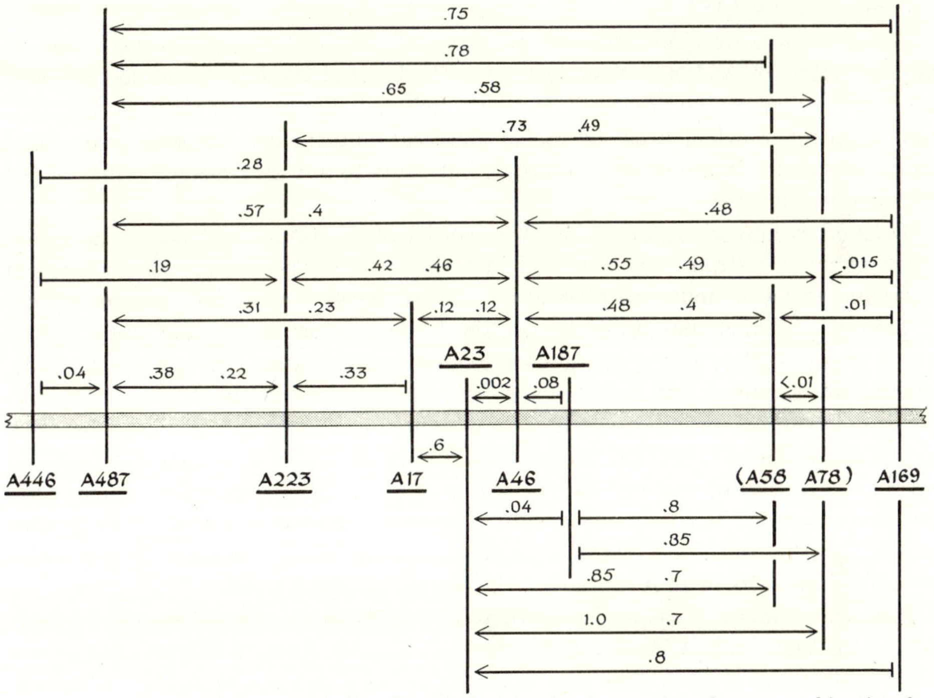

Fig. 1.—The order of mutationally altered sites in the A gene based on recombination frequencies, deletion mapping, and three-point genetic tests. The head of each arrow points to the recipient in each transduction cross. If two values are given, the cross was performed in both directions. In such cases each value is placed near the recipient in the cross.

Recombination frequency data: Recombination frequency data alone establishes close linkage of three groups of mutants: (1) A446 and A487; (2) A46, A23, and A187; and (3) A58, A78, and A169. The recombination values obtained also suggest that the order of these groups is as indicated in the figure, but they do not permit ordering within any of the groups.[21] In most cases recombination experiments were performed in both directions, i.e., each strain served as donor and recipient, and fair agreement was observed between the two values. However, it is evident from the data in Figure 1 that there were some exceptions. The exceptions generally involved leaky mutants, and it is possible that the scoring difficulties encountered with these strains were responsible.

Two of the mutants examined, A23 and A187, gave recombination values considerably higher than those obtained with mutant A46. Other mutants which were independent isolates resembling A23 (A27, A28, A36) and A46 (A95, A178) exhibited the same recombination behavior as the strain they resembled. Extensive mapping experiments with other A mutants and the strains mentioned above indicate that the A23 and A187 recombination values are probably exceptionally high rather than the A46 values exceptionally low. Since the mutational alterations in strains of the A23 and A46 type were probably single nucleotide changes,[22] it would appear that differences of single nucleotide pairs between donor DNA and recipient DNA can influence the frequency of recombinational events. An alternative explanation, that all A23 and A187 double mutants are prototrophic, seems very unlikely on the basis of numerous tests with recombinants from crosses involving A23 and A187.

Recombinants were not obtained in crosses of A23 × A27, A28, or A36; A46 × A95 or A178; or A58 × A90 or A94. On the basis of these tests and the primary structure studies to be described, it was concluded that each of these groups consists of members which arose by repeat identical mutations at the same site. Mutants A58 and A78 are clearly different, but they do not give recombinants at the 0.01 per cent level.

Deletion mapping: Information on the order of the mutational alterations in the relevant mutant strains was obtained in transduction experiments with a series of V_1^R-*tryp*⁻ deletion mutants (Table 1). These latter strains had regions of the A gene deleted, including, in each case, the V_1 locus which is situated at one end of the A gene.[15] Thus all the deletions extended into the A gene from the same side, and are overlapping, but may have different end points in the A gene. The recovery of tryptophan-independent recombinants from a transduction cross between any A mutant and a deletion mutant would indicate that the mutationally altered site in the A mutant was outside the region of the A gene that was missing in the deletion mutant. Recombination values were determined in transduction crosses with some of the deletion mutants to approximate the relative distance from a mutationally altered site in an A mutant to the end point of a deletion.

It is clear from the crosses with T^-70 and T^-689 that the altered sites in mutants A23, A46, A187, A58, A78, and A169—but not those in A446, A487, and A223—are within the region of the A gene that is missing in these deletion mutants. The rela-

TABLE 1

RECOMBINATION TESTS WITH VARIOUS TRYP⁻ DELETION MUTANTS

	his⁻V_1^R *tryp*⁻ Deletion Mutant Recipient					
Donor	T^-201	T^-5	T^-70	T^-689	T^-211	T^-226
A446	0	0	0.08‡	0.08‡		
A487	0	0	0.19*	0.5		
A223	0	0	0.17	0.06	+	+
A23	0	0	0	0	1.3	+
A46	0	0	0	0	0.86	0.77
A187	0	0	0		1.3	+
A58	0	0	0	0	0.09	0.1
A78	0	0	0		0.1	0.1
A169			0		0.1	0.1
A38	+	+	+	+	+	+

+, 0 = recombinants or no recombinants, respectively, in qualitative transduction experiments.
* = each recombination value represents the uncorrected observed ratio of *tryp*⁺ to *his*⁺ transductants.
‡ A446 gives unusually low recombination values in all experiments.

tive order of the altered sites in mutants A446, A487, and A223 could not be established by quantitative transduction experiments with the same deletion mutants. The quantitative transduction experiments with deletion mutants $T^{-}211$ and $T^{-}226$ divide the other mutants into two groups: (A46, A23, and A187) and (A58, A78, and A169). The combined results presented suggest the following order of altered sites: A38......(A446, A487, A223)......(A46, A23, A187)......(A58, A78, A169)......V_1.

Three-point genetic tests: Genetic tests employing outside markers were performed to determine relative order within each group of closely linked mutants. These tests are summarized in Table 2. The results obtained support the conclusions concerning group order that were arrived at by the previous methods and indicate relative orders within each closely linked group. Crosses 1–7 establish the order anth......A34......A446......A487......A223......A46, crosses 8 and 9 the order anth ......A46......A78......A169, and crosses 10 and 11 the order anth......(A23, A46) ......A187. The order anth......A23-A46 had tentatively been assigned on the basis of other data.[22] Crosses 12–15 confirm orders established by other methods. The combined genetic analyses with the mutants examined suggest the sequence of mutational alterations shown in Figure 1—A446-A487-A223-A23-A46-A187-(A58, A78)-A169.

TABLE 2

OUTSIDE-MARKER ORDERING OF MUTATIONALLY ALTERED SITES

Transduction cross	Nonselective markers	Recombinants detected*		*anth*$^+$ (%)	Order
(1) 34 → *anth*$_2^-$-223	*anth*$_2$	72 *anth*$^+$;	332 *anth*$^-$	18	*anth*-34-223
(2) 46 → *anth*$_2^-$-223	*anth*$_2$	180 *anth*$^+$;	196 *anth*$^-$	48	*anth*-223-46
(3) 446 → *anth*$_2^-$-223	*anth*$_2$	14 *anth*$^+$;	112 *anth*$^-$	11	*anth*-446-223
(4) 487 → *anth*$_2^-$-223	*anth*$_2$	29 *anth*$^+$;	118 *anth*$^-$	20	*anth*-487-223
(5) 34 → *anth*$_2^-$-487	*anth*$_2$	86 *anth*$^+$;	368 *anth*$^-$	19	*anth*-34-487
(6) 46 → *anth*$_2^-$-487	*anth*$_2$	27 *anth*$^+$;	28 *anth*$^-$	49	*anth*-487-46
(7) 446 → *anth*$_2^-$-487	*anth*$_2$	7 *anth*$^+$;	31 *anth*$^-$	23	*anth*-446-487
(8) 46 → *anth*$_2^-$-78	*anth*$_2$	52 *anth*$^+$;	220 anth$^-$	19	*anth*-46-78
(9) 169 → *anth*$_2^-$-78	*anth*$_2$	22 *anth*$^+$;	24 *anth*$^-$	48	*anth*-78-169
(10) 187 → *anth*$_1^-$-23	*anth*$_1$	9 *anth*$^-$;	42 *anth*$^+$	83	*anth*-23-187
(11) 187 → *anth*$_1^-$-46	*anth*$_1$	14 *anth*$^-$;	116 *anth*$^+$	89	*anth*-46-187
(12) 46 V_1^R→ *anth*$_1^-$-58 V_1^S	*anth*$_1$; V_1	22 *anth*$^-$ V_1^R;	1 *anth*$^+$ V_1^R	4	*anth*-46-58-V_1
(13) 58 V_1^R→ *anth*$_1^-$-46 V_1^S	*anth*$_1$; V_1	10 *anth*$^+$ V_1^S;	1 *anth*$^+$ V_1^R;	73	*anth*-46-58-V_1
			4 *anth*$^-$ V_1^S		
(14) 169 V_1^R→ *anth*$_1^-$-46 V_1^S	*anth*$_1$; V_1	6 *anth*$^+$ V_1^S;	1 *anth*$^-$ V_1^S	90	*anth*-46-169-V_1
(15) 446 → *anth*$_1^-$-46	*anth*$_1$	81 *anth*$^-$;	22 *anth*$^+$	21	*anth*-446-46
		20 *anth*$^-$-446-46;	65 *anth*$^+$ 446$^-$-46	76	

Order of markers: *anth*-A34 .... (A446, etc.) --- V_1^R.

* In crosses with the outside marker *anth*$_2$ the percentage of *anth*$^+$ recombinants is approximately 20% if the order is $\frac{+\quad x}{anth^-\quad y}$ and approximately 50% if the order of x and y is reversed. With the marker *anth*$_1$ different values are obtained but the order of x and y relative to *anth* can be clearly established. The explanation for the different values obtained with the two outside markers is not known.

Primary Structure Studies.—Amino acid substitutions have been detected in primary structure studies with each of the 16 A mutants.[6, 7, 14, 17, 18] The substitutions observed and the peptides in which they are present are shown in Figure 2. The conclusions from the genetics studies are also included in the figure (i.e., order of alterations and approximation of recombination distances between alterations). It is apparent that mutants with alterations extremely close to one another in the A gene have amino acid substitutions close to one another in the A protein. Primary structure studies with the A protein[17, 18] have established the linear sequence

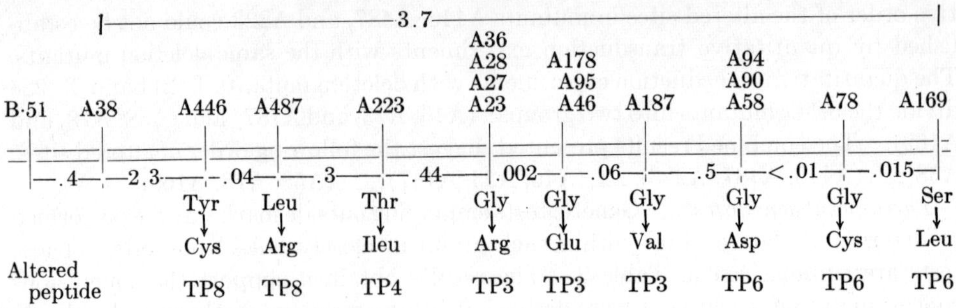

FIG. 2.—Amino acid substitutions in the A proteins of various mutants.[6, 7, 14, 17, 18] The A58, A78, A90, A94, and A169 substitutions will be described in detail elsewhere.[18]

of peptides TP-11, TP-8, TP-4, TP-18, TP-3, and TP-6, constituting a 75-residue segment of the A protein. This sequence and the sequence of the amino acids of most of these peptides are presented in Figure 3. This segment accounts for approximately one fourth of the residues in the A protein and is not at the amino or car-

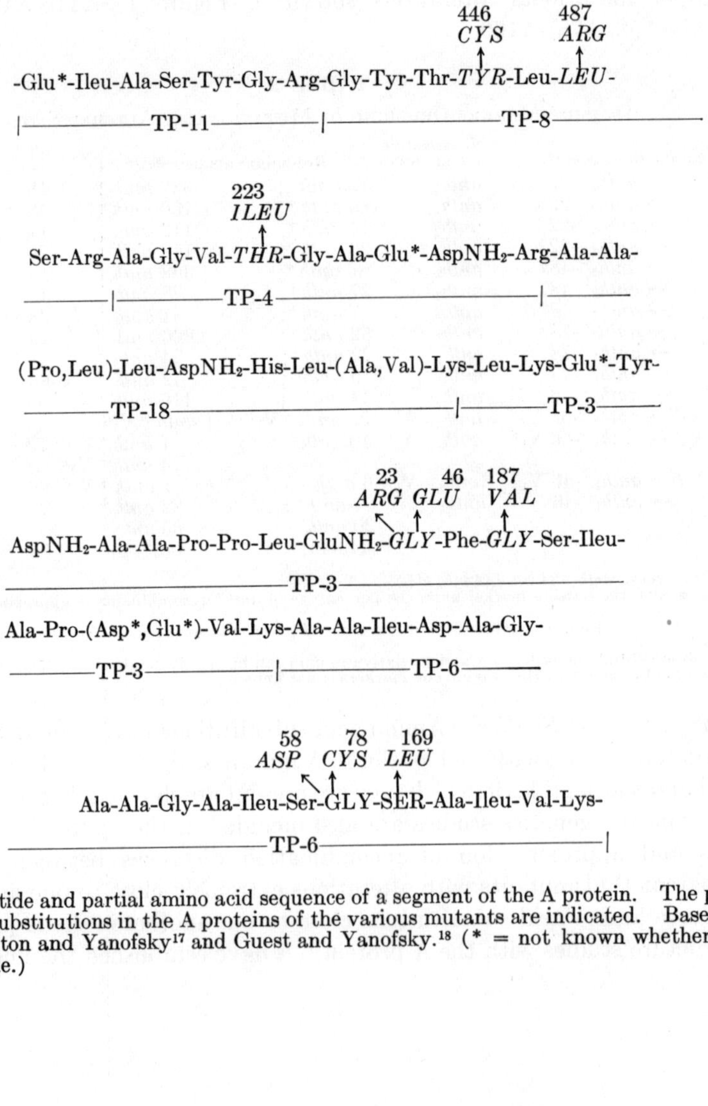

FIG. 3.—Peptide and partial amino acid sequence of a segment of the A protein. The positions of amino acid substitutions in the A proteins of the various mutants are indicated. Based on the studies by Carlton and Yanofsky[17] and Guest and Yanofsky.[18] (* = not known whether present as acid or amide.)

boxyl end of the protein. The positions of the amino acid replacements in the A proteins of the various mutants are also shown. It is clear that the positions of the amino acid replacements in the segment of the A protein are in the same relative order as the order of the mutationally altered sites of the corresponding mutants in the A gene. These findings convincingly demonstrate a colinear relationship between gene structure and protein structure.

The relationship between the map and residue distances observed is also of interest and can be seen most clearly from the representative values summarized in Table 3. Although the map distance/residue distance ratio varies between 0.01 and 0.05, in most cases this value is approximately 0.02. It would appear, therefore, that distances on the genetic map are representative of distances between amino acid residues in the corresponding protein.

TABLE 3

RELATIONSHIP BETWEEN MAP DISTANCES AND RESIDUE DISTANCES

Mutant pair	Map distance	Amino acid residue distance	Map distance / Residue distance
A58-A78	<0.01	0	—
A46-A23	0.002	0	—
A58-A169	0.01	1	0.01
A78-A169	0.015	1	0.015
A46-A187	0.08	2	0.04
A23-A187	0.04	2	0.02
A446-A487	0.04	2	0.02
A487-A223	0.3	6	0.05
A446-A223	0.19	8	0.02
A46-A58	0.44	23	0.02
A46-A78	0.52	23	0.02
A23-A58	0.78	23	0.03
A23-A78	0.85	23	0.04
A46-A169	0.48	24	0.02
A23-A169	0.8	24	0.03
A223-A46	0.44	28	0.02
A487-A46	0.48	34	0.01
A446-A46	0.28	36	0 01
A78-A223	0.61	51	0.01
A169-A487	0.75	58	0.01

Summary.—The concept of colinearity of gene structure and protein structure was examined with 16 mutants with alterations in one segment of the A gene and the A protein of tryptophan synthetase. The results obtained demonstrate a linear correspondence between the two structures and further show that genetic recombination values are representative of the distances between amino acid residues in the corresponding protein.

Note added in proof: Dr. Francis Crick has recently forwarded a manuscript which deals with a study of colinearity in another system (Sarabhai *et al.*, *Nature*, in press).

The authors are indebted to Virginia Horn for performing the genetic analyses described in this paper. They are also indebted to Deanna Thorpe, Patricia Schroeder, Donald Vinicor, and John Horan for their excellent technical assistance.

* This investigation was supported by grants from the National Science Foundation and the U.S. Public Health Service.

† Present address: B. C. Carlton, Department of Biology, Yale University, New Haven, Connecticut; D. R. Helinski, Department of Biology, Princeton University, Princeton, New Jersey; U. Henning, Max-Planck Institut für Zellchemie, Munich, Germany.

‡ Guinness Research Fellow, on leave from Department of Biochemistry, Oxford, England.

[1] Beadle, G. W., and E. L. Tatum, these PROCEEDINGS, **27,** 499 (1941).

[2] Sanger, F., and H. Tuppy, *Biochem. J.*, **49,** 481 (1951).

[3] Watson, J. D., and F. H. C. Crick, in *Viruses*, Cold Spring Harbor Symposia on Quantitative Biology, vol. 18 (1953), p. 123.

[4] Crick, F. H. C., *Symp. Soc. Exptl. Biol.*, **12,** 138 (1958).

[5] Yanofsky, C., D. R. Helinski, and B. D. Maling, in *Cellular Regulatory Mechanisms*, Cold Spring Harbor Symposia on Quantitative Biology, vol. 26 (1961), p. 11.

[6] Helinski, D. R., and C. Yanofsky, these PROCEEDINGS, **48,** 173 (1962).

[7] Henning, U., and C. Yanofsky, these PROCEEDINGS, **48,** 183 (1962).

[8] *Ibid.*, 1497 (1962).

[9] Helinski, D. R., and C. Yanofsky, *Biochim. Biophys. Acta*, **63,** 10 (1962).

[10] Carlton, B. C., and C. Yanofsky, *J. Biol. Chem.*, **238,** 2390 (1963).

[11] Kaiser, A. D., *J. Mol. Biol.*, **4,** 275 (1962).

[12] Yanofsky, C., in *Synthesis and Structure of Macromolecules*, Cold Spring Harbor Symposia on Quantitative Biology, vol. 28 (1963), in press.

[13] Yanofsky, C., in *The Bacteria*, ed. I. C. Gunsalus and R. Y. Stanier (Academic Press), vol. 5, in press.

[14] Helinski, D. R., and C. Yanofsky, *J. Biol. Chem.*, **238,** 1043 (1963).

[15] Somerville, R., and C. Yanofsky, unpublished observations.

[16] Drexler, H., and J. R. Christensen, *Virology*, **13,** 31 (1961).

[17] Carlton, B. C., and C. Yanofsky, in preparation.

[18] Guest, J., and C. Yanofsky, in preparation.

[19] Lennox, E. S., *Virology*, **1,** 190 (1955).

[20] Yanofsky, C., and E. S. Lennox, *Virology*, **8,** 425 (1959).

[21] On the basis of preliminary two-point mapping data with mutant A2, it earlier had tentatively been concluded that the alterations in mutants A58, A78, A90, and A94 were to the left of the alterations in A17 and A46.[5] The more extensive data presented in Figure 1 and the deletion and three-point data indicate that the correct order is as shown in Figure 1.

[22] Yanofsky, C., in *Synthesis and Structure of Macromolecules*, Cold Spring Harbor Symposia on Quantitative Biology, vol. 28 (1963), in press.

Reprinted from NATURE, Vol. 201, 1964

CO-LINEARITY OF THE GENE WITH THE POLYPEPTIDE CHAIN

By Dr. A. S. SARABHAI, Dr. A. O. W. STRETTON and Dr. S. BRENNER
Medical Research Council, Laboratory of Molecular Biology, Cambridge
AND
Dr. A. BOLLE
Institut de Biologie Moleculaire, Université de Genève

THE 'sequence hypothesis' states that the amino-acid sequence of a protein is specified by the nucleotide sequence of the gene determining that protein[1]. It has always been assumed that a simple congruence exists between these two sequences such that the order of the codons in the gene is the same as the order of the corresponding amino-acids in the polypeptide chain. There is no direct evidence for this co-linearity; but it is known that mutations close together in the gene affect the same amino-acid in the protein[2]. In this article we show that a class of suppressible mutations affecting the head protein of bacteriophage *T4D* produce fragments of the polypeptide chain. This property allows us to prove that the gene is co-linear with the polypeptide chain. Dr. C.

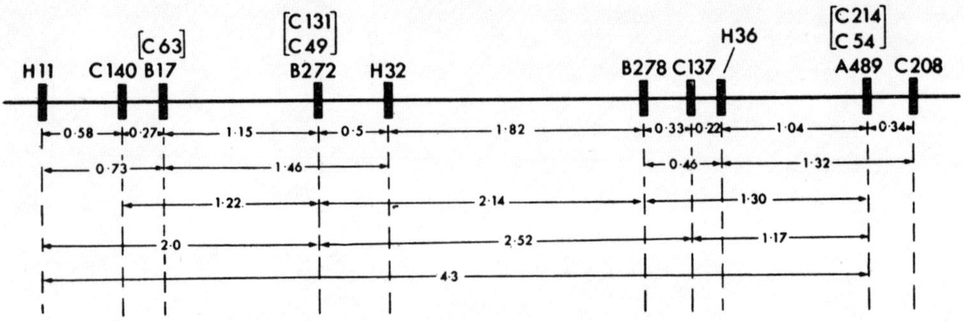

Fig. 1. Genetic map of *amber* mutants of the head protein gene. In many cases the recombination frequencies represent the average of several experiments and may differ slightly from those shown in Table 1

Yanofsky of Stanford University has informed us that he has shown co-linearity in an examination of the tryptophan synthetase of *E. coli*.

Suppressible Mutants

Benzer and Champe[3] have found that certain strains of *Escherichia coli* permit the growth of some phage mutants which do not grow on other strains. Bacteria with this permissive property are said to suppress the effect of the mutation. These suppressible mutants could be divided into different classes depending on the permissive host strain used. Benzer and Champe[4] later showed that one class of these mutants (subset 1), when grown in non-permissive bacteria, appeared to prevent the reading of the genetic message. They suggested that these mutants contained a nonsense codon and that permissive hosts could translate this codon into an amino-acid.

Mutations have been found in other genes in bacteria and bacteriophages which are suppressed by strains suppressing subset 1 mutants[5]. In particular, a large number of suppressible mutants of this class, called *amber* mutants, have been isolated in bacteriophage *T4D* (ref. 6). These are distributed throughout the genetic map of the phage, and it appears that any gene is susceptible to this type of mutation.

In order to characterize the molecular consequences of these suppressible mutations and their mechanism of suppression, we have examined *amber* mutants affecting the head protein of the bacteriophage. About 90 per cent of the protein of the phage particle is head protein[7], and Koch and Hershey[8] have shown that 60–70 per cent of the proteins synthesized late during infection are ultimately incorporated into mature phage particles. Thus more than one-half of the late protein synthesis of the infected

Table 1. Ordering of Mutants by 3-Factor Crosses

E. coli CR63, at a concentration of 2 × 10[8] cells/ml. in buffer, was infected with a mixture of the parental phages, each at a multiplicity of 5. After 8 min at 37° C to allow adsorption, the complexes were diluted 1 : 200 into tryptone broth and lysed with chloroform after 1 h at 37° C. Appropriate dilutions were plated by preadsorption on *E. coli CR63* to measure total progeny yield and on *E. coli B* to measure wild-type recombinants

Cross	Recombination (per cent)	Order deduced									
*B*17 × *H*11	0·80										
*B*17 × *H*11 + *B*278	0·14	*H*11		*B*17			*B*278				
*B*278 × *A*489	1·30										
*B*278 × *B*17 + *A*489	0·34			*B*17			*B*278			*A*489	
*B*272 × *B*278	2·20										
*B*272 × *H*11 + *B*278	0·32	*H*11			*B*272		*B*278				
*B*272 × *B*17	1·30										
*B*272 × *B*17 + *A*489	0·24			*B*17	*B*272					*A*489	
*C*140 × *B*17	0·27										
*C*140 × *H*11 + *B*17	0·04										
*C*140 × *B*17 + *B*272	0·20	*H*11	*C*140	*B*17	*B*272						
*H*32 × *B*272	0·5										
*H*32 × *B*17 + *B*272	0·36										
*H*32 × *B*272 + *B*278	0·05			*B*17	*B*272	*H*32	*B*278				
*C*137 × *B*278	0·33										
*C*137 × *B*272 + *B*278	0·28										
*C*137 × *B*278 + *A*489	0·06										
*H*36 × *B*278	0·46										
*H*36 × *B*272 + *B*278	0·40										
*H*36 × *B*278 + *A*489	0·06				*B*272		*B*278	(*C*137	*H*36)	*A*489	
*H*36 × *C*137	0·22										
*H*36 × *C*137 + *A*489	0·05										
*C*137 × *H*36 + *A*489	0·20							*C*137	*H*36	*A*489	
*C*208 × *A*489	0·34										
*C*208 × *B*17	4·0										
*C*208 × *B*17 + *A*489	0·33										
*C*208 × *B*278 + *A*489	0·29			*B*17			*B*278			*A*489	*C*208
Order:		*H*11	*C*140	*B*17	*B*272	*H*32	*B*278	*C*137	*H*36	*A*489	*C*208

Table 2. Distribution of Peptides in Amber-infected Cells

Segment	*H*11	*C*140	*B*17	*B*272	*H*32	*B*278	*C*137	*H*36	*A*489	*C*208	Wild	No. of peptides found: Chymo-tryptic	No. of peptides found: Tryptic	Example (see Fig. 2)
I	+	+	+	+	+	+	+	+	+	+	+	13	11	—
II	–	+	+	+	+	+	+	+	+	+	+	0	0	—
III	–	–	+	+	+	+	+	+	+	+	+	0	0	—
IV	–	–	–	+	+	+	+	+	+	+	+	8	3	Cys
V	–	–	–	–	+	+	+	+	+	+	+	0	1	His *T7c*
VI	–	–	–	–	–	+	+	+	+	+	+	10	2	Tyr *C12b*
VII	–	–	–	–	–	–	+	+	+	+	+	2	2	Try *T6*
VIII	–	–	–	–	–	–	–	+	+	+	+	5	1	Pro *T2a*
IX	–	–	–	–	–	–	–	–	+	+	+	8	11	Try *T2*
X	–	–	–	–	–	–	–	–	–	+	+	1	1	Tyr *C2*
XI	–	–	–	–	–	–	–	–	–	–	+	7	10	His *C6*

Note: The number of peptides found in a segment is not a measure of the length of the segment, since in some cases the same peptide sequence may be scored more than once, either because of being labelled by two different amino-acids, or because of partial enzymatic degradation products.

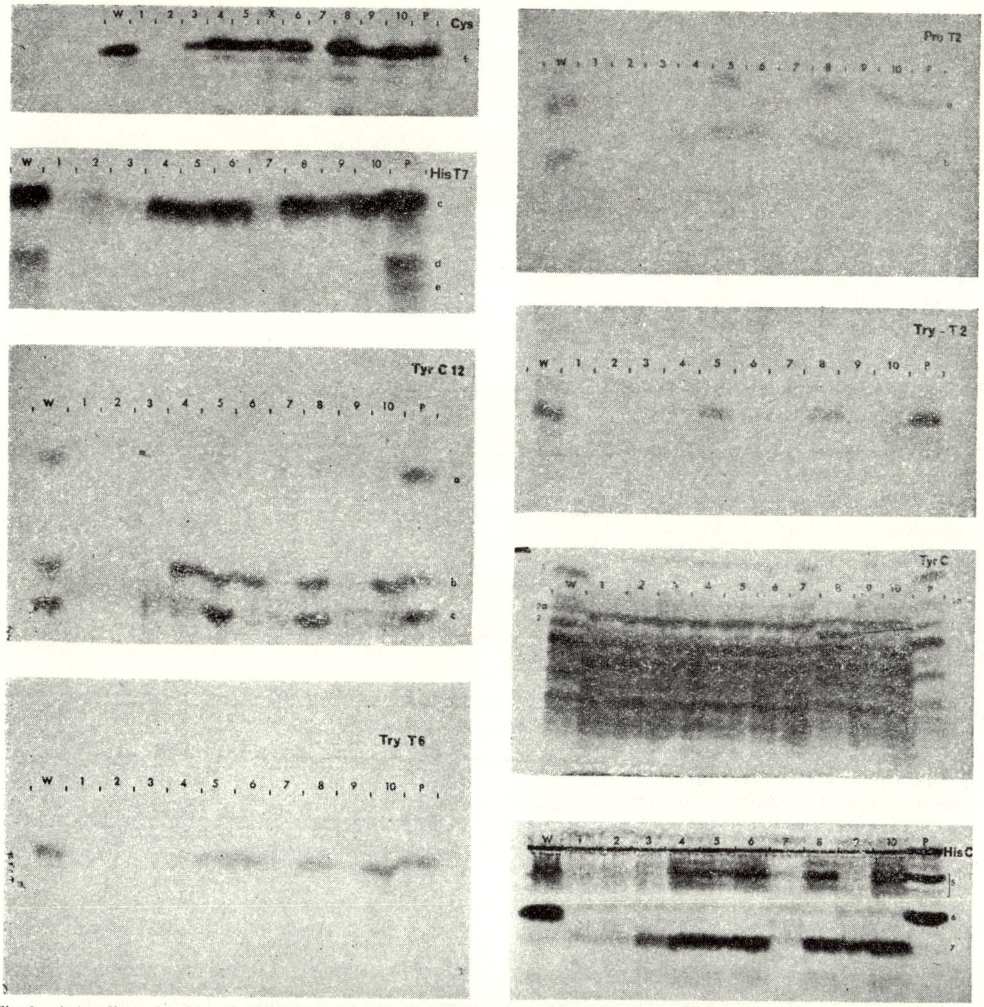

Fig. 2. Autoradiographs of tryptic (T) and chymotryptic (C) peptides, labelled with radioactive amino-acids. P is wild-type phage, W (and X in one case) is wild-type infected cells, 1–10 are *amber* mutant infected cells, as follows: 1 = *H*11, 2 = *B*17, 3 = *B*272, 4 = *B*278, 5 = *A*489, 6 = *C*137, 7 = *C*140, 8 = *C*208, 9 = *H*32, and 10 = *H*36.

10 ml. cultures of *E.* coli B (5×10^8 cells/ml.) in minimal medium were infected with phage (multiplicity = 5) and, 5 min later, superinfected to produce lysis inhibition. For labelling with sulphur-35 the minimal medium used contained no sulphur at the time of infection. 10 min after the initial infection, 5 μc. of carbon-14-labelled amino-acid, or 50 μc. of $^{35}SO_4^{--}$, was added. After 1–1·5 h protein synthesis was terminated by the addition of chloramphenicol (50 μg/ml.). The cells were removed by centrifugation, suspended in water, and heated at 100° C for 10 min. Lysozyme (50 μg/ml.), DNase (10 μg/ml.) and RNase (10 μg/ml.) were added, and, after digestion for 14 h at 37° C, the suspensions were dialysed against 2,500 volumes of distilled water for 18 h. Each preparation was made 1 per cent in ammonium bicarbonate and divided into two parts, one part was digested with chymotrypsin (10 μg/ml.), the other with trypsin (10 μg/ml.). After 14 h, digestion was stopped by the addition of acetic acid. For each radioactive amino-acid, all ten *amber* mutants and a wild-type were treated in parallel as described here. At the same time, labelled wild-type phage was prepared by allowing 20 ml. cultures of *T4D*-infected cells to assimilate radioactive amino-acid for 4 h. After lysis with chloroform, the phage was purified by differential centrifugation. Suspensions were heated, the released DNA digested with DNase, and the denatured ghosts removed by centrifugation. These were washed, suspended in 1 per cent ammonium bicarbonate, and proteolytic digestion and subsequent treatment carried out as described here.

The same amount of radioactivity of each digest was applied to Whatman 3 *MM* paper, and after high-voltage ionophoresis at *p*H 6·4 the peptide zones located by autoradiography. Selected zones were purified by ionophoresis at *p*H 3·6

cell is devoted to the manufacture of head protein; this has allowed us to detect fragments of the head polypeptide chain without any prior purification.

Isolation and Genetic Mapping of *Amber* Mutants

A previous examination of the *amber* mutants of bacteriophage *T4D* suggested that seven cistrons might be involved in the synthesis of the head protein[6]. We first proved that the cistron characterized by one of these (*am B*17) controlled the head protein. This was done by a special complementation test between *am B*17 and a known peptide difference[9] in the head protein.

Fourteen mutants, induced in phage *T4D* by base analogue mutagens (5-bromouracil, 2-aminopurine and hydroxylamine) were located in the head protein cistron by their failure to complement *am B*17 on the non-permissive strain (*E. coli B*). Some of these were shown to be recurrences by spot test crosses. In all, ten different sites were found, and one representative of each site was chosen. Double mutants were constructed by crossing two mutants and were tested by back-crossing progeny to the two parents. In most cases, the double mutants are less well suppressed by the permissive strain (*E. coli CR*63) than are the parents.

Standard crosses were carried out, and the frequency of wild-type recombinants was measured by plating on *E. coli B*. This was multiplied by 2 to give the recombination frequency. The results are summarized in

Fig. 1 in the form of a genetic map of the region, showing relevant 2-factor recombination frequencies. The additivity of recombination frequencies is good, but we have not relied on this to establish the order of the mutants. This has been done by 3-factor crosses (Table 1), and although such crosses are complicated by high negative interference[10] they uniquely determine the genetic order shown in Fig. 1.

Head Protein Peptides in *Amber* Mutants

The proteins being synthesized in the phage-infected cells were labelled with radioactive amino-acids. The cells were then lysed, the nucleic acid removed and without further purification the mixture was digested with either trypsin or chymotrypsin. The peptides thus produced were characterized by high-voltage ionophoresis on paper, their position being located by autoradiography. Labelled proteins from purified wild-type phage were also examined in the same way. The technical details are given in the legend to Fig. 2.

The peptides characteristic of the head protein of the phage can easily be recognized in digests of cells infected with the wild-type phage. This shows that the contamination introduced by other proteins synthesized in the infected cells is not serious. We have only scored peptides present in digests of both wild-type phage and the wild-type infected cells. The first general result is that while some of the peptides of the head protein are present in cells infected with a given *amber* mutant, others are absent, and the pattern differs for each *amber* mutant. Hence peptides can be classified as to whether or not they are included in the polypeptide synthesized by an *amber* mutant.

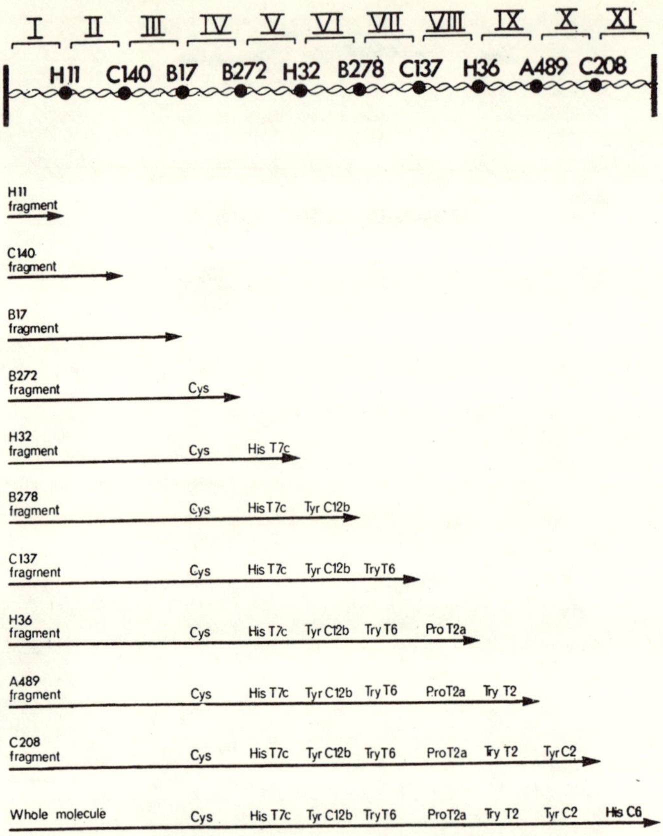

Fig. 3. Diagram showing hierarchical ordering of the fragments and the correlation of the segments with the genetic map. The peptides are those shown in Fig. 2

For example, in Fig. 2 it can be seen that the Cys peptide is clearly absent in cells infected with the mutants *H*11, *C*140 and *B*17 but is present in all other cases. This suggests that the mutants produce fragments of the head polypeptide chain such that some of the fragments include the Cys peptide while others do not. The peptide His *T*7*c* is synthesized in cells infected with *H*32, *B*278, *C*137, *H*36, *A*489 and *C*208 but not in cells infected with *H*11, *C*140, *B*17 and *B*272. This peptide would therefore be absent in the fragments made by the last four mutants. These two peptides distinguish the fragment made by *B*272 from the other fragments, since only in this case is the Cys peptide present and the His *T*7*c* peptide absent. By extension of the experimental data it should be possible to characterize each fragment uniquely in this way. Some 96 tryptic and chymotryptic peptides, detected by separate labelling experiments with ^{14}C-tyrosine, phenylalanine, histidine, tryptophan, proline and arginine, and $^{35}SO_4^{--}$ (methionine + cysteine), have been scored, and examples are given in Fig. 2.

The number of peptides found in any one mutant is a measure of the length of the fragment produced. It is therefore possible to arrange the fragments in order of increasing size: each fragment should include all the peptides present in smaller fragments, together with extra ones. The difference between two fragments defines a segment of the polypeptide chain, and the arrangement of the fragments by increasing size orders these segments. If the gene and the polypeptide chain are co-linear, then the order of the polypeptide segments should be identical with the order of the segments into which the genetic map is divided by the mutants.

The hierarchical order of the peptides is shown in Table 2 and in Fig. 3. No peptides scored as head protein peptides have been found to distinguish *H*11, *C*140 and *B*17 from each other. Thus only nine of the eleven segments can be defined. However, we note that since peptides defining segment I are present in all *amber* mutants, there is no internal evidence that they are peptides of the head protein. This difficulty could be resolved by extensive purification, but until this is carried out we have chosen to leave segment I out of account. The order deduced from the fragments is therefore: (*H*11, *C*140, *B*17) *B*272, *H*32, *B*278, *C*137, *H*36, *A*489, *C*208 with eight segments uniquely defined. The order is the same as the genetic order (compare Fig. 1) and hence to a high degree of confidence these experiments show that the two structures are co-linear.

No exception to this order has been found for peptides which are present in the wild-type phage and absent from at least one *amber* mutant. However, peptides are found in some *amber* mutants which are not present in the wild-type phage. We might expect such peptides for two reasons:

(1) The point of termination of a fragment may cut a wild-type peptide, and thus produce a new peptide which occurs in that *amber* mutant only. We appear to have found an example of this in one case.

(2) Some wild-type peptides never show on the paper because they form part of the insoluble 'core'. However, when fragments of the protein are digested such peptides may become soluble, either because they have become accessible to the proteolytic enzyme or because the region of the protein which co-precipitated them is now absent.

Such peptides can also be assigned to segments. We have been able to score ten of these peptides. They agree with the order of segments previously found and in addition there is one peptide which defines segment II; that is, it is present in all *amber* mutants except *H*11.

The remarkable property of the *amber* mutations of the head protein cistron and, by inference, of all other suppressible mutations of this class, is that they lead to the synthesis of a fragment of the polypeptide chain. Qualitatively, fragments appear to be produced in amounts comparable with that of the wild-type protein. Although we do not as yet have any direct evidence, it is likely that the fragments include the N-terminus of the head protein, since proteins are synthesized sequentially from their N-terminal ends[11]. If this is so, then the nonsense codon introduced by the *amber* mutation results in polypeptide chain termination, without permitting re-initiation of synthesis beyond that point. The exact details of this mechanism are unknown, as indeed is the mechanism of suppression itself. However, it is hoped that further analysis of the *amber* mutants should throw more light on these problems.

To summarize: we have shown that suppressible *amber* mutations of the head protein lead to the production of fragments of the polypeptide chain. Analysis of the fragments allows us to define eight segments of the polypeptide chain which are in the same order as the segments defined on a genetic map. We conclude that the gene is co-linear with the polypeptide chain.

We thank Dr. R. H. Epstein for providing some of the mutants used, and Dr. F. H. C. Crick for interesting discussions. We also thank Mrs. R. M. Fishpool, Miss M. I. Wigby and Mr. P. A. Wright for their assistance.

[1] Crick, F. H. C., *Symp. Soc. Exp. Biol.*, **12**, 138 (1958).
[2] Henning, U., and Yanofsky, C., *Proc. U.S. Nat. Acad. Sci.*, **48**, 1497 (1962). See also Rothman, F., *Cold Spring Harbor Symp. Quant. Biol.*, **26**, 23 (1961).
[3] Benzer, S., and Champe, S. P., *Proc. U.S. Nat. Acad. Sci.*, **47**, 1025 (1961).
[4] Benzer, S., and Champe, S. P., *Proc. U.S. Nat. Acad. Sci.*, **48**, 1114 (1962).
[5] Campbell, A., *Virology*, **14**, 22 (1961). Garen, A., and Siddiqi, O., *Proc. U.S. Nat. Acad. Sci.*, **48**, 1121 (1962).
[6] Epstein, R. H., Bolle, A., Steinberg, C. M., Kellenberger, E., Boy de la Tour, E., Chevalley, R., Edgar, R. S., Susman, M., Denhardt, G. H., and Lielausis, A., *Cold Spring Harbor Symp. Quant. Biol.*, **28** (in the press).
[7] Van Vunakis, H., Baker, W. H., and Brown, R. K., *Virology*, **5**, 327 (1958). Brenner, S., Streisinger, G., Horne, R. W., Champe, S. P., Barnett, L., Benzer, S., and Rees, M. W., *J. Mol. Biol.*, **1**, 281 (1959).
[8] Koch, G., and Hershey, A. D., *J. Mol. Biol.*, **1**, 260 (1959).
[9] Brenner, S., and Barnett, L., *Brookhaven Symp. Biol.*, **12**, 86 (1959).
[10] Chase, M., and Doermann, A. H., *Genetics*, **43**, 332 (1958).
[11] Dintzis, H. M., *Proc. U.S. Nat. Acad. Sci.*, **47**, 247 (1961). Goldstein, A., and Brown, B. J., *Biochim. Biophys. Acta*, **53**, 438 (1961).

Reprinted from PROCEEDINGS OF THE NATIONAL ACADEMY OF SCIENCES, Vol. 47, 1961

*ASSEMBLY OF THE PEPTIDE CHAINS OF HEMOGLOBIN**

BY HOWARD M. DINTZIS

DEPARTMENT OF BIOLOGY, MASSACHUSETTS INSTITUTE OF TECHNOLOGY

Communicated by John T. Edsall, January 16, 1961

The mechanism by which proteins are synthesized has been a matter of intense speculation in recent years.[1, 2] Some published speculative models propose simultaneous bond formation between all neighboring activated amino acids on a preloaded template (a sort of stamping machine operation). Others suggest various forms of sequential addition of amino acids to a steadily growing polypeptide chain. In addition there have been hypothesized all degrees of exchange between amino acids already incorporated into growing peptide chains on the template and various classes of precursor "activated" amino acids in solution.[3]

A common concept of how peptide chains may grow is based on their linear chemical nature and assumes serial addition of amino acids, starting at one end of the chain and progressing steadily to the other end. A less orderly picture involves peptide sections growing randomly here and there on the template and finally coalescing into a single chain. Since we know very little about the geometric nature of the templates upon which protein synthesis occurs,[4] we cannot a priori rule out all manner of complex growth mechanisms. For example, if the substructure of the template is folded or coiled in a regular manner it is possible that short, evenly spaced bits of peptide chain are made first on those parts of the template most accessible to the external solution and that the intervening bits are added later at a slower rate. Also since nothing is known about the type of bonds holding the activated amino acids to the template just prior to peptide bond formation, we cannot assume that chain growth is necessarily unidirectional. It is possible that chain growth is initiated at both the amino end and the carboxyl end and progresses towards the middle, or conversely, begins in the middle and proceeds toward both ends.

It is apparent that there exists no shortage of hypothetical models of protein chain growth. The difficulty lies in finding an analytical technique capable of yielding enough information to eliminate conclusively most wrong models and, if possible, to narrow the choice to a single correct one.

Data concerning the actual mechanism of protein assembly should in principle be obtainable by studying both the newly formed protein molecules and the ribosome templates on which they are supposedly formed. However, no method exists for fractionating from a cellular extract all ribosomes engaged in the production of a single type of protein molecule. If a type of cell could be found which is engaged solely in the synthesis of a single kind of protein molecule, then presumably all ribosomes in such a cell would contain incomplete bits of that kind of protein molecule and no others.

Fortunately, a close approximation to this highly desirable situation exists in the case of immature mammalian blood cells producing hemoglobin. These cells, reticulocytes, account for 80 to 90 per cent of the red cells present in the blood of rabbits made anemic by daily injection of phenylhydrazine. The cells may be isolated from the blood and placed in an incubation medium where they will continue producing hemoglobin for many hours.[5, 6] During such an incubation over 90 per cent of the soluble protein produced appears as hemoglobin. It is therefore reasonable to expect most of the growing peptide chains present in the ribosome fraction of such cells to represent incomplete hemoglobin molecules.

If we have available a technique for splitting the peptide chains of both completed and incompleted hemoglobin molecules at a definite number of specific sites, we should be in a position to test which one, if any, of the above hypothetical mechanisms of protein assembly is correct. This is so because each model of protein assembly leads to a definite prediction as to the time and space distribution of newly added amino acids in short sections of peptide chain, both in finished hemoglobin and in the ribosomal particles. Amino acids labeled with radioactive isotopes provide a means of detecting newly added amino acids. In living reticulocyte cells there exist a very large number of finished hemoglobin molecules (10–20% of the cell by weight) and in addition a large number of ribosomal particles, supposedly containing unfinished hemoglobin molecules in different stages of completion. If, at a given moment, we add a radioactively labeled amino acid to the incubation medium containing reticulocytes, then we expect polypeptide produced thereafter to be labeled with radioactive amino acid.

The data to be presented in this paper strongly support a model of protein synthesis involving growth by some kind of sequential addition of amino acids. In Figure 1 are shown some of the predicted consequences of this type of model. For the purposes of illustration we have chosen a model involving chain initiation at one end of the polypeptide followed by sequential addition of one amino acid after another until the other end is reached. We shall assume that some digestion technique can be used to split each polypeptide chain at a definite number of specific sites, yielding the set of peptides *a*, *b*, *c*, . . . *g*, and that furthermore, the set can be separated and the amount of newly added amino acid present in each member *a*, . . . *g* determined quantitatively.

In the finished hemoglobin at short times, we would then expect a steep gradient of radioactive label through the peptides, with only a few peptides labeled at very short times. At longer times the gradient of radioactivity along the peptide chain should become shallower as more and more completely labeled molecules are produced. At all times, the peptide *g*, closest to the finish line, should have the most radioactive label, and the peptide *a*, closest to the starting line, should have the least radioactive label.

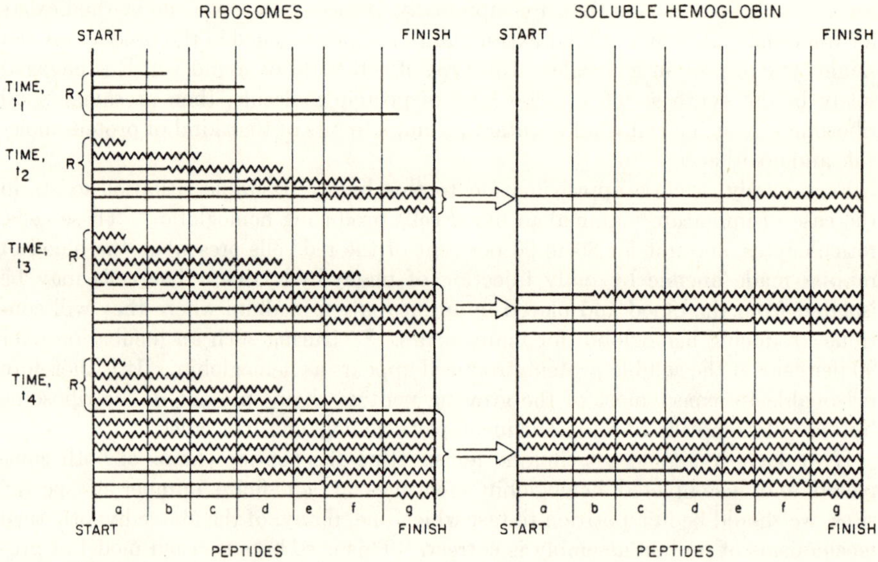

FIG. 1.—Model of sequential chain growth. The straight lines represent unlabeled polypeptide chain. The zigzag lines represent radioactively labeled polypeptide chain formed after the addition of radioactive amino acid at time t_1. The groups of peptides labelled *R* are those unfinished bits attached to the ribosomes at each time; the rest, having reached the finish line, are assumed to be present in the soluble hemoglobin. In the ribosome at time t_2, the top two completely zigzag lines represent peptide chains formed completely from amino acids during the time interval between t_1 and t_2. The middle two lines represent chains which have grown during the time interval but have not reached the finish line and are therefore still attached to the ribosomes. The bottom two chains represent those which have crossed the finish line, left the ribosomes, and are to be found mixed with other molecules of soluble hemoglobin.

On the other hand, in the ribosomes at very short times we would expect an almost uniform distribution of total label among the various peptides since each growing chain will have added only a small radioactive section (Fig. 1). After times long enough to flush out the nonradioactive bits of growing chain, there should be a gradient of total radioactivity from the initial peptide *a*, with the most, to the final peptide *g*, with the least. For this model this is so because there exist in a population of ribosomes at any moment, more sections of peptide *a* than *b*, more *b* than *c*, and so on. Thus, the expected gradient of label in the ribosomes is opposite to that in the finished hemoglobin, both in space direction and in time development.

General Experimental Considerations.—The technique used for forming and separating a reproducible set of peptides was a modification of the method involving a combination of paper electrophoresis and chromatography, at right angles, used by Ingram for human hemoglobin,[7] and termed "fingerprinting" by him. The enzyme trypsin, which splits polypeptides with high specificity wherever the amino acids lysine and arginine occur, furnishes the means of splitting at a definite number of sites. For various reasons, many details of Ingram's procedure for tryptic digestion, paper electrophoresis, and chromatography were modified in the present study.

The problem of obtaining quantitative data on the amount of radioactivity in each peptide was solved by the use of two different isotopic labels. Short incubations were done with H^3-leucine, and very long incubations with C^{14}-leucine. The very long incubations were assumed to give hemoglobin of uniform specific activity in each leucine position. The H^3- and C^{14}-labeled preparations were mixed and carried through the stages of digestion, electrophoresis and chromatography together. The ratio of H^3 to C^{14} was taken as a measure of the amount of label in each peptide obtained from the short time incubations. This method gave an internal standardization automatically correcting both for the differential losses and for the different number of leucine residues in the peptides.

In order to slow the rate of hemoglobin synthesis to the point where samples could be handled with convenience, incubations were tried at various temperatures below body temperature. It was found that the rate of incorporation of C^{14}-leucine into hemoglobin fell slowly with temperature until a point about 10° was reached, whereupon incorporation abruptly stopped. Incorporation of labeled amino acid was found to proceed smoothly at 15° at approximately $^1/_4$ of the rate at 37° (Table 1) and this temperature was routinely used for all short-time experiments.

TABLE 1

INCORPORATION OF C^{14} LEUCINE INTO RABBIT HEMOGLOBIN AT VARIOUS TEMPERATURES OF INCUBATION

Temperature of incubation	Experiment 1		Experiment 2	
	Cpm/mg	% of 37° value	Cpm/mg	% of 37° value
0	0	0.00	...	...
5	14	0.22	...	...
10	280	4.3	...	...
15	2,230	34	8,150	17
20	...	...	17,700	38
25	...	...	52,600	110
30	...	...	45,000	95
37	6,500	100	47,000	100

Hemoglobin was dialyzed for 5 days against water, precipitated with trichloroacetic acid, dissolved in dilute NaOH, reprecipitated with trichloroacetic acid, washed with acetone and ether, and then plated in thin layers containing approximately 20 mg. Counting was done using a Nuclear Chicago end window gas flow counter, the results corrected to zero thickness.

It has been previously shown that the structural protein of ribosomes is not appreciably labeled at short times of incubation.[8] It was therefore assumed that the labeled peptides resulting from a digest of ribosomes with ribonuclease and trypsin represent growing hemoglobin chains and not ribosomal structural proteins. On tryptic digestion the ribosome structural protein did yield a large number of ninhydrin staining peptides which were distinct from those of hemoglobin but, as expected, they did not contain radioactive label.

Incubation of cells: Rabbit reticulocytes prepared from phenylhydrazine-treated animals were washed and incubated according to the procedures of Borsook *et al.*[6] The cells were incubated at 37° for 15 min to allow them to renew metabolites, then at 15° for 5 min. To 1.8 ml cells in a total volume of 4 ml incubation mixture was added 0.24 mg 4, 5 H^3-DL-leucine (5 mc, New England Nuclear Corporation, 3.6 mc per μmole) and the incubation continued at 15°. At various intervals aliquots of approximately 1 ml were removed with a pipet and quickly placed in precooled vials surrounded by solid carbon dioxide.

Uniformly labeled C^{14} leucine hemoglobin was prepared in a similar manner from approximately 1 mg of L-leucine, uniform C^{14}, (50μc Nuclear Chicago), which was incubated with 10 ml of sterile whole blood at 37° for 5 or 24 hr. During 24-hr incubations a significant amount of cell lysis occurred, partly offsetting the approximately 50 per cent higher specific activity obtainable. Typical incubations with C^{14} L-leucine of specific activity 6–8 c/millimole gave hemoglobin of approximate activity 1×10^5 dpm/mg.

Preparation of hemoglobin and ribosomes for tryptic digestion: Samples containing approximately 0.45 ml cells were thawed and the broken cells diluted to a volume of 7 ml with cold solution containing 0.14 *M* KCl, 0.001 *M* $MgCl_2$ and 0.01 *M* Tris-Cl pH 7.3. Solution of this composition had been previously shown to stabilize rabbit reticulocyte ribosomes[3] and was used in all operations where ribosomes were present. The solutions were then centrifuged at 20,000 *g* for 10 min to remove cell walls and debris, and then at 130,000 *g* for 1½ hr to remove ribosomes.

Hemoglobin: The ribosome-free supernatant was dialyzed for 5–7 days in the cold against daily changes of 5×10^{-4} *M* KH_2PO_4, 5×10^{-4} *M* K_2HPO_4 saturated with toluene to prevent bacterial growth. The slight precipitate which formed was centrifuged off and the supernatant hemoglobin frozen until used.

Short time labeled H^3-leucine hemoglobin (4–60 min at 15°) solution was mixed with long time labeled C^{14}-leucine hemoglobin (5–24 hr at 37°) solution in a ratio such that both the H^3 and C^{14} could be counted with good accuracy. This ratio was usually near 10 dpm H^3 per dpm C^{14}. The combined hemoglobin solution was then used to prepare globin by acid acetone precipitation.[4]

Ribosomes: The ribosome pellets were dissolved in 7 ml stabilizing buffer at 0°C, centrifuged for 5 min at 20,000 *g* to remove denatured protein and then reprecipitated by centrifuging at 130,000 *g* for 1½ hr. The ribosomes were redissolved and recentrifuged three times to remove

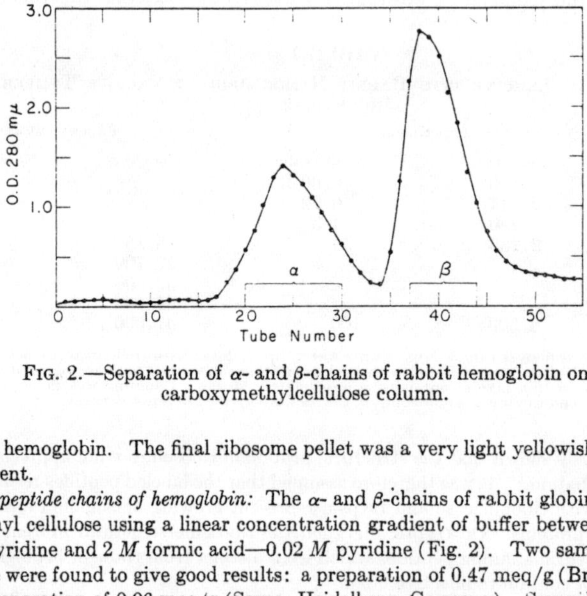

FIG. 2.—Separation of α- and β-chains of rabbit hemoglobin on carboxymethylcellulose column.

free leucine and hemoglobin. The final ribosome pellet was a very light yellowish color and completely transparent.

Separation of peptide chains of hemoglobin: The α- and β-chains of rabbit globin were separated on carboxymethyl cellulose using a linear concentration gradient of buffer between 0.2 *M* formic acid—0.02 *M* pyridine and 2 *M* formic acid—0.02 *M* pyridine (Fig. 2). Two samples of carboxymethyl cellulose were found to give good results: a preparation of 0.47 meq/g (Brown Co., Berlin, N. H.) and a preparation of 0.06 meq/g (Serva, Heidelberg, Germany). Several preparations of higher capacity from various companies did not give as good results. Solutions of separated chains were dried under vacuum in the presence of sulfuric acid and soda lime.

Tryptic digestion of hemoglobin samples: Autotitrator: Dried samples were dissolved in water to a concentration of 10–20 mg/ml. The pH was adjusted to 9.5 with 0.10 *N* NaOH from an initial value between 3 and 4. Dense precipitation occurred near neutral pH but the solution became clear again at pH 9.5. 0.01 ml of 1% trypsin (Worthington 2x crystallized, salt-free in 10^{-3} *M* HCl) was added for each ml of solution and the digestion was allowed to proceed at 37° until definite evidence of a plateau in base uptake was obtained (approximately 1½ hr).

Buffer: 10 mg of dried sample was dissolved in 0.5 ml water, 0.015 ml 0.5 *M* NH_4OH was added, followed by 0.01 ml 1% trypsin and 0.025 ml buffer made of 1 *M* NH_4OAc + NH_4OH to pH 9.75. Digestion proceeded for 4 hours at room temperature.

In all cases digestion was stopped by the addition of several drops of glacial acetic acid. The samples were then dried under high vacuum in the presence of sulfuric acid and soda lime and then dissolved at a concentration of 100 mg/ml in 0.4% acetic acid-0.1% pyridine, giving a preparation which was often clear, but sometimes had slight to medium turbidity.

Tryptic digestion of ribosome samples: The ribosome pellet from 0.45 ml cells, approximately 3 mg dry weight, was dissolved in 1 ml water. 10–20 mg uniformly labeled C^{14} leucine globin was dissolved in 1 ml water. The two solutions were mixed and adjusted to pH 8.5 in an autotitrator at 37°. 0.02 ml 1% ribonuclease (Worthington crystalline) was added, followed, after 15 min, by 0.02 ml 1% trypsin. The digestion was followed in the autotitrator for 15 min, then the pH was raised to 9.5 and the digestion followed for approximately 1$^1/_2$ hr until a plateau was reached. The samples were acidified and dried as in the case of hemoglobin digestion.

Paper electrophoresis: Electrophoresis was carried out on a water-cooled metal plate insulated with a thin sheet of polyethylene. Strips of Whatman No. 3MM paper 12 in. wide and 37 in. long were wet with buffer of pH 4.5 (2.5% pyridine, 2.5% acetic acid, 5% n-butanol, all concentrations v/v) and blotted. Eight-inch wicks made of 4 thicknesses of the same paper were overlapped at each end, and 0.02 ml of solution containing 2 mg of sample was applied at the origin. The paper was then covered with polyethylene sheeting pressed flat by weights applied over a sponge rubber pad. Electrophoresis was carried out at 2,000 volts and approximately 100 ma, for 16 hr, after which the paper was dried.

Chromatography: The dried papers were trimmed to a length of 33 in. and stapled into cylinders 12 in. high. Chromatography was then conducted at room temperature in glass jars 12 in. wide and 24 in. high, using a mixture of 42.5 vols n-butanol, 27.5 vols pyridine, 30 vols water. Occasionally it was necessary to increase the chromatographic resolution by sewing a 4-in. strip of paper to the top of the sample sheet before stapling into a circle.

Isolation and counting of peptides: The dried chromatograms were dipped in 0.25 per cent ninhydrin in acetone, dried, and heated at 90° for 5 min. The resulting blue paper spots were cut out, placed in 20 ml counting vials and 5 ml of water was added to each. The vials were then heated in an oven at 90° for 30 min to extract the peptides from the paper, after which time the paper was removed from the vial with a tweezer and the solution evaporated to dryness overnight in an oven at 90°. 0.20 ml of 0.01 HCl was added to each vial, followed by 20 ml of scintillator solution made up of three parts toluene, one part absolute ethanol, and containing 1% phenylbiphenylyloxadiazole-1,3,4(PBD) and 0.05% p-bis [2-(5-phenyloxazolyl)]-benzene (POPOP). The resulting solutions were measured for C^{14} and H^3 activity simultaneously using a TriCarb scintillation counter equipped with split channel operation so that the lower voltage channel counted both C^{14} and H^3 while the upper voltage channel counted mainly C^{14}. The recovery of radioactivity from eluted peptides of hemoglobin amounted to approximately 50 per cent of the amount applied at the origin spot for paper electrophoresis.

The TriCarb scintillation counter was run with 1040 volts on the photomultiplier tubes. The lower pulse height discriminator was set to register pulses between 10 and 50 volts, giving an efficiency of 6.5% for H^3 and 20% for C^{14} with a background of 40 cpm. The upper pulse height discriminator was set to register pulses of 100 volts or higher, giving an efficiency of 0.14% for H^3 and 37% for C^{14} with a background of 60 cpm. Mixtures of isotopes ranging from 2 dpm H^3/dpm C^{14} to 40 dpm H^3/dpm C^{14} were used.

Figure 3 shows a rather typical peptide separation. To improve photographic reproduction, the ninhydrin staining was done with twice the usual concentration of ninhydrin. The result shows more clearly than usual the presence of "ghost" spots, which are defined as weak spots sometimes present but usually absent or barely detectable. The spots which are always or almost always present have been numbered arbitrarily from left to right.

Peptide 31 is the leucine-containing peptide farthest from the origin as determined by radioactivity count on peptides made from uniformly labeled hemoglobin. There are approximately four ninhydrin staining spots farther from the origin than peptide 31, but since they were not labeled by leucine, they were routinely removed from the paper by electrophoresis, to increase the separation of the remaining peptides. The total number of peptides found with reasonable reproducibility is thus about 35, appreciably above the number 26 reported in human hemoglobin by Ingram.[7] It should be noted that a number of peptides, e.g., 2, 7, 16, 19, 23, stain quite weakly and may represent products of incomplete tryptic digestion, or partial digestion by other enzymes such as chymotrypsin which may be present as trace impurities in the trypsin.

The separation and identification of peptides was not uniformly good. In some runs spots were either missing or badly smeared into other spots. Consequently it was necessary to eliminate

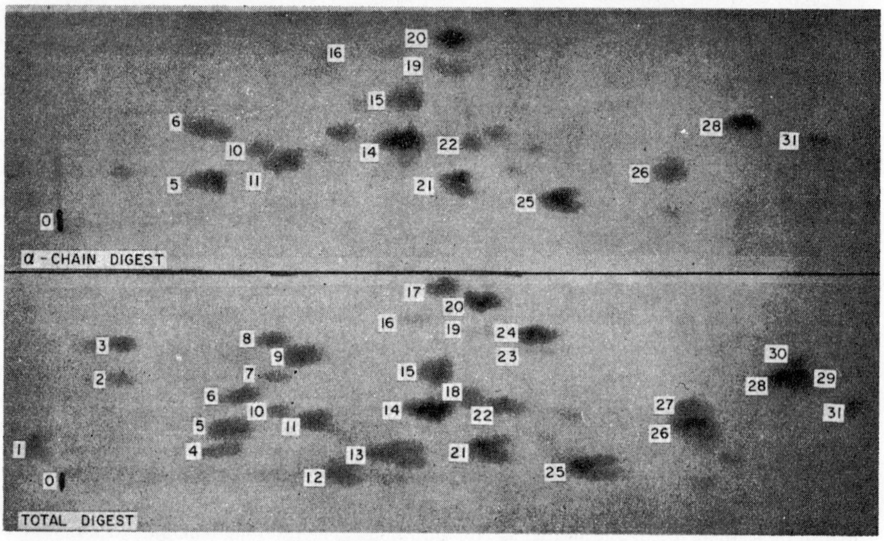

FIG. 3.—Peptide maps of tryptic digest of total rabbit globin (bottom) and column separated α-chain (top). The point of application of digest to the paper is marked by O. The positive electrode is to the left.

from elution and counting in each run those peptides which could not be identified with certainty or which badly overlapped with neighbors known to contain leucine.

A total of 18 peptides, 9 from the α-chain and 9 from the β-chain, were found reproducibly to contain leucine in significant quantity. The average relative yield of C^{14} in these peptides, obtained from digests of uniformly labeled single chains, is shown in Table 2. In addition, smaller

TABLE 2

RELATIVE YIELD OF C^{14} LEUCINE FROM TRYPTIC PEPTIDES

α-Chain Peptides		β-Chain Peptides	
Peptide number	Average relative yield	Peptide number	Average relative yield
10	0.4	1	1.2
11	0.7	3	0.8
14	4.5	9	2.6
16	0.3	12	0.5
20	1.0	13	1.0
21	0.4	17	1.2
22	0.2	18	0.5
25	0.6	24	1.0
31	0.6	27	0.6

amounts of label were found in peptides 2 (β), 6 (α), 19 (α), and 28 (β), but since their yields were quite small and variable, no attempt was made to do quantitative measurements on them (except for a few studies on peptide 28, reported below). This study, which is dependent on the use of radioactive leucine, is therefore based on slightly over half of the total number of recognizable peptides produced from hemoglobin by trypsin digestion. Extension to the remaining peptides awaits the availability of other amino acids, preferably lysine and arginine, of very high H^3 specific activity.

Results.—After 7 min of incubation at 15° in the presence of H^3 leucine, a marked difference in the relative amount of tritium contained could be found in the peptides of both the α- and β-chains. The peptides could be arranged in a more or less definite order of increasing tritium content (Fig. 4), such that only the relative order of nearest neighbors was in doubt.

At different times of incubation the same relative order of the peptides was maintained (Fig. 5). The shape of the curves indicates extreme nonuniformity of labeling at 4 min of incubation, with a number of peptides containing no detectable H^3 leucine. By 60 min of incubation the gradient of radioactivity has been largely, but not entirely, eliminated.

To check the significance of the varying amounts of H^3 leucine found in different peptides two types of control experiments were made (Table 3). First, hemoglobins made by incubation for 5 hr at 37° with H^3 leucine and with C^{14} leucine were mixed, digested and counted for H^3 and C^{14}. These samples gave a uniform ratio within experimental error; see Table 3, column (*a*). Next, samples from a 7-min incubation at 15° giving marked nonuniform labeling with H^3 leucine (Table 3, column (*b*)) were checked to see if any systematic counting error was involved. To each sample

TABLE 3

CONTROTS ON COUNTING ACCURACY

Peptide number	(*a*) Long-time incubation	(*b*) Short-time incubation	(*c*) Increment ratio
α-Chain			
21	1.01	0.08	1.02
10	0.94	0.08	0.98
20	1.06	...	...
25	1.04	0.36	1.03
11	1.04	0.38	1.05
14	1.07	0.69	1.05
31	1.02	...	...
22	1.02	0.84	1.00
16	0.88	1.06	1.02
β-Chain			
13	...	0.05	0.98
24	0.93	0.10	1.02
1	1.01	0.16	1.03
17	0.94	0.23	1.02
3	0.94	0.34	0.88
9	0.99	0.54	1.02
18	0.97	0.59	0.89
12	1.05	0.70	1.00
27	0.86	...	...

(*a*) 5-hr incubation H^3 leucine, 5-hr incubation C^{14} leucine, relative amount of tritium.
(*b*) 7-min incubation H^3 leucine, 30-hr incubation C^{14} leucine, relative amount of tritium.
(*c*) Ratio of increases in H^3 to C^{14} after adding H^3 leucine and C^{14} leucine to each counting vial of (*b*).

vial a constant amount of H^3 standard and C^{14} standard were added, and the radioactivity redetermined. The measured increments in H^3 and C^{14} activity were constant within experimental error and the normalized ratio of increments $\Delta H^3/\Delta C^{14}$ was also constant (Table 3, column (c)).

The results obtained by digesting ribosomes were less reproducible for a number of reasons. First, the α- and β-chains could not be separated, since by definition we were looking for incomplete chain fragments in the ribosomes, and thus did not dare lose fragments in an attempt at fractionation. Secondly, the over-all background of radioactivity between ninhydrin staining spots was much higher in the ribosomes. This is perhaps to be expected from the model in Figure 1, where there

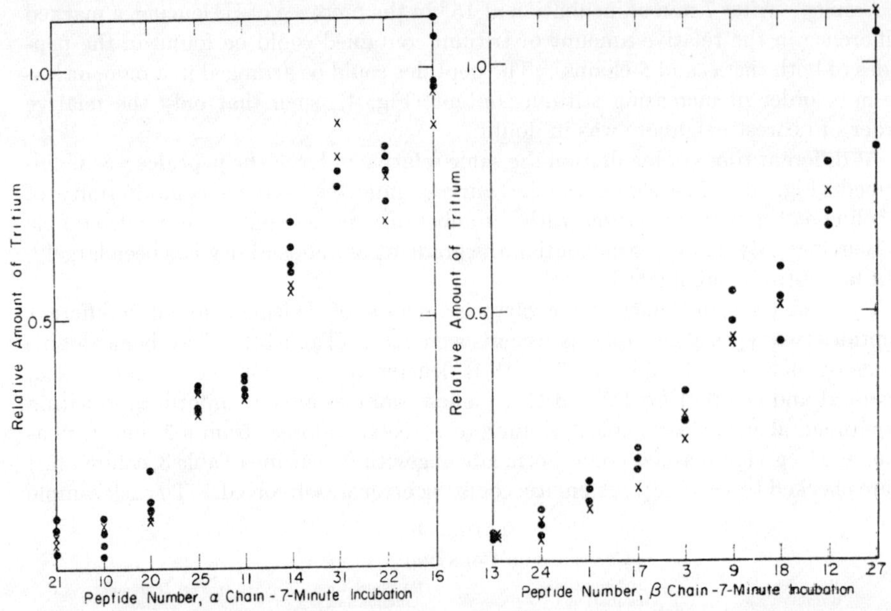

FIG. 4.—Distribution of H^3-leucine among tryptic peptides of soluble rabbit hemoglobin. Peptides produced by tryptic digestion in an autotitrater are indicated by ●. Peptides produced from a separate incubation by tryptic digestion in buffer are indicated by X.

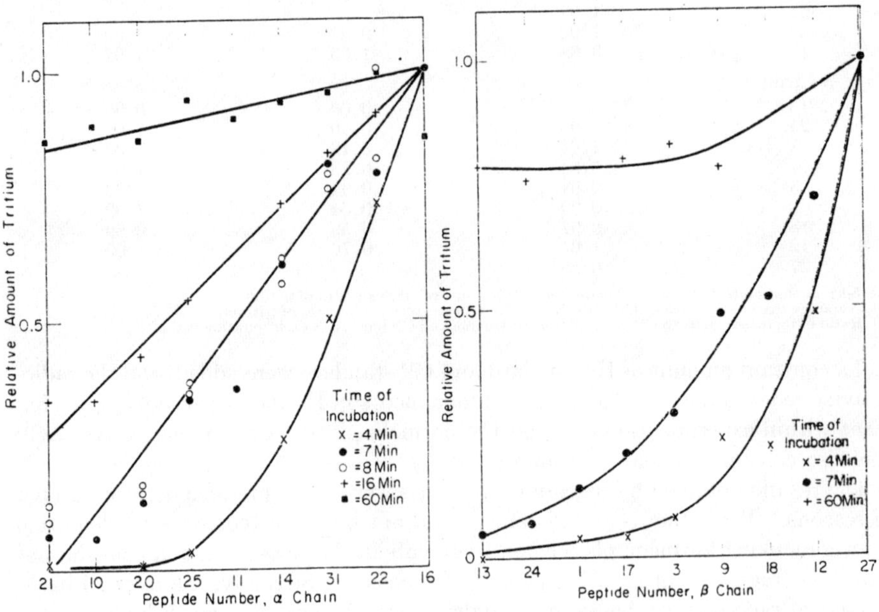

FIG. 5.—Distribution of H^3-leucine among tryptic peptides of soluble rabbit hemoglobin after various times of incubation at 15°C. The points indicated for 7 minutes (●) are the result of averaging all points shown in Figure 4.

are shown end bits of growing chain which do not span vertical lines. Such bits would not correspond to tryptic peptides from hemoglobin and would not be expected to separate with the known peptides; hence they would contribute to the background of radioactivity.

Figure 6 shows the data obtained from ribosomes of cells which had been incubated for short periods (4 to 7 min) at 15°. It is hard to see any significant trend to the data, with the possible exception that the terminal peptides (16 and 27) seem lower than the rest. It thus appears that at these short times of incubation the hemoglobin peptides in ribosomes are labeled almost uniformly.

Figure 7 shows results from ribosomes of cells which had been incubated 60 min at 15°. In this case there is a clear trend visible in the peptides of the α-chain with a less definite result in the case of the β-chain. The gradient of radioactivity is opposite to that in Figure 4.

After 7 min of incubation with H^3 leucine at 15° the hemoglobin peptides isolated from soluble hemoglobin (Fig. 4) had an average specific activity of 1.2×10^5 dpm H^3 per mg. The average specific activity of the hemoglobin peptides prepared from the ribosomes isolated from the same cells may be calculated if one can make an estimate of the weight fraction of ribosomal particles which is present as growing hemoglobin chains. If we make the extreme assumption that the purified ribosomes are pure hemoglobin, then the specific activity of the average peptide in the ribosome (Fig. 6) is 7×10^6 dpm H^3 per mg, or 60 times that of the average peptide in soluble hemoglobin. If we take as more likely the previously reported[8] estimate that growing peptide chains amount to approximately 0.1 per cent of the ribosomal mass, then the ratio of peptide specific activity in ribosomes to that in soluble hemoglobin becomes 60,000. This latter assumption also leads to the conclusion that the specific activity of the H^3-leucine in the ribosomal hemoglobin peptides is approximately 1.5 times that of the H^3-leucine used for the incubation, a result obviously too high but within the combined errors of experiment and assumptions. These results indicate conclusively that the tryptic peptide fragments of hemoglobin isolated from ribosomal particles are precursors of finished hemoglobin molecules and do not represent contamination of the ribosomal particles by completed molecules from the soluble pool.

The results given in Figures 4, 5, 6, and 7 are in agreement with the model shown in Figure 1 in all particulars. The predicted gradient of radioactivity in the peptides of soluble hemoglobin, becoming less pronounced with time, and the inverse gradient in the peptides of the ribosomes, becoming more pronounced with time, are both found. The development of a gradient of radioactivity in the ribosomal hemoglobin peptides at long times is perhaps the most direct proof to date that ribosomes contain incomplete growing peptides. A gradient might be expected at short times in the ribosomes due to contamination from nonuniformly labeled molecules produced elsewhere, but it is hard to see how a gradient could develop with increasing time except by means of the mechanism shown in Figure 1.

It must be stressed that the data given thus far do not constitute proof of the correctness of the particular model in Figure 1, although they are in complete agreement with it. This is the case because all the data presented above are limited to time measurements. To test the model completely, the sequence of amino acid residues along the peptide chain must also be known. Specifically it must be

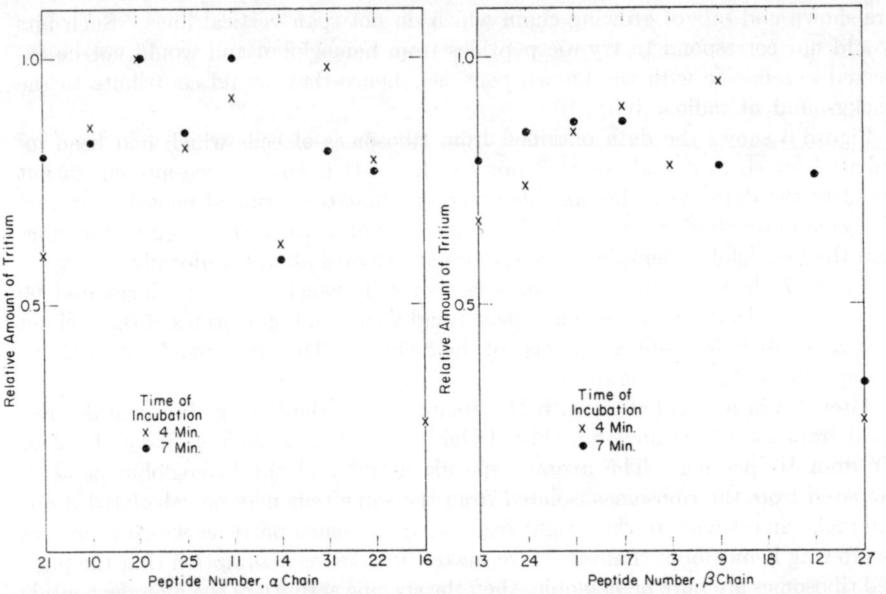

FIG. 6.—Distribution of H^3-leucine among tryptic peptides of ribosomes after short incubations.

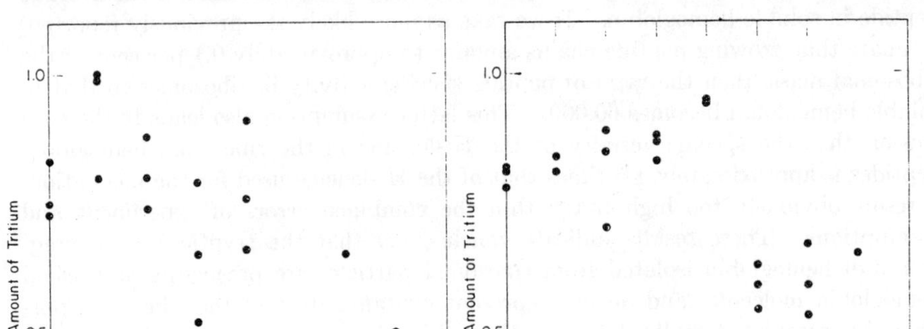

FIG. 7.—Distribution of H^3-leucine among tryptic peptides of ribosomes after long incubations.

shown that peptides at the end of the time sequence, e.g., 21 and 31, are nearest to the end of the polypeptide chain, and furthermore that peptides which are neighbors in the time sequence of labeling with H^3 leucine, e.g., 14 and 31, are also neighbors on the polypeptide chain.

An attempt has been made to identify the tryptic peptide nearest to the free carboxyl end of each hemoglobin chain. Guidotti has reported that by using a mixture of carboxypeptidases A and B he was able to remove sequentially approximately a dozen amino acids from the carboxyl end of the α- and β-chains of human globin.[10] In the case of human globin a leucine residue was one of those removed from each chain. It was therefore reasonable to assume that a similar operation on rabbit globin could remove a leucine residue from the peptide nearest to the carboxyl end.

Uniformly labeled C^{14} leucine globin was incubated according to the conditions of Guidotti with carboxypeptidase A (Worthington, DFP treated) and carboxypeptidase B (kindly donated by Dr. Martha Ludwig). After digestion it was heated at 100° for 15 min to denature the enzyme, and then mixed with undigested uniformly labeled H^3 leucine globin which had received the same treatment, except that no carboxypeptidase had been added to it. The mixture of C^{14} leucine globin and H^3 leucine globin was then digested with trypsin and the peptides were separated and counted as described above. If the carboxypeptidase had no effect on any recognizable leucine-containing peptide, then we would expect to obtain a constant ratio of H^3 to C^{14} in each resulting tryptic peptide (e.g., see Table 3, column (*a*)). If, however, C^{14}-leucine were removed from a peptide by the action of carboxypeptidase, we would expect a decrease in C^{14} leucine content in that peptide with a corresponding increase in the ratio of H^3 to C^{14}. The experiment was also done in reverse, with the H^3 leucine globin being digested with carboxypeptidase. As a final control both C^{14} leucine globin and H^3 leucine globin were carried through all operations except that no carboxypeptidase was added to either. The results are indicated in Table 4.

TABLE 4

RELATIVE TRITIUM CONTENT OF TRYPTIC PEPTIDES FOLLOWING CARBOXYPEPTIDASE ACTION

Peptide number, α-chain	(*a*) C^{14} Globin Digested with Carboxypeptidases			(*b*) H^3 Globin Digested with Carboxypeptidases		(*c*) No Digestion with Carboxypeptidases	
	Digest 1		Digest 2				
10	1.0	1.0	1.2	1.0	0.8	1.3	1.1
11	1.0	1.0	1.2	...	1.0	1.1	1.3
14	1.0	1.0	1.1	1.1	1.0	1.0	1.1
16	11.0	20.0	2.7	0.07	0.0	0.3	0.4
20	1.0	1.0	0.7	1.0	1.0	1.0	1.0
21	1.0	1.0	1.0	1.0	1.1	1.0	1.0
22	0.9	0.9	0.9	1.2	1.6	1.0	1.1
25	1.0	1.0	1.2	1.1	1.1	1.0	1.1
31	1.0	1.1	1.0	1.2	1.0	1.2	1.0
Peptide number, β-chain							
1	1.0	1.0	1.0	0.9	1.0	0.9	1.0
3	0.9	1.0	1.0	0.9	1.0	1.0	1.0
9	0.9	0.9	0.9	0.9	1.0	0.9	0.9
12	1.4	0.8	1.2	1.1	1.1	1.0	0.9
13	1.2	0.8	1.0	0.9	0.9	0.9	1.2
17	1.0	1.0	1.0	1.0	1.0	1.0	1.0
18	0.8	0.9	1.0	0.9	0.9	2.0	0.9
24	1.0	1.0	1.1	0.9	0.9	1.0	1.0
27	1.6	1.4	0.9	0.8	0.8	0.8	0.9
28	30.0	8.0	13.0	0.07	0.01	0.8	0.9

The only α-chain peptide which shows significant deviation from constant ratio in the expected direction is peptide 16. Unfortunately, the control experiment shows some ratio deviation in peptide 16 (column (*c*)) but not enough to upset the conclusion that peptide 16 is near the carboxyl end of the α-chain. The ratio variation of peptide 16 in the control experiment (Table 4, column (*c*)) may be due to the fact that the proteins were heated at 100° for 15 min to inactivate carboxypeptidase, whereas in previous experiments (Table 3, column (*a*)) this was not done.

In the β-chain none of the major yield peptides showed a significant ratio change, but peptide 28, which was previously sometimes present in minor yield, was present in good yield and clearly showed the behavior expected of a peptide near the carboxyl end of the β-chain. On re-examination of the data from 4- and 7-min H^3-leucine incubations, four clear cases were found where peptide 28 had been present but in low yield. The average tritium content of peptide 28 in these four runs was found to be 1.06 ± 0.27 times the tritium content of peptide 27. Although the average yield of peptide 28 was only 0.18 ± 0.04 times the yield of peptide 27, it is tempting to conclude that peptide 28 is closely related to peptide 27 in the time sequence of labeling with H^3 leucine.

It would thus appear that in both the α- and β-chains those leucine-containing peptides which are the first to be labeled with H^3-leucine in the soluble hemoglobin are nearest to the free carboxyl end of the chain. According to the model shown in Figure 1, this implies that chain growth terminates at or near the free carboxyl end of the molecule.

Discussion.—The NH_2-terminal amino acid of both the α- and β-chains of rabbit hemoglobin is valine.[11] Attempts have been reported to find the rate of short time radioactive labeling of the NH_2-terminal valine relative to the average of all other valines in the hemoglobin molecule. Using whole rabbit reticulocytes, Loftfield[2] reported results indicating that the NH_2-terminal valine is labeled last. On the other hand, Bishop *et al.*,[12] using a cell-free system from rabbit reticulocytes, reported results indicating that the NH_2-terminal valine is labeled first. Reports on other protein-synthesizing systems are equally conflicting. Thus the work of Yoshida and Tobita[13] on bacterial amylase indicates that synthesis proceeds from the amino-terminal toward the carboxyl-terminal end. Complications are present in the interpretation of their work because of the very long times of incubation involved and the presence of various protein precursor pools. Shimura *et al.*[14] using the fibroin synthesizing gland of the silk worm obtained results indicating that the NH_2-terminal glycine is added last.

Muir *et al.*[15] reported finding uniform labeling in hemoglobin labeled *in vivo*. This is to be expected from the results reported in this paper. Thus Figure 5 shows that labeling is uniform within 20 per cent after 60 min of incubation at 15°, corresponding to 15 min of incubation at 37°. Kruh *et al.*[16] have reported nonuniform labeling in hemoglobin after very long *in vivo* experiments. This result is not consistent with the data reported in this paper and possibly represents phenomena different from the original synthesis.

A different approach was used by Loftfield and Eigner who reported kinetics of amino acid incorporation into ferritin[17] and hemoglobin[2] after short times of incubation. Their data indicate that for the first few minutes of labeling the specific activity of newly formed protein increases as the square of the time, becoming linear only after several minutes. From these data they concluded that a scheme essentially the same as that of Figure 1 is indicated. However, this result cannot distinguish between a random and a sequential process of attaching amino acids to the template.

It is perhaps worth noting that if the model shown in Figure 1 is finally proved to be correct, then the experimental technique described in this paper could be useful for structure determination. Thus, it should be possible to determine the spatial sequence of tryptic peptides in proteins of unknown structure by determining the time order of labeling.

It has previously been reported[8] that to account for the production of new hemoglobin in living rabbit reticulocytes, each ribosomal particle must, on the average, make one polypeptide chain of hemoglobin in 1.5 min. That result was obtained by dividing the total rate of hemoglobin synthesis by the total number of ribosomal particles. From Figure 5 it may be seen that the last peptide on each chain to be labeled receives its label at some time between 4 and 7 min of incubation at 15°. Since the rate of labeling was found to be approximately $^1/_4$ as great at 15° as at 37° (Table 1), this implies that the total time of assembly of each polypeptide chain at 37° is approximately 1.5 min. The agreement between the average rate of synthesis, 1.5 min, and the individual rate of chain synthesis, also 1.5 min, strongly implies that most of the ribosomal particles present in rabbit reticulocytes are, in fact, producing hemoglobin. Since there are approximately 150 amino acid residues in each chain, the average rate of growth is close to two amino acids added per second.

In all of the above discussion a number of possible complications have been ignored because of insufficient data to evaluate their effects. Thus we have ignored the effects of both delay time and dilution of specific activity suffered by labeled leucine during its passage into the cells and subsequent reactions prior to actual peptide bond formation. The fact that we have not needed to invoke these processes to explain the results suggests that the effects are small. Likewise we have ignored the possible existence of hemoglobin in transitory forms between completed polypeptide chains and final soluble hemoglobins. We might imagine, for example, that α-chains and β-chains are produced on separate ribosomal particles and that furthermore single α- and β-chains are insoluble and stay on the ribosomes, while α_2 and β_2 dimers are soluble. This leads to the notion of a small pool of completed chain attached to the ribosome, which would change slightly the results expected in Figure 1, and would lead to a less steep predicted slope in Figure 7.

The figures for this paper have been drawn with uniform spacing between adjacent peptides. This, of course, does not imply that the labelled amino acids are uniformly spaced along the actual polypeptide chain. When the actual sequence of the peptide chains is determined we shall be in a position to plot the relative amount of labeling in each amino acid against its position in the chain. Only when that is done will it be worthwhile to consider the detailed shape of the curves for evidence concerning uniformity of growth rate along the polypeptide chain.

In summary it may be concluded that the growth of the peptide chains of hemoglobin is not a random process but a steady sequential addition of amino acids to growing chains at the rate of approximately two amino acids per second. The number of initiation points per chain is, at most, very small and most likely only one.

The chain growth terminates near or at the free carboxyl end. Taken together, these conclusions indicate that chain growth proceeds steadily from the free amino end toward the free carboxyl end in rabbit hemoglobin.

The author wishes to acknowledge the expert technical assistance of Miss Judith Karossa and Mrs. Ruth Langridge in the early and later parts, respectively, of this investigation.

* This work was supported by a grant from the National Institutes of Health.

[1] Steinberg, D., M. Vaughan, and G. B. Anfinsen, *Science*, **124**, 389 (1956).
[2] Loftfield, R., *Proc. 4th Intern. Congr. Biochem.*, **8**, 222 (1960).
[3] Borsook, H., *Proc. 3rd Intern. Congr. Biochem.*, **92** (1956).
[4] Dibble, W. E., and H. M. Dintzis, *Biochim. et. Biophys. Acta*, **37**, 152 (1960).
[5] Kruh, J., and H. Borsook, *J. Biol. Chem.*, **220**, 905 (1956).
[6] Borsook, H., E. H. Fischer, and G. Keighley, *J. Biol. Chem.*, **229**, 1059 (1957).
[7] Ingram, V. M., *Biochim. et Biophys. Acta*, **28**, 539 (1958).
[8] Dintzis, H., H. Borsook, and J. Vinograd, in *Microsomal Particles and Protein Synthesis*, ed. R. B. Roberts (New York: Pergamon Press, 1958), p. 95.
[9] Wilson, S., and D. B. Smith, *Can. J. Biochem. and Physiol.*, **37**, 405 (1959).
[10] Guidotti, G., *Biochim. et Biophys. Acta.* **42**, 177 (1960).
[11] Osawa, H., and K. Satake, *J. Biochem. (Tokyo)*, **42**, 905 (1956).
[12] Bishop, J., J. Leaky, and R. Schweet, these PROCEEDINGS, **46**, 1030 (1960).
[13] Yoshida, A., and T. Tobita, *Biochim. et Biophys. Acta*, **37**, 513 (1960).
[14] Shimura, K., H. Fukai, J. Sato, and R. Saeki, *J. Biochem. (Tokyo)*, **43**, 101 (1956).
[15] Muir, H., A. Neuberger, and J. Perrone, *Biochem. J.*, **52**, 87 (1952).
[16] Kruh, J., J. Dreyfus, and G. Schapira, *J. Biol. Chem.*, **235**, 1075 (1960).
[17] Loftfield, R. B., and E. A. Eigner, *J. Biol. Chem.*, **231**, 925 (1958).

Reprinted from PROCEEDINGS OF THE NATIONAL ACADEMY OF SCIENCES, Vol. 48, 1962

*ON THE ROLE OF SOLUBLE RIBONUCLEIC ACID IN CODING FOR AMINO ACIDS**,†

BY FRANÇOIS CHAPEVILLE,§ FRITZ LIPMANN,‡ GÜNTER VON EHRENSTEIN,** BERNARD WEISBLUM, WILLIAM J. RAY, JR., AND SEYMOUR BENZER††

THE ROCKEFELLER INSTITUTE, JOHNS HOPKINS UNIVERSITY, AND PURDUE UNIVERSITY

Communicated April 25, 1962

In protein synthesis, each amino acid is first joined specifically with a corresponding sRNA through the mediation of an activating enzyme. These aminoacyl-sRNA's, by reaction with a ribosomal preparation,[1, 2] form proteins with specific amino acid sequences.[3] According to the "adaptor" hypothesis of Crick[4] and Hoagland,[5] the position of a particular amino acid would be determined not by the amino acid itself, but by hydrogen bonding between the RNA template and a complementary nucleotide sequence in the sRNA carrying the amino acid. The experiment described in this paper was designed as a direct test of the adaptor hypothesis, by attaching an amino acid to its normal sRNA and then, without breaking the bond, converting the amino acid to another one of the natural amino acids. It is then possible to determine whether the coding properties of this hybrid are determined by the sRNA or the amino acid. We have made use of the fact that cysteine can be altered by reductive desulfhydration with Raney Nickel to alanine.

This reaction can be performed while the cysteine is attached to its sRNA, producing the hybrid Ala-sRNA^{CySH}. A superscript denotes the normal amino acid accepting specificity of an sRNA, the actual amino acid attached being indicated as a prefix. Figure 1 illustrates this procedure.

To determine the coding properties of the hybrid molecule, we have utilized the finding by Nirenberg and Matthaei,[6] Lengyel, Speyer, and Ochoa,[7] Speyer, Lengyel, Basilio, and Ochoa,[8] and Martin, Matthaei, Jones, and Nirenberg,[9] that polyuridylic-guanylic acid will stimulate ribosomal incorporation into polypeptides of certain amino acids, including cysteine, but not alanine. As shown below, this difference between cysteine and alanine also applies when they are attached to their normal acceptors, i.e., CySH-sRNA^{CySH} is reactive with poly UG but Ala-sRNA^{Ala} is not. The hybrid molecule Ala-sRNA^{CySH} proved to be just as reactive as CySH-sRNA^{CySH}, leading to the conclusion that the sRNA moiety indeed determines the coding specificity.

Materials and Methods.—Preparation of C^{14}-L-cysteine: L-C^{14}-cysteine of high specific activity was prepared from L-C^{14}-serine. The yeast serine sulfhydrase described by Schlossmann and Lynen[10] catalyzes the reaction:

$$CH_2OH{\cdot}CHNH_2{\cdot}COOH + H_2S \rightarrow CH_2SH{\cdot}CHNH_2{\cdot}COOH + H_2O.$$

The enzyme was prepared from National bakers' yeast which had been frozen in liquid nitrogen and stored at $-20°$; 50 gm of frozen yeast were extracted by stirring at 3° for 5 hr with 100 ml of 0.05 M K_2HPO_4 and 0.05 M EDTA. From then on, the procedure of Schlossmann and Lynen was followed up to the ammonium sulfate step. The ammonium sulfate precipitate between 40 and 65 per cent saturation was dissolved in 15 ml of 0.05 M Tris HCl, pH 7.5, and dialyzed against 0.02 M Tris HCl buffer. The dialysate was stored at $-20°$. 0.75 μmole of C^{14}-L-serine (65 μC/μmole, New England Nuclear Corporation) was dissolved in a total volume of 1.5 ml containing: 3 μmoles of EDTA; 0.5 ml of 0.5 M Tris HCl, pH 8.5, saturated with H_2S; 0.3 μmole of pyridoxal phosphate; and 2 mg of enzyme. After incubation at 37° for 3 hr in nitrogen, 20 ml of ethanol containing 0.2 ml of 2 N HCl were added to the cysteine. After centrifugation, the supernatant was evaporated and the residue re-extracted with 10 ml ethanol plus 0.2 ml 2 N HCl, and the extraction and evaporation were repeated once more.

Before the ethanol addition, 75 μmoles of C^{12}-serine, 20 μmoles of C^{12}-alanine, and 30 μmoles of C^{12}-glycine were added to dilute residual radioactivity of the serine and possible degradation products. The repeated extraction-evaporation procedure was used, since cysteine is more soluble in ethanol than the other amino acids. The final product contained approximately 40 per cent of the input radioactivity. On paper electrophoresis at pH 1.85 in 7.8 per cent acetic acid and 2.5 per cent formic acid (70 volts/cm, 60 min), no detectable radioactivity was found except with cysteine. The dry product was stored at $-20°$.

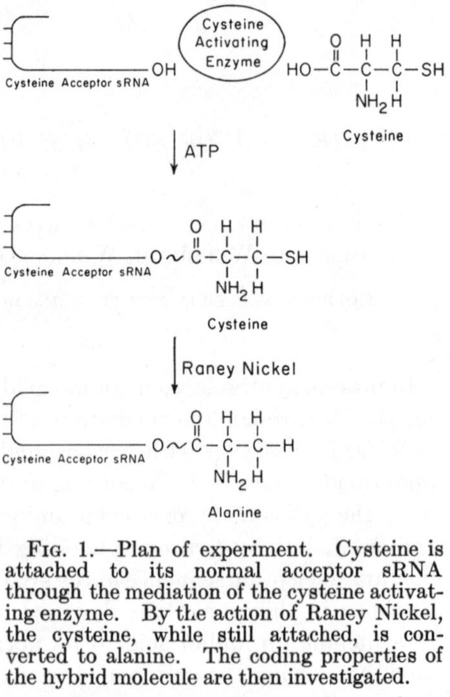

FIG. 1.—Plan of experiment. Cysteine is attached to its normal acceptor sRNA through the mediation of the cysteine activating enzyme. By the action of Raney Nickel, the cysteine, while still attached, is converted to alanine. The coding properties of the hybrid molecule are then investigated.

Preparation of C^{14}-CySH-sRNA^{CySH}: *E. coli*-sRNA was prepared as described,[11] and a 105,000

× *g* supernatant of *E. coli* extract, passed through a Sephadex G-25 column to remove amino acids, was used for charging. While charging with C^{14}-cysteine, β-mercaptoethanol was added. After incubation and phenol extraction, a 100-fold excess of C^{12}-cysteine was added. The alcohol precipitate of the phenol extract was dissolved in 0.01 *M* mercaptoethanol and dialyzed at 3° for 16 hr against 10^{-4} *M* mercaptoethanol solution. To test the charged sRNA, a sample was electrophorized at pH 1.85 as described above for the identification of cysteine. Radioactivity was detected only in the ultraviolet-absorbing, nonmoving position. However, when the charged sRNA was hydrolyzed, with addition of C^{12}-cysteine, for 30 min at pH 11, the radioactivity now appeared in the pH 1.85 electropherogram in the region common, under these conditions, for both cysteine and cystine. Frequently in this procedure, cysteine yielded some material that moved more slowly, presumably due to oxidation. With different batches of sRNA, variable amounts of C^{14}-cysteine were charged, for example, 4,000 and 7,200 cpm/mg RNA, corresponding to 0.067 and 0.12 mμmole/mg respectively. When C^{14}-phenylalanine was charged on these preparations, a variation of ratios of cpm/mg in the same sense was observed.

Reduction of CySH-sRNA^{CySH} to Ala-sRNA^{CySH}: Raney Nickel No. 28, suspended in water, was obtained from Raney Catalyst Company, Chattanooga, Tennessee, and stored at 0°. The preparation was washed with water before use. Organic sulfur compounds are easily reduced by Raney Nickel.[12, 13] It acts at room temperature, and, in the case of cysteine, noncatalytically (cf. for example, Mozingo *et al.*[13]):

$$RCH_2SH + Ni(H)_x \rightarrow RCH_3 + H_2S + Ni.$$

Generally, 1.5 ml of a solution containing 8.5 mg of C^{14}-CySH-sRNA^{CySH} (53,000 cpm), 0.4 *M* acetate buffer, pH 5, 0.12 ml of saturated EDTA, and 170 mg of Raney Nickel was shaken at room temperature for 30 min. After adding another 0.12 ml of saturated EDTA solution, the Nickel was centrifuged off. To the clear supernatant, 2 vol of cold ethanol were added and the mixture kept at −20° for 2 hr. After centrifugation, the precipitate was taken up in 0.5 ml of water and dialyzed against water overnight at 3°; 18,000 cpm on 5.25 mg. of RNA were nondialyzable, and on paper electrophoresis at pH 1.85 the radioactivity and the ultraviolet-absorbing material were found in the nonmoving portion. After hydrolysis for 30 min at pH 11 and electrophoresis at pH 1.85 in the presence of 20 μg each of C^{12}-cysteine and alanine, the radioactivity moved away from the ultraviolet spot and approximately 70 per cent was found in the region corresponding to alanine and 30 per cent in that of cysteine. Reduction of CySH-sRNA^{CySH} varied under these conditions between 30 and 70 per cent. During the reduction procedure about half the sRNA-bound amino acid was lost.

Preparation of polynucleotide phosphorylase: Polynucleotide phosphorylase was prepared from lyophilized *Micrococcus lysodeikticus* (Miles Chemical Co., Clifton, N.J.) by the method of Singer and Guss.[14]

Preparation of poly UG: The reaction mixture contained in a total volume of 1.2 ml: 25 μmoles of UDP and 5 μmoles of GDP neutralized at pH 8.2; 5 μmoles of $MgCl_2$; 1 μmole of EDTA; 100 μmoles of Tris HCl buffer pH 8.2; and approximately 5 units[14] of the enzyme. Incubation was for 85 min at 35°. After addition of phenol as in the preparation of sRNA, the poly UG was isolated, redissolved, dialyzed against water, and lyophilized. Approximately 3 mg of material were obtained; when dissolved in 2 ml, it gave a viscous solution containing 46 optical density units per ml at 260 mμ. The ratio of 280 to 260 absorption was 0.38.

In the experiments with Raney Nickel-treated Phe-sRNA^{Phe}, polyuridylic acid was used that had been prepared as described for poly UG, except that UDP only was added.

E. coli ribosomes were prepared from frozen bacterial paste by the method described.[11] Crude 105,000 × *g E. coli* supernatant was prepared by centrifugation for 2 hr of 20 gm of alumina-ground cell paste extracted with 60 ml of Tris Mg acetate.

Results.—Speyer *et al.*[8] and Martin *et al.*[9] showed that, in the presence of phenylalanine, poly UG stimulates the ribosomal incorporation of cysteine but not of alanine. It had been shown that such transfer can occur directly from aminoacyl-sRNA, at least in the case of phenylalanine.[15] The time curve in Figure 2 shows the transfer of cysteine from CySH-sRNA^{CySH} in the presence of poly UG. In the absence of poly UG, the incorporation in such an experiment is typically about

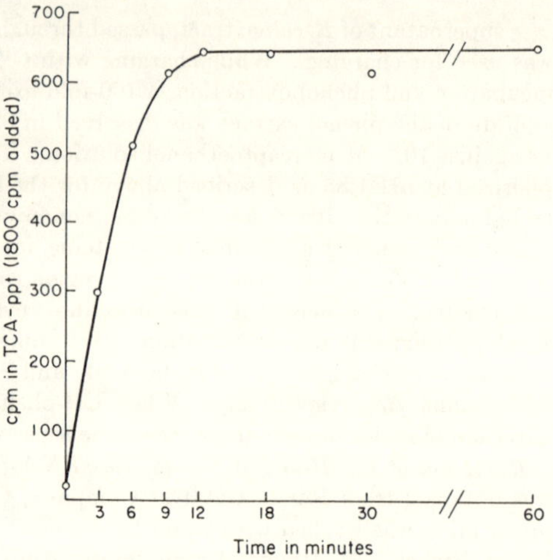

FIG. 2.—Time curve of the transfer of cysteine from sRNA into TCA-insoluble product. Each sample contained, in a total volume of 0.25 ml: 2 mg of *E. coli* ribosomes, 0.25 mg of *E. coli* 105,000 × *g* crude supernatant (expressed in mg of protein), 0.4 mg of sRNA charged with C^{12}-phenylalanine, 0.25 mg of sRNA charged with C^{14}-cysteine (1,800 cpm), 0.01 *M* PEP, 8 μg of PEP-kinase, 0.01 *M* GTP, 0.012 *M* reduced glutathione, 0.012 *M* $MgCl_2$, 0.03 *M* KCl, 10 μg of poly UG, and 0.05 *M* Tris HCl pH 7.4. The incubation was at 35° and the reaction was stopped in each tube by the addition of 5 ml of 5 per cent TCA at the indicated time. The precipitate was extracted with this TCA at 90° for 15 min, washed twice with TCA and once with 1:1 cold ether/alcohol. It was then dissolved in formic acid, plated, and counted in the Nuclear Chicago windowless gas flow counter.

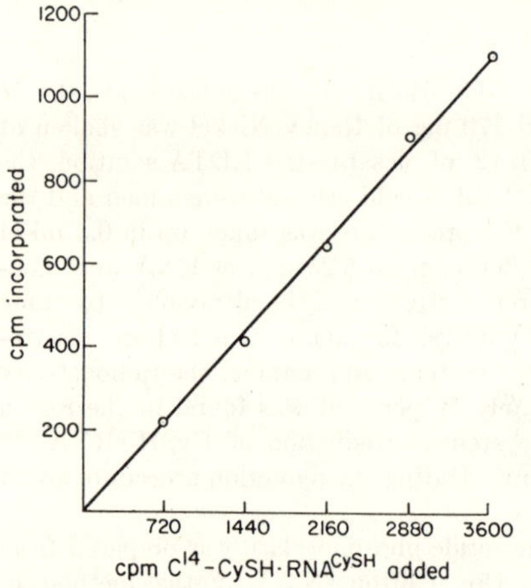

FIG. 3.—Amount of C^{14}-cysteine transferred versus input. Incubation was for 30 min. The conditions were the same as in Figure 2, except that 1.5 mg of ribosomes were used. The specific activity was 7,200 cpm/mg RNA.

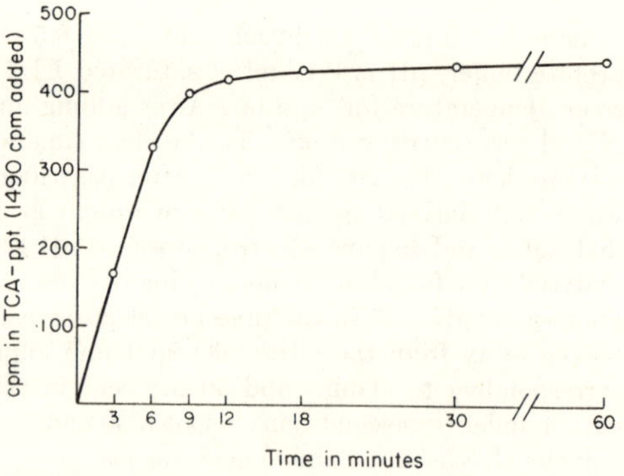

FIG. 4.—Time curve with Raney Nickel-treated C^{14}-CySH-sRNA^{CySH} containing 60 per cent of the radioactivity as alanine and 40 per cent as cysteine. Conditions were the same as in Figure 2, except that 0.5 mg of sRNA (1,490 cpm) was used for each incubation.

200 cpm. As shown in Figure 3, the incorporation with different amounts of CySH-sRNA^{CySH} is proportional to the input, about one third being incorporated. Incorporation in the absence of poly UG was found generally to amount to about 20 per cent of that with poly UG.

When CySH-sRNA^{CySH} was treated with Raney Nickel and about 60 per cent of the radioactivity had been converted to Ala-sRNA^{CySH}, about one third of the total counts were incorporated into the insoluble fraction, as seen in Figure 4, a result comparable to that obtained with the unreduced preparation. The similarity of the time curves for incorporation with reduced and unreduced preparations indicates that exposure to Raney Nickel did not affect the sRNA except that some amino acid was lost. To confirm this further, Phe-sRNA^{Phe} was subjected to a Raney Nickel treatment analogous to that used for the reduction of CySH-

TABLE 1

LACK OF EFFECT OF RANEY NICKEL TREATMENT ON POLY-U-DEPENDENT TRANSFER OF PHENYLALANINE FROM C^{14}-PHE-sRNA^{Phe}

	C^{14}-Phe-sRNA^{Phe}	TCA precipitate Without poly-U	TCA precipitate With poly-U
	(cpm)	(cpm)	(cpm)
Untreated	1,295	108	775
Raney Nickel-treated	1,280	111	782

C^{14}-Phe-sRNA^{Phe} (70 O.D. units at 260 mμ, 18,500 cpm) was treated with Raney Nickel as described for C^{14}-CySH-sRNA^{CySH}. After this treatment, 49.5 O.D. units containing 9,200 cpm were recovered. The C^{14}-Phe-sRNA^{Phe} was incubated with 0.5 μg of poly-U in 0.5 ml under the conditions described in Figure 2.

sRNA^{CySH} and, as shown in Table 1, incorporation of phenylalanine in the presence and absence of poly-U is unchanged by the treatment. It was also found that the acceptor activity of sRNA, at least for arginine, is not damaged by Raney Nickel.

As shown in Table 2, poly UG does not stimulate incorporation with authentic

TABLE 2

POLY UG-STIMULATED INCORPORATION FROM RANEY NICKEL-TREATED C^{14}-CYSH-sRNA^{CySH}

	Aminoacyl-sRNA added			Incorporation into TCA Insoluble Product	
	CySH-sRNA^{CySH}	Ala-sRNA^{Ala}	Ala-sRNA^{CySH}	Without poly UG	With poly UG
	(cpm)	(cpm)	(cpm)	(cpm)	(cpm)
Untreated	1800	—	—	162	508
Untreated	—	2800	—	158	169
Raney Nickel-treated CySH-sRNA^{CySH}	650	—	840	67	527

Conditions were as in Figure 2, except that the volume of the medium was 0.5 ml and 4 mg of ribosomes were used. Incubation was for 30 min.

Ala-sRNA^{Ala}. Therefore, the result described in Figure 4 indicates that reduction of the cysteine on sRNA to alanine does not alter its incorporation. To prove the incorporation from Ala-sRNA^{CySH} unambiguously, the TCA-extracted residue was hydrolyzed and analyzed. To simplify the analysis, the cysteine was oxidized with performic acid to give cysteic acid. After evaporation, the oxidized hydrolysate was dissolved in 0.1 *N* HCl and run over a small column of Dowex 50; this retains alanine but lets cysteic acid pass through. The alanine was then eluted with 4 *N* HCl. An experiment of this type is presented in Table 3. In this case, large

TABLE 3

ANALYSIS OF OXIDIZED HYDROLYSATE OF TCA-PRECIPITATE SEPARATION ON DOWEX 50

	Input		Incorporated—Dowex 50 analysis		
	CySH-sRNA^{CySH} (cpm)	Ala-sRNA^{CySH} (cpm)	Total (cpm)	Filtrate Cysteic acid (cpm)	Eluate alanine (cpm)
Untreated CySH-sRNA^{CySH}	9,000	—	3,050	2,835	83
Raney Nickel-treated CySH-sRNA^{CySH}	3,500	6,500	3,290	990	2,020

To the 2 ml reaction mixture, incubated at 37° for 30 min, 5 ml of 7 per cent TCA were added. After heating for 15 min at 90°, the insoluble material was centrifuged, resuspended, and washed 3 times with 5 per cent TCA at room temperature. It was then washed with 5 ml of cold alcohol/ether solution and spun down at 0°. To the final precipitate, 5 μmoles of C^{12}-cysteine, 2.5 μmoles of C^{12}-alanine, 0.1 ml of pure formic acid, and 1.5 ml of 6 *N* HCl were added. The mixture was hydrolyzed in the autoclave in sealed tubes for 28 hr at 105°. The HCl was evaporated in a rotary evaporator, and the dry material dissolved in 0.5 ml of performic acid and heated at 50° for 1 hr. The performic acid was evaporated in an air current at 50°, and the residue taken up with water and put on a Dowex 50 column in the acid form (0.6 × 4.5 cm). The column was washed with 5 ml of 0.1 *N* HCl. The filtrate and acid-washing fluid containing cysteic acid were combined and kept. Alanine and other amino acids were eluted from the column with 4 ml of 6 *N* HCl. Filtrate and eluate were evaporated and redissolved in water, and an aliquot of each used for counting radioactivity and for paper electrophoresis.

TABLE 4

ANALYSIS OF OXIDIZED HYDROLYSATE OF TCA-PRECIPITATE BY ELECTROPHORETIC SEPARATION

	Input		Incorporated-Electrophoretic Analysis		
	CySH-sRNA^{CySH} (cpm)	Ala-sRNA^{CySH} (cpm)	Total (cpm)	Cysteic acid spot (cpm)	Alanine spot (cpm)
Untreated CySH-sRNA^{CySH}	2,870	—	520	225	3
Raney Nickel-treated CySY-sRNA^{CySH}	660	1,540	395	66	141

A small-scale incubation was made, as described in Table 3, except that before hydrolysis only 0.15 μmole of C^{12}-cysteine and 0.15 μmole of C^{12}-alanine were added. The hydrolysate was oxidized with performic acid and electrophorized under the same conditions as in Figure 5. The alanine and cysteic acid spots were eluted and counted.

amounts of incubate were analyzed, and it may be seen that again about one third of the counts present in either the reduced or unreduced input were incorporated into the TCA-insoluble material. With the treated preparation containing 65 per cent of its counts as Ala-sRNA^{CySH}, two thirds of the counts were retained on the column and appeared in the Dowex eluate as expected for alanine.

As a further check, the Dowex filtrates and eluates from the experiment in Table 3 were compared by electrophoresis followed by radioautography. As shown in Figure 5, radioactivity was found in the position corresponding to alanine, whereas no radioactivity appeared in this area with the untreated CySH-sRNA^{CySH}. Cysteic acid moved slightly in the opposite direction and was proportionally weaker with the reduced sample (No. 2) than with the untreated (No. 1). Quantitative data are given in Table 4 for an experiment in which an aliquot of the TCA precipitate was first counted for total counts, the oxidized hydrolysate was then electrophorized on paper as described in Figure 5, and the spots for cysteic acid and alanine were eluted from the paper and counted.

FIG. 5.—Analysis of the radioactivity transferred into TCA-precipitable material. Each sample is the combined Dowex filtrate and eluate from Table 3. The samples were electrophorized on Whatman 3 MM paper at pH 1.85 (acetic acid:formic acid:water, 7.8:2.5:89.7) at 70 volts/cm for 60 min. The electropherogram was stapled to X-ray film and exposed for 25 days. (1) untreated C^{14}-CySH-sRNA^{CySH} (2) Raney Nickel-treated C^{14}CySH-sRNA^{CySH}.

Thus, alanine is accepted by the poly UG system from Ala-sRNA^{CySH}, but not from Ala-sRNA^{Ala}. We conclude, then, that poly UG recognizes the specific sRNA molecule rather than the amino acid.

Discussion.—The results give direct support to the hypothesis that sRNA functions as an adaptor in specifying the fit of an amino acid on a template. Once attached to sRNA, the cysteine could be altered to another one of the natural amino acids, but the coding properties of the hybrid remained unchanged. In other experiments,[16] it has been shown that attached cysteine may also be oxidized to cysteic acid, a rather drastic change in the chemical properties of the amino acid, without loss of reactivity with poly UG. It remains to be shown that the result obtained with the artificial UG polymer carries over to the synthesis of a natural protein. This is currently being tested in the hemoglobin synthesizing system, utilizing reticulocyte ribosomes.

Summary.—A direct test is made of the hypothesis that sRNA functions as an adaptor in specifying the fit of amino acids on a template. Cysteine, attached to its normal acceptor sRNA, is converted, while still attached, to alanine, by reduction with Raney Nickel. The coding properties of the resultant hybrid are examined by the use of poly UG which stimulates incorporation into polypeptide of cysteine but not alanine. Alanine, attached to the cysteine acceptor, is incorporated, indicating that the adaptor hypothesis is correct.

Since an amino acid, once attached, no longer participates in coding, it follows that the code (i.e., the correspondence between nucleotide sequence in template RNA and amino acid sequence in protein) is embodied in the precise structures and interrelationships of the set of sRNA adaptors and activating enzymes.

* Supported by grants from the National Science Foundation and from the National Cancer Institute, National Institutes of Health, U.S. Public Health Service.

† The following abbreviations are used: sRNA, soluble ribonucleic acid; poly-U, polyuridylic acid; poly UG, polyuridylic-guanylic acid; PEP, phosphoenolpyruvate; TCA, trichloroacetic acid.

§ Helen Hay Whitney Foundation Fellow. Present address: Département de Biologie, Centre d'Études Nucléaires, Scalay, France.

‡ The Rockefeller Institute, New York 21, N. Y.

** Johns Hopkins University, Baltimore 18, Md.

†† Department of Biological Sciences, Purdue University, Lafayette, Ind.

[1] Hoagland, M. B., M. L. Stephenson, J. F. Scott, L. I. Hecht, and P. C. Zamecnik, *J. Bicl. Chem.*, **231**, 241 (1958).

[2] Nathans, D., G. von Ehrenstein, R. Monro, and F. Lipmann, *Fed. Prcc.*, **21**, 127 (1962).

[3] von Ehrenstein, G., and F. Lipmann, these PROCEEDINGS, **47**, 941 (1961).

[4] Crick, F. H. C., *Symposium Soc. Exp. Bicl.*, **12**, 138 (1958).

[5] Hoagland, M. B., in *Structure and Function of Genetic Elements*, Brookhaven Symposia in Biology, No. 12 (1959), p. 40.

[6] Nirenberg, M. W., and J. H. Matthaei, these PROCEEDINGS, **47**, 1588 (1961).

[7] Lengyel, P., J. F. Speyer, and S. Ochoa, these PROCEEDINGS, **47**, 1936 (1961); **48**, 282 (1962).

[8] Speyer, J. F., P. Lengyel, C. Basilio, and S. Ochoa, these PROCEEDINGS, **48**, 441 (1962).

[9] Martin, R. G., J. H. Matthaei, O. W. Jones, and M. W. Nirenberg, *Biophys. Biochem. Res. Comm.*, **6**, 410 (1962).

[10] Schlossmann, K., and F. Lynen, *Biochem. Z.*, **328**, 591 (1957).

[11] Nathans, D., and F. Lipmann, these PROCEEDINGS, **47**, 497 (1961).

[12] Schröter, R., in *Newer Methods of Preparative Organic Chemistry* (New York: Interscience Publishers, Inc., 1948), p. 72.

[13] Mozingo, R., D. E. Wolf, S. A. Harris, and K. Folkers, *J. Am. Chem. Scc.*, **65**, 1013 (1943).

[14] Singer, M. F., and J. K. Guss, *J. Biol. Chem.*, **237**, 182 (1962).

[15] Nirenberg, M. W., J. H. Matthaei, and O. W. Jones, these PROCEEDINGS, **48**, 104 (1962).

[16] Chapeville, F., *Fed. Proc.*, **21**, 414d (1962).

Reprinted from PROCEEDINGS OF THE NATIONAL ACADEMY OF SCIENCES, Vol. 48, 1962

A PHSYICAL BASIS FOR DEGENERACY IN THE AMINO ACID CODE

BY BERNARD WEISBLUM,* SEYMOUR BENZER,* AND ROBERT W. HOLLEY†

PURDUE UNIVERSITY AND U. S. PLANT, SOIL AND NUTRITION LABORATORY

Communicated June 25, 1962

If two or more nucleotide sequences specify the same amino acid, the code is said to be degenerate for that amino acid. Evidence that such degeneracy exists in *E. coli* comes from experiments on the stimulation of ribosomal incorporation of amino acids by synthetic polyribonucleotides. For instance, Martin *et al.*[1] and Speyer *et al.*[2] found that either poly UC or poly UG stimulates incorporation of leucine, so that at least two different combinations code for the same amino acid. Since the specificity of coding seems to reside in sRNA molecules functioning as adaptors,[3, 4] such degeneracy might be due to the existence of two sRNA acceptors for leucine, each having different coding properties. Two peaks of leucine-acceptor activity were demonstrated in yeast sRNA by Doctor *et al.*[5] using countercurrent distribution. Berg and Lagerkvist[6] have shown that *E. coli* sRNA also contains two leucine acceptors with different acceptor-terminal nucleotide sequences. In this paper, two leucine acceptors in *E. coli*, separated by countercurrent distribution, are shown to have different coding properties. One responds preferentially to poly UC, the other to poly UG.

Materials and Methods.—Separation of two leucine acceptors: sRNA from *E. coli*, strain KB, was prepared by phenol extraction and passage over DEAE cellulose as described by Holley *et al.*,[7] then stripped of amino acids, dialyzed, and lyophilized, as described by von Ehrenstein and Lipmann.[8] Countercurrent distribution was done exactly as described by Apgar *et al.*[9] 200 mg of *E. coli* sRNA was added to the first 5 tubes of a 200-tube apparatus, and the final fractions were combined into 40 sets of 5 tubes each, dialyzed, recovered by evaporation, and each dissolved in 2 ml of deionized water. A 100,000 × *g* supernatant of an *E. coli* extract was used as the source of amino acyl RNA synthetase activity. For assays of acceptor activity of the sRNA fractions, the reaction mixture contained in a volume of 0.5 ml: Tris-HCl, pH 7.2, 50 μM; $MgCl_2$, 5 μM; KCl, 5 μM; adenosine triphosphate, Na salt (ATP), 0.5 μM; 0.2 μc of C^{14}-amino acid, either DL-leucine, 10 μc/μM, or L-arginine, 5 μc/μM (Calbiochem); reduced glutathione (GSH), 2 μM; bovine serum albumin (BSA), 100 μg; *E. coli* 100,000 × *g* supernatant, 750 μg protein, 0.05 ml of the countercurrent fraction to be tested. After incubation for 25 min at 30°C, 3 ml cold salt-ethanol (0.5 *M* NaCl, 67% ethanol) and 1 mg of crude yeast RNA were added and the tubes chilled in an ice bath for 5 min. After 4 centrifugations at 4°C (resuspending in salt-ethanol), the precipitate was dissolved in 1.5 *N* NH_4OH, dried, and assayed for radioactivity using a thin-window, gas-flow counter (Nuclear Chicago Corporation), with a C^{14}-counting efficiency of 19%.

Attachment of C^{14}-leucine to peak I and peak II sRNA: The reaction mixture for attaching C^{14}-leucine to the sRNA fractions contained in a total volume of 1 ml: Tris-HCl, pH 7.2, 100 μM; $MgCl_2$, 10 μM; KCl, 10 μM; GSH, 2 μM; ATP, 2.5 μM; C^{14}-leucine, 138 μc/μM (New England Nuclear Corporation), 0.024 μM; 19 C^{12}-amino acids (leucine omitted), 0.2 μM each; 100,000 × *g* supernatant protein, 1.5 mg; sRNA from peak I or peak II (Fig. 1), 0.38 or 0.19 mg, respectively.

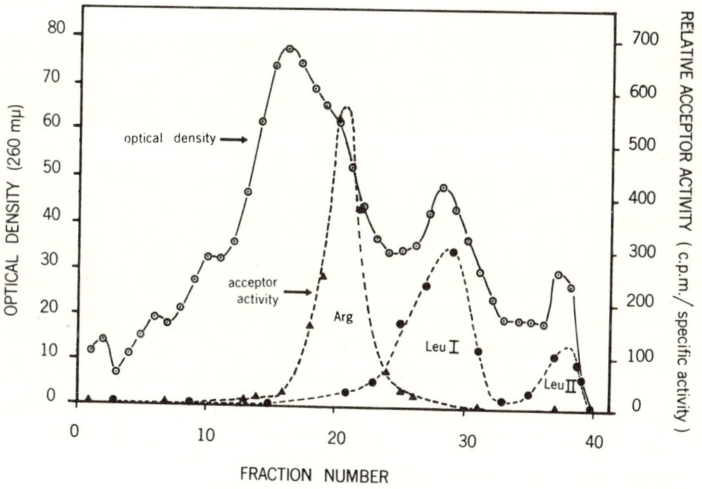

FIG. 1.—Resolution of two leucine acceptors by countercurrent distribution of *E. coli* sRNA. Arginine acceptor activity is shown for comparison. The optical density refers to that of each countercurrent fraction dissolved in 2 ml of water. For the assay of acceptor activity, 0.05 ml was taken from each tube.

After incubation for 30 min at 35°C, the reaction was stopped by addition of 1 ml phenol, and the charged sRNA was isolated as described by von Ehrenstein and Lipmann.[8] The amino acyl-sRNA obtained contained 570,000 cpm per mg sRNA in the case of peak I, and 520,000 cpm per mg sRNA in the case of peak II.

Attachment of C^{14}-leucine to unfractionated sRNA: The reaction mixture was similar to above with the addition of an ATP-generating system. It contained in a total volume of 2 ml: Tris-HCl, pH 7.2, 200 μM; $MgCl_2$, 20 μM; KCl 20 μM; GSH, 4 μM; ATP, 5 μM; phosphoenolpyruvate (PEP), Na salt (Calbiochem), 25 μM; PEP kinase (Boehringer), 40 μg; C^{14}-leucine, 10 μc/μM (Calbiochem), 0.1 μM; sRNA, 20 mg; 100,000 × *g* supernatant, 3 mg protein. In this particular preparation, no mixture of C^{12}-amino acids was included. After incubation for 30 min at 35°C, the charged sRNA was isolated by the phenol procedure as above. The product was dialyzed for 18 hr against deionized water at 4°C, and stored frozen at −20°C. It contained 8,600 cpm per mg sRNA.

For attachment of the C^{12}-amino acid mixture to sRNA, the same proportions were used, except that the C^{14}-leucine was replaced by 2 μM each of 19 amino acids (leucine omitted).

Ribosomal preparation: Preincubated 30,000 × *g* supernatant (iS-30) of *E. coli* extract was prepared as described by Nirenberg and Matthaei,[11] omitting DNase. The preparation used had an OD_{260} of 223 and contained 20 mg protein per ml. When tested for activity as described,[10, 11] 180-fold stimulation of incorporation of free C^{14}-phenylalanine was produced by 10 μg of poly U. Under conditions similar to those used in the leucine transfer experiments below, 75% of sRNA-bound C^{14}-phenylalanine was found to be incorporated into the hot trichloroacetic acid (TCA)-insoluble product in the presence of 10 μg of poly U, in agreement with Nirenberg *et al.*[11]

Ribosomal incorporation of C^{14}-leucine bound to sRNA: The reaction mixture contained, in a total volume of 1 ml: Tris-HCl, pH 7.6, 100 μM; $MgCl_2$, 14 μM; KCl, 50 μM; GTP, 0.03 μM; PEP, 5 μM; PEP kinase, 20 μg; β-mercaptoethanol, 5 μM; C^{14}-leucyl-sRNA (unfractionated, or from peak I or peak II), as stated in the legends of Figures 2, 3, and 4; C^{12}-amino acyl-sRNA (leucine omitted), 0.1 mg; C^{12}-leucine, 2 μM; 5 μg of either poly U, poly UC, or poly UG, as stated; iS-30, 0.1 ml. Under the conditions used, the amount of C^{14}-leucine incorporated into the acid-insoluble product in 10 min was proportional to the input C^{14}-leucyl-sRNA.

After incubation for the various stated times, the reaction was stopped by addition of 5 ml of 5% TCA and the resulting precipitate washed, digested, and plated as described by Siekevitz.[12]

Synthetic polyribonucleotides: Poly UC (input ratio 1:1) and poly UG (input ratio 5:1) were kindly donated by Dr. S. Ochoa and Dr. R. F. Beers, respectively. Poly U was purchased from the Miles Laboratories, Clifton, New Jersey.

Results.—Separation of two leucine acceptors: Figure 1 shows the results of a countercurrent fractionation of *E. coli* sRNA. The assay of leucine-acceptor activity shows two clearly separated peaks. Samples taken from these will be subsequently referred to as "peak I" and "peak II" sRNA.

Coding properties of peak I and peak II sRNA: Samples of peak I and peak II sRNA were charged with C^{14}-leucine and their response to specific polymers was determined. Nirenberg *et al.*[11] have shown that poly U-stimulated incorporation of C^{14}-phenylalanine can take place by direct transfer of the amino acid from the amino acyl-sRNA complex. The same applies to sRNA-bound cysteine in response to poly UG, as shown by Chapeville *et al.*[4] While the observations that both poly UC and poly UG stimulate ribosomal incorporation of leucine were made using the free amino acid, it can be seen below that the same result applies to leucine bound to sRNA.

Figure 2 shows the kinetics of transfer of C^{14}-leucine from peak I sRNA in response to 3 different polymers. Poly UC produces the highest level of stimulation. A strikingly different result is observed if leucine is transferred from peak II sRNA, as shown in Figure 3. In that case, poly UG is the most active, although poly U also has some effect. It is concluded therefore that peak I sRNA and peak II sRNA have different coding properties.

Note in Figure 3 that peak II responds somewhat to Poly U. It is unlikely that this effect is due to contamination of the poly U with G, since base analysis of the polymer showed that the G content, if any, was less than 0.5%.

Figure 4 gives, for comparison, the results for unfractionated sRNA (not taken through the countercurrent procedure). The pattern of response to the various polymers corresponds roughly to the sum of both peaks. In making the comparison, one must take into account that peak I is about four times as abundant as peak II and also that the input of counts in Figure 2 is about twice that in Figure 3. Thus, there is no indication that the properties of the sRNA were significantly altered by the countercurrent fractionation. A reservation must be noted here, however. In preparing the C^{14}-leucyl-sRNA used in Figure 4, no mixture of C^{12}-amino acids was added. Therefore, it cannot be excluded that minor impurity amino acids in the C^{14}-leucine could have made some contribution to the transfer observed in Figure 4.

Discussion.—Degeneracy, in the case of leucine, can apparently be accounted for by the different coding properties of two acceptors. Berg and Lagerkvist[6]

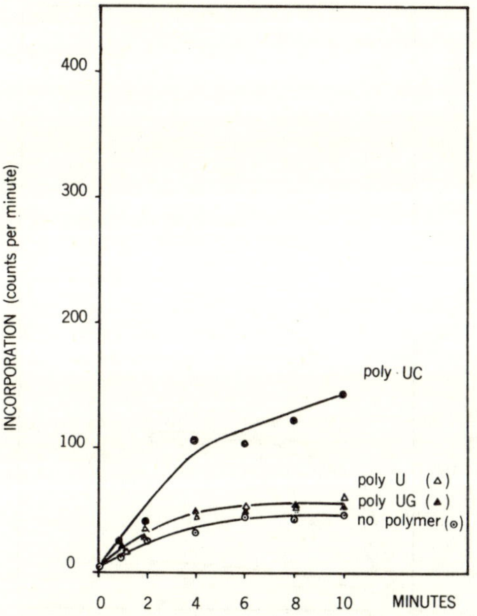

FIG. 2.—Polymer-dependent incorporation of leucine bound to peak I sRNA. The input of C^{14}-leucyl peak I-sRNA was 0.012 mg, containing 6,700 cpm. 0.1 mg C^{12}-amino acyl-sRNA (minus leucine) was also added. The procedure is described in *Methods*.

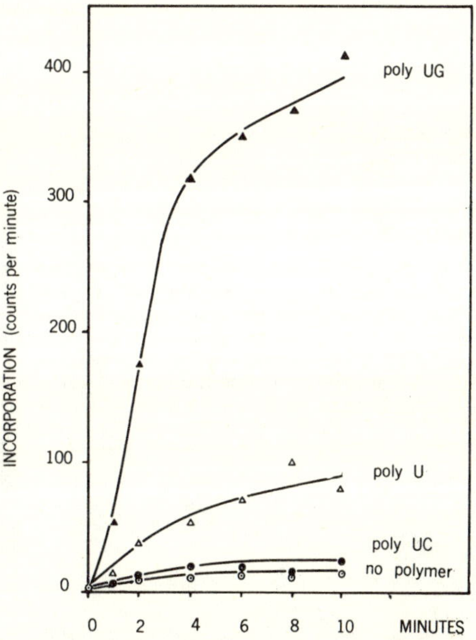

FIG. 3.—Polymer-dependent incorporation of leucine bound to peak II sRNA. The reaction mixture and procedure were as in Fig. 2, except that the labeled input was 0.006 mg of C^{14}-leucyl peak II-sRNA, containing 3,100 cpm.

found two nucleotide sequences at the acceptor terminus for the leucine-acceptor sRNA in *E. coli*—GCACCA and GUACCA in proportion four to one. This relative abundance agrees with the sizes of peak I and peak II. However, there is no obvious complementary relation between these sequences and the responses of peak I and peak II sRNA's to poly UC and poly UG. This suggests that the coding properties of the sRNA molecule are not determined by these residues alone.

Two separable acceptors have also been demonstrated by countercurrent distribution for histidine,[13] threonine, and leucine[5] in yeast, as well as lysine[14] in rat liver. Sueoka and Yamane,[15] using a different technique, have demonstrated multiple peaks in *E. coli* sRNA for various amino acids. However, the existence of two sRNA's for a given amino acid does not necessarily imply degeneracy in coding, since modification of an sRNA molecule may be possible without altering its coding specificity.

Whether degeneracy is widespread in nature and plays an important part in protein synthesis remains to be determined. Some degree of degeneracy may confer increased survival value upon a cell, by reducing the number of nonsense coding units, i.e., those that correspond to no amino acid. A mutation that changes one amino acid in a polypeptide chain to another amino acid is less likely to be deleterious than one which completely impedes the completion of the growing chain.[16]

It should be pointed out that two kinds of degeneracy are possible. The first kind is illustrated by the difference between peak I and peak II, a given amino acid

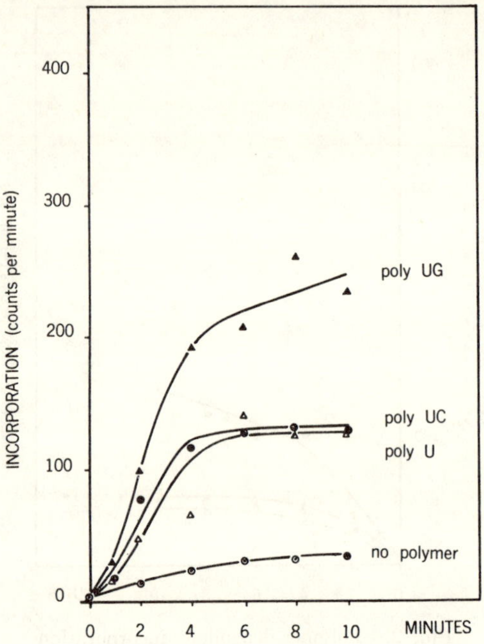

FIG. 4.—Polymer-dependent incorporation of leucine bound to unfractionated *E. coli* sRNA. The reaction mixture and procedure were as in Fig. 2, except that 0.5 mg of C^{14}-leucyl-sRNA, containing 4,300 cpm, was used. The specific activity of the leucine was lower (10 $\mu c/\mu M$) than that used in Figs. 2 and 3.

corresponding to two different coding units. The existence of degeneracy of the second kind, where two different amino acids correspond to the same coding unit, is suggested by the observations that both leucine and phenylalanine respond to poly U.

The present experiments also provide independent proof for the role of sRNA as an adaptor in coding. As shown by Chapeville *et al.*[4] two different amino acids (cysteine and alanine) attached to the same sRNA possess the same coding specificity. In the present experiment, two sRNA's were found to have different coding properties, even though loaded with the same amino acid, showing that the coding specificity resides in the sRNA.

Summary.—*E. coli* contains two leucine-acceptor sRNA's separable by countercurrent distribution. Ribosomal incorporation of leucine attached to one of these is stimulated by poly UC, while leucine attached to the other responds more readily to poly UG. This provides an explanation for the degeneracy observed in coding experiments with leucine. The experiments also confirm that the coding specificity is carried by the sRNA.

This research was supported by grants from the National Science Foundation and the National Institutes of Health. One of us (B. W.) is the holder of a National Institutes of Health Postdoctoral Fellowship.

* Department of Biological Sciences, Purdue University, Lafayette, Indiana.

† U. S. Plant, Soil and Nutrition Laboratory, Ithaca, New York.

[1] Martin, R. G., J. H. Matthaei, O. W. Jones, and M. W. Nirenberg, *Biochem. Biophys. Res. Comm.*, **6,** 410 (1962).

[2] Speyer, J. F., P. Lengyel, C. Basilio, and S. Ochoa, these PROCEEDINGS, **48,** 441 (1962).

[3] Crick, F. H. C., in *The Biological Replication of Macromolecules,* Symposia of the Society for Experimental Biology (London: Cambridge University Press, 1959) vol. 12, p. 138.

[4] Chapeville, F., F. Lipmann, G. von Ehrenstein, B. Weisblum, W. J. Ray, Jr., and S. Benzer, these PROCEEDINGS, **48,** 1086 (1962).

[5] Doctor, B. P., J. Apgar, and R. W. Holley, *J. Biol. Chem.*, **236,** 1117 (1961).

[6] Berg, P., and U. Lagerkvist, *Acides Ribonucléiques et Polyphosphates*, C.N.R.S., Paris, 1962.

[7] Holley, R. W., J. Apgar, B. P. Doctor, J. Farrow, M. A. Marini, and S. H. Merrill, *J. Biol. Chem.*, **236,** 200 (1961).

[8] von Ehrenstein, G., and F. Lipmann, these PROCEEDINGS, **47,** 941 (1961).

[9] Apgar, J., R. W. Holley, and S. H. Merrill, *J. Biol. Chem.*, **237,** 796 (1962).

[10] Nirenberg, M. W., and J. H. Matthaei, these PROCEEDINGS, **47,** 1588 (1961).

[11] Nirenberg, M. W., J. H. Matthaei, and O. W. Jones, these PROCEEDINGS, **48,** 104 (1962).

[12] Siekevitz, P., *J. Biol. Chem.*, **195,** 549 (1952).

[13] Apgar, J., R. W. Holley, and S. H. Merrill, *Biochim. Biophys. Acta*, **53,** 220 (1961).

[14] Apgar, J., and R. W. Holley, in preparation.

[15] Sueoka, N., and T. Yamane, these PROCEEDINGS, **48,** 1454 (1962).

[16] Benzer, S., and S. P. Champe, these PROCEEDNGS, **48,** 1114 (1962).

Reprinted from PROCEEDINGS OF THE NATIONAL ACADEMY OF SCIENCES, Vol. 47, 1961

THE DEPENDENCE OF CELL- FREE PROTEIN SYNTHESIS IN E. COLI UPON NATURALLY OCCURRING OR SYNTHETIC POLYRIBONUCLEOTIDES

BY MARSHALL W. NIRENBERG AND J. HEINRICH MATTHAEI*

NATIONAL INSTITUTES OF HEALTH, BETHESDA, MARYLAND

Communicated by Joseph E. Smadel, August 3, 1961

A stable cell-free system has been obtained from *E. coli* which incorporates C^{14}-valine into protein at a rapid rate. It was shown that this apparent protein synthesis was energy-dependent, was stimulated by a mixture of L-amino acids, and was markedly inhibited by RNAase, puromycin, and chloramphenicol.[1] The present communication describes a novel characteristic of the system, that is, a requirement for template RNA, needed for amino acid incorporation even in the presence of soluble RNA and ribosomes. It will also be shown that the amino acid incorporation stimulated by the addition of template RNA has many properties expected of *de novo* protein synthesis. Naturally occurring RNA as well as a synthetic polynucleotide were active in this system. The synthetic polynucleotide appears to contain the code for the synthesis of a "protein" containing only one amino acid. Part of these data have been presented in preliminary reports.[2, 3]

Methods and Materials.—The preparation of enzyme extracts was modified in certain respects from the procedure previously presented.[1] *E. coli* W3100 cells harvested in early log phase were washed and were disrupted by grinding with alumina (twice the weight of washed cells) at 5° for 5 min as described previously.[1] The alumina was extracted with an equivalent weight of buffer containing 0.01 *M* Tris(hydroxymethyl)aminomethane, pH 7.8, 0.01 *M* magnesium acetate, 0.06 *M* KCl, 0.006 *M* mercaptoethanol (standard buffer). Alumina and intact cells were removed by centrifugation at 20,000 × *g* for 20 min. The supernatant fluid was decanted, and 3 μg DNAase per ml (Worthington Biochemical Co.) were added, rapidly reducing the viscosity of the suspension, which was then centrifuged again at 20,000 × *g* for 20 min. The supernatant fluid was aspirated and was centrifuged at 30,000 × *g* for 30 min to clear the extract of remaining debris. The liquid layer was aspirated (S-30) and was centrifuged at 105,000 × *g* for 2 hr to sediment the ribosomes. The supernatant solution (S-100) was aspirated, and the solution just above the pellet was decanted and discarded. The ribosomes were washed by resuspension in the standard buffer and centrifugation again at 105,000 × *g* for 2 hr. Supernatant fluid was discarded and the ribosomes were suspended in standard buffer (W-Rib). Fractions S-30, S-100, and W-Rib were dialyzed against 60 volumes of standard buffer overnight at 5° and were divided into aliquots for storage at −15°.

In some cases, fresh S-30 was incubated for 40 min at 35°. The reaction mixture components in μmoles per ml were as follows: 80 Tris, pH 7.8; 8 magnesium acetate; 50 KCl; 9 mercaptoethanol; 0.075 each of 20 amino acids;[1] 2.5 ATP, K salt; 2.5 PEP, K salt; 15 μg PEP kinase (Boehringen & Sons, Mannheim, Germany). After incubation, the reaction mixture was dialyzed at 5° for 10 hr against 60 volumes of standard buffer, changed once during the course of dialysis. The incubated S-30 fraction was stored in aliquots at −15° until needed (Incubated-S-30).

RNA fractions were prepared by phenol extraction using freshly distilled phenol. Ribosomal RNA was prepared from fresh, washed ribosomes obtained by the method given above. In later RNA preparations, a 0.2% solution of sodium dodecyl sulfate recrystallized by the method of Crestfield *et al.*[4] was added to the suspension of ribosomes before phenol treatment. The suspension was shaken at room temperature for 5 min. Higher yields of RNA appeared to be obtained when the sodium dodecyl sulfate step was used; however, good RNA preparations were also obtained when this step was omitted. An equal volume of H_2O-saturated phenol was added to ribosomes suspended in standard buffer after treatment with sodium dodecyl sulfate, and the suspension was shaken vigorously at room temperature for 8–10 min. The aqueous phase was aspirated from the phenol phase after centrifugation at 1,450 × *g* for 15 min. The aqueous layer was extracted two more times in the same manner, using $^1/_2$ volume of H_2O-saturated phenol in each case. The final aqueous phase was chilled to 5° and NaCl was added to a final concentration of 0.1%. Two volumes of ethyl alcohol at −20° were added with stirring to precipitate the RNA. The suspension was centrifuged at 20,000 × *g* for 15 min and the supernatant solution was decanted and discarded. The RNA pellet was dissolved in minimal concentrations of standard buffer (minus mercaptoethanol) by gentle homogenization in a glass Potter-Elvehjem homogenizer (usually the volume of buffer used was about $^1/_3$ the volume of the original ribosome suspension). The opalescent solution of RNA was dialyzed for 18 hr against 100 volumes of standard buffer (minus mercaptoethanol) at 5°. The dialyzing buffer was changed once. After dialysis, the RNA solution was centrifuged at 20,000 × *g* for 15 min and the pellet was discarded. The RNA solution, which contained less than 1% protein, was divided into aliquots and was stored at −15° until needed.

Soluble RNA was prepared from 105,000 × *g* supernatant solution by the phenol extraction method described above. Soluble RNA was also stored at −15°. Alkali-degraded RNA was prepared by incubating RNA samples with 0.3 *M* KOH at 35° for 18 hr. The solutions then were neutralized and dialyzed against standard buffer (minus mercaptoethanol). RNAase-digested samples of RNA were prepared by incubating RNA with 2 μg per ml of crystalline

RNAase (Worthington Biochemical Company) at 35° for 60 min. RNAase was destroyed by four phenol extractions performed as given above. After the last phenol extraction, the samples were dialyzed against standard buffer minus mercaptoethanol. RNA samples were treated with trypsin by incubation with 20 μg per ml of twice recrystallized trypsin (Worthington Biochemical Company) at 35° for 60 min. The solution was treated four times with phenol and was dialyzed in the same manner.

The radioactive amino acids used, their source, and their respective specific activities are as follows: U-C^{14}-glycine, U-C^{14}-L-isoleucine, U-C^{14}-L-tyrosine, U-C^{14}-L-leucine, U-C^{14}-L-proline, L-histidine-2(ring)-C^{14}, U-C^{14}-L-phenylalanine, U-C^{14}-L-threonine, L-methionine (methyl-C^{14}), U-C^{14}-L-arginine, and U-C^{14}-L-lysine obtained from Nuclear-Chicago Corporation, 5.8, 6.2, 5.95, 6.25, 10.5, 3.96, 10.3, 3.9, 6.5, 5.8, 8.3 mC/mM, respectively; C^{14}-L-aspartic acid, C^{14}-Lglutamic acid, C^{14}-L-alanine, obtained from Volk, 1.04, 1.18, 0.75 m C/mM, respectively; D-L-tryptophan-3 C^{14}, obtained from New England Nuclear Corporation, 2.5 mC/mM; S^{35}-L-cystine obtained from the Abbott Laboratories, 2.4 mC/mM; U-C^{14}-L-serine obtained from the Nuclear-Chicago Corporation, 0.2 mC/mM. Other materials and methods used in this study are described in the accompanying paper.[1] All assays were performed in duplicate.

Results.—Stimulation by ribosomal RNA: In the previous paper,[1] it was shown that DNAase markedly decreased amino acid incorporation in this system after 20 min. For the purpose of this investigation, 30,000 × *g* supernatant fluid fractions previously incubated with DNAase and other components of the reaction mixtures (Incubated-S-30 fractions) were used for many of the experiments.

Figure 1 shows that incorporation of C^{14}-L-valine into protein by Incubated-S-30 fraction was stimulated by the addition of purified *E. coli* soluble RNA. Maximal stimulation was obtained with approximately 1 mg soluble RNA. In some experiments, increasing the concentration 5-fold did not further stimulate the system. Soluble RNA was added to all reaction mixtures unless otherwise specified.

Figure 2 demonstrates that *E. coli* ribosomal RNA preparations markedly stimu-

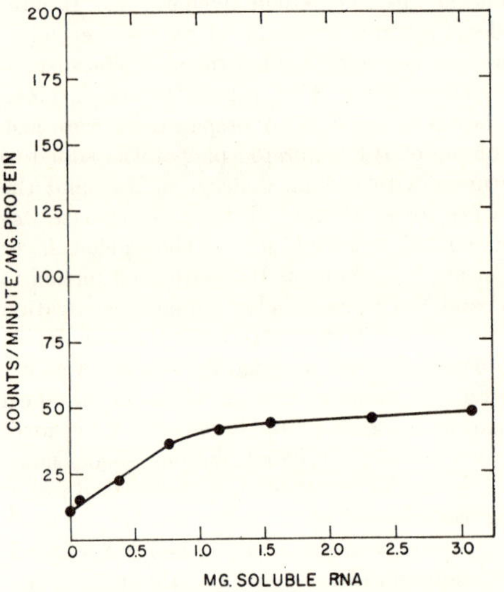

FIG 1.—Stimulation of amino acid incorporation into protein by *E. coli* soluble RNA. Composition of reaction mixtures is specified in Table 1. Samples were incubated at 35° for 20 min. Reaction mixtures contained 4.4 mg. of Incubated-S-30 protein.

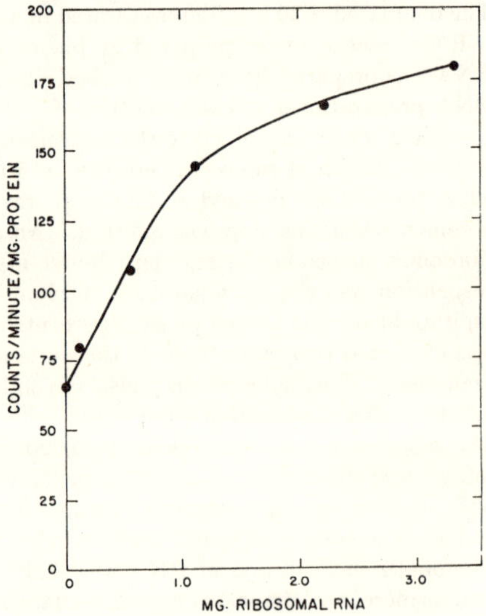

FIG. 2.—Stimulation of amino acid incorporation into protein by *E. coli* ribosomal RNA in the presence of soluble RNA. Composition of reaction mixtures is specified in Table 1. Samples were incubated at 35° for 20 min. Reaction mixtures contained 4.4 mg of Incubated-S-30 protein and 1.0 mg *E. coli* soluble RNA.

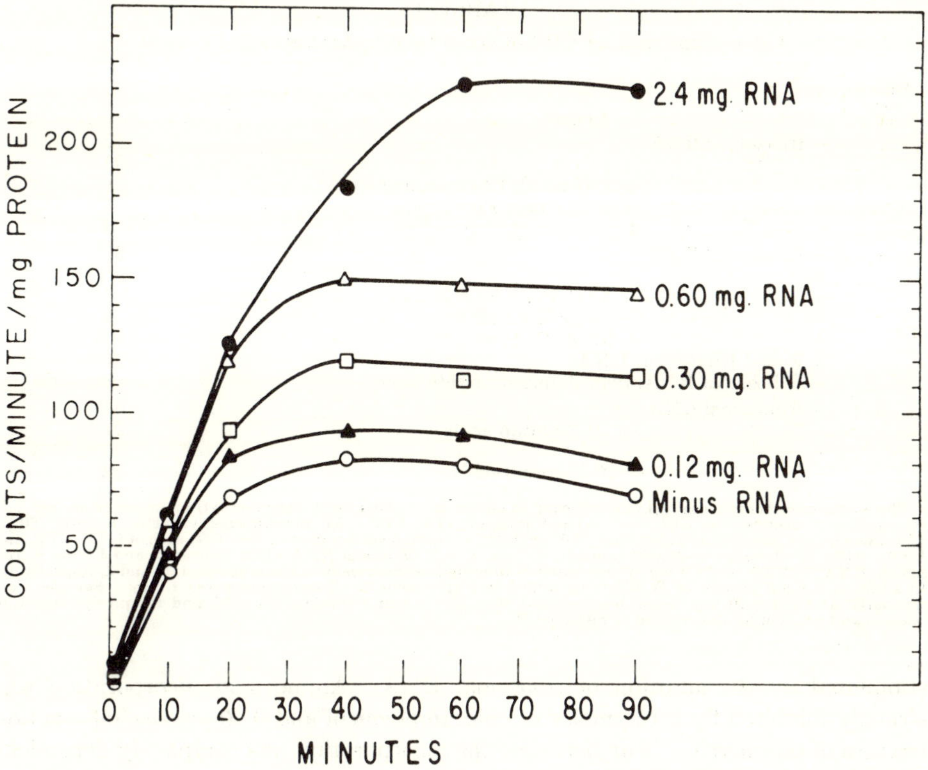

FIG. 3.—Dependence of C^{14}-L-valine incorporation into protein upon ribosomal RNA. The composition of the reaction mixtures and the incubation conditions are presented in Table 1. Reaction mixtures contained 0.98 mg of *E. coli* soluble RNA and 4.4 mg of Incubated-S-30-protein.

lated incorporation of C^{14}-valine into protein even though maximally stimulating concentrations of soluble RNA were present in the reaction mixtures. A linear relationship between the concentration of ribosomal RNA and C^{14}-valine incorporation into protein was obtained when low concentrations of ribosomal RNA were used. Increasing the soluble RNA concentration up to 3-fold did not replace the effect observed when ribosomal RNA was added.

The effect of ribosomal RNA in stimulating incorporation of C^{14}-valine into protein is presented in more detail in Figure 3. In the absence of ribosomal RNA, incorporation of C^{14}-valine into protein by the incubated-S-30 fraction was quite low when compared with S-30 (not incubated before storage at $-15°$) and stopped almost completely after 30 min. At low concentrations of ribosomal RNA, maximum amino acid incorporation into protein was proportional to the amount of ribosomal RNA added, suggesting stoichiometric rather than catalytic action of ribosomal RNA. Total incorporation of C^{14}-valine into protein was increased more than 3-fold by ribosomal RNA in this experiment even in the presence of maximally stimulating concentrations of soluble RNA. Ribosomal RNA may be added at any time during the course of the reaction, and, after further incubation, an increase in incorporation of C^{14}-valine into protein will result.

Characteristics of amino acid incorporation stimulated by ribosomal RNA: In Table 1 are presented the characteristics of C^{14}-L-valine incorporation into protein

TABLE 1

CHARACTERISTICS OF C^{14}-L-VALINE INCORPORATION INTO PROTEIN

Experiment no.	Addition	Counts/min/mg protein
1	− Ribosomal RNA	42
	+ " "	204
	+ " " + 0.15 μmole Chloramphenicol	58
	+ " " + 0.20 μmole Puromycin	7
	+ " " deproteinized at zero time	8
2	− Ribosomal RNA	35
	+ " "	101
	+ " " − ATP, PEP, PEP kinase	7
	+ " " + 10 μg RNAase	6
	+ " " + 10 μg DNAase	110
	+ Boiled Ribosomal RNA	127
	+ Ribosomal RNA, deproteinized at zero time	8
3	− Ribosomal RNA	34
	− " " − 20 L amino acids	21
	+ " "	99
	+ " " − 20-L-amino acids	52

The reaction mixtures contained the following in μmole/ml: 100 Tris(hydroxymethyl) aminomethane. pH 7.8; 10 magnesium acetate; 50 KCl; 6.0 mercaptoethanol; 1.0 ATP; 5.0 phosphoenolpyruvate, K salt; 20 μg phosphoenolpyruvate kinase, crystalline; 0.05 each of 20 L-amino acids minus valine; 0.03 each of GTP, CTP, and UTP; 0.015 C^{14}-L-valine (~70,000 counts); 3.1 mg. *E. coli* ribosomal RNA where indicated, and 1.0 mg *E. coli* soluble RNA; 3.2, 3.2, and 1.4 mg of incubated-S-30 protein were present in Experiments 1, 2, and 3, respectively. In addition 4.4 mg protein of W-Rib were added in Experiment 3. Total volume was 1.0 ml. Samples were incubated at 35° for 20 min, were deproteinized with 10 per cent trichloroacetic acid, and the precipitates were washed and counted by the method of Siekevitz.[22]

stimulated by the addition of ribosomal RNA. Amino acid incorporation was strongly inhibited by 0.15 μmoles of chloramphenicol and 0.20 μmoles/ml reaction mixture of puromycin. Furthermore, the incorporation was completely dependent upon the addition of ATP and an ATP-generating system and was totally inhibited by 10 μg/ml RNAase. Equivalent amounts of DNAase had no effect upon the incorporation stimulated by the addition of ribosomal RNA. Placing a ribosomal RNA preparation in a boiling water bath for 10 min did not destroy its C^{14}-valine incorporation activity; instead, a slight increase in activity was consistently observed. However, when these RNA preparations were placed in a boiling water bath, a copious, white precipitate resulted. Upon cooling the suspension in an ice bath, the precipitate immediately dissolved.

The data of Table 1 also demonstrate that the incorporation of amino acids into protein in the presence of ribosomal RNA was further stimulated by the addition of a mixture of 20 L-amino acids, suggesting cell-free protein synthesis.

C- and N-terminal analyses of the ribosomal RNA-dependent product of the reaction were performed with carboxypeptidase and 1-fluoro-2,4-dinitrobenzene respectively (Dr. Frank Tietze kindly performed these analyses). Four per cent of the radioactivity was released from the C-terminal end and 1% was associated with the N-terminal end. The remainder of the C^{14}-label was internal. Similar results were obtained when reactions were performed using S-30 enzyme fractions which had not been treated with DNAase. Protein precipitates isolated from reaction mixtures after incubation were completely hydrolyzed with HCl, and the C^{14}-label incorporated into protein was demonstrated to be valine by paper chromatography.

Many of the experiments presented in this paper were performed with enzyme fractions prepared with DNAase added to reduce their viscosity. Ribosomal

RNA also stimulated C^{14}-valine incorporation when enzyme extracts prepared in the absence of DNAase were used.

To be effective in stimulating amino acid incorporation into protein, the ribosomal RNA *required the presence of washed ribosomes.* The data of Table 2 show

TABLE 2

THE INEFFECTIVENESS OF RIBOSOMAL RNA IN STIMULATING C^{14}-L-VALINE INCORPORATION INTO PROTEIN IN THE PRESENCE OF RIBOSOMES OR 105,000 × *g* SUPERNATANT SOLUTIONS ALONE

Additions	Counts/min
Complete	51
" + 2.1 mg Ribosomal RNA	202
" − Ribosomes	17
" − Ribosomes + 2.1 mg Ribosomal RNA	20
" − Supernatant solution	36
" − Supernatant solution + 2.1 mg Ribosomal RNA	45
" Deproteinized at zero time	25

The components of the reaction mixtures and the incubation conditions are presented in Table 1. 0.86 and 3.3 mg protein were present in the ribosome (W-Rib) and 105,000 × *g* supernatant (S-100) fractions, respectively.

that both ribosomes and 105,000 × *g* supernatant solution were necessary for ribosomal RNA-dependent amino acid incorporation. No incorporation of amino acids into protein occurred when the 105,000 × *g* supernatant solution alone was added to ribosomal RNA preparations, demonstrating that ribosomal RNA preparations were not contaminated with intact ribosomes. This conclusion also was substantiated by showing that the activities of ribosomal RNA preparations were not destroyed by boiling, although the activities of the ribosomes were destroyed by such treatment.

The effect of ribosomal RNA upon the incorporation of seven different amino acids is presented in Table 3. The addition of ribosomal RNA increased the incorporation of every amino acid tested.

The effect shown by ribosomal RNA was not observed when other polyanions were used, such as polyadenylic acid, highly polymerized salmon sperm DNA, or a high-molecular-weight polymer of glucose carboxylic acid (Table 4). Pretreatment of ribosomal RNA with trypsin did not affect its biological activity. However, treatment of the ribosomal RNA with either RNAase or alkali resulted in a complete loss of stimulating activity. The active principle, therefore, appears to be RNA.

The sedimentation characteristics of the ribosomal RNA preparations were examined in the Spinco Model E ultracentrifuge (Fig. 4*A*). Particles having the characteristics of S-30, S-50, or S-70 ribosomes were not observed in these preparations. The S_{20}^{W} of the first peak was 23, that of the second peak 16, and that of the third, small peak, 4. Pretreatment with trypsin did not affect the S_{20}^{W} values of the peaks appreciably (Fig. 4*C*); however, treatment with RNAase completely destroyed the peaks (Fig. 4*B*), confirming the ancillary evidence which had suggested that the major component was high-molecular-weight RNA.

Preliminary attempts at fractionation of the ribosomal RNA were performed by means of density-gradient centrifugation employing a linear sucrose gradient. The results of one such experiment are presented in Figure 5. Amino acid incorporation activity of the RNA did not follow absorbancy at 260 mμ; instead, the activity seemed to be concentrated around fraction No. 5, which was approximately one-third of the distance from the bottom of the tube. These results again

TABLE 3

SPECIFICITY OF AMINO ACID INCORPORATION STIMULATED BY RIBOSOMAL RNA

C^{14}-Amino Acid	Addition	Counts/min/mg protein
C^{14}-L-Valine	Complete	25
"	" + Ribosomal RNA	137
C^{14}-L-Threonine	"	31
"	" + Ribosomal RNA	121
C^{14}-L-Methionine	"	121
"	" + Ribosomal RNA	177
C^{14}-L-Arginine	"	49
"	" + Ribosomal RNA	224
C^{14}-L-Phenylalanine	"	77
"	" + Ribosomal RNA	147
C^{14}-L-Lysine	"	36
"	" + Ribosomal RNA	175
C^{14}-L-Leucine	"	134
"	" + Ribosomal RNA	272
"	" Deproteinized at zero time	6

The composition of the reaction mixtures are presented in Table 1. The mixture of 20 L-amino acids included all amino acids except the C^{14}-amino acid added to one reaction mixture. Reaction mixtures contained 4.4 mg Incubated-S-30 protein. Samples were incubated at 35° for 60 min. 2.1 mg ribosomal RNA were added where indicated.

TABLE 4

RIBOSOMAL RNA CONTROL EXPERIMENTS DESCRIBED IN TEXT

Experiment No.	Addition	Counts/min/mg protein
1	Complete	54
	" + 2.4 mg Ribosomal RNA	144
	" + 2.0 mg Polyadenylic acid	10
	" + 2.0 mg Salmon sperm DNA	41
	" + 2.0 mg Polyglucose carboxylic acid	49
	" + 2.4 mg Ribosomal RNA, deproteinized at zero time	7
2	Complete	39
	" + 2.0 mg Ribosomal RNA*	150
	" + 2.1 mg Ribosomal RNA preincubated with trypsin*	166
	" + 2.0 mg Ribosomal RNA preincubated with RNAase*,†	47
	" Deproteinized at zero time	8
3	Complete	20
	" + 1.2 mg Ribosomal RNA	82
	" + 1.2 mg Alkali degraded ribosomal RNA†	21
	" Deproteinized at zero time	7

The composition of the reaction mixtures and the incubation conditions are given in Table 1. 4.4, 3.2, and 4.4 mg Incubated-S-30 protein were present in Experiments 1, 2, and 3, respectively. 2.4, 0.98, and 0 mg *E. coli* soluble RNA were present in Experiments 1, 2, and 3, respectively.

* Ribosomal RNA preparations were deproteinized by phenol extraction after enzymatic digestion as specified under *Methods and Materials*.

† mg Ribosomal RNA refers to RNA concentration before digestion.

demonstrate that the activity was not associated with a soluble RNA fraction, present in maximum concentration in fraction No. 11, near the top of the tube. In addition, all amino acid incorporation analyses were performed in the presence of added soluble RNA, and the addition of more soluble RNA would not stimulate C^{14}-L-valine incorporation into protein.

Effects of RNA obtained from different species: The data of Table 5 demonstrate that RNA from different sources stimulates C^{14}-valine incorporation into protein. Yeast ribosomal RNA prepared by the method of Crestfield *et al.*[4] was considerably more effective in stimulating incorporation than equivalent amounts of *E. coli* ribosomal RNA. Yeast ribosomal RNA prepared by this method has little or no amino acid acceptor activity and has a molecular weight of about 29,000.[7] Tobacco mosaic virus RNA prepared by phenol extraction and having a molecular weight of

TABLE 5

STIMULATION OF AMINO ACID INCORPORATION BY RNA FRACTIONS PREPARED FROM DIFFERENT SPECIES

Additions	Counts/min/mg protein
None	42
+ 0.5 mg *E. coli* ribosomal RNA	75
+ 0.5 mg Yeast ribosomal RNA	430
+ 0.5 mg Tobacco mosaic virus RNA	872
+ 0.5 mg Ehrlich ascites tumor microsomal RNA	65

The components of the reaction mixtures and the incubation conditions are presented in Table 1. Reaction samples contained 1.9 mg Incubated-S-30 protein.

approximately 1,700,000† stimulated amino acid incorporation strongly. Marked stimulation due to tobacco mosaic virus RNA was observed also with *E. coli* enzyme extracts which had not been treated with DNAase. More complete details of this work will be presented in a later publication.

Stimulation of amino acid incorporation by synthetic polynucleotides: The data of Figure 6 show that the addition of 10 μg of polyuridylic acid‡ per ml of reaction mixture resulted in a remarkable stimulation of C^{14}-L-phenylalanine incorporation. Phenylalanine incorporation was almost completely dependent upon the addition of polyuridylic acid, and incorporation proceeded, after a slight lag period, at a linear rate for approximately 30 min.

The data of Table 6 demonstrate that no other polynucleotide tested could replace polyuridylic acid. The absolute specificity of polyuridylic acid was con-

TABLE 6

POLYNUCLEOTIDE SPECIFICITY FOR PHENYLALANINE INCORPORATION

Experiment no.	Additions	Counts/min/mg protein
1	None	44
	+ 10 μg Polyuridylic acid	39,800
	+ 10 μg Polyadenylic acid	50
	+ 10 μg Polycytidylic acid	38
	+ 10 μg Polyinosinic acid	57
	+ 10 μg Polyadenylic-uridylic acid (2/1 ratio)	53
	+ 10 μg Polyuridylic acid + 20 μg polyadenylic acid	60
	Deproteinized at zero time	17
2	None	75
	+ 10 μg UMP	81
	+ 10 μg UDP	77
	+ 10 μg UTP	72
	Deproteinized at zero time	6

Components of the reaction mixtures are presented in Table 1. Reaction mixtures contained 2.3 mg Incubated-S-30 protein. 0.02 μmoles U-C^{14}-L-phenylalanine (~125,000 counts/minute) was added to each reaction mixture. Samples were incubated at 35° for 60 min.

firmed by demonstrating that randomly mixed polymers of adenylic and uridylic acid‡ (Poly A-U, 2/1 ratio and 4/1 ratio) were inactive in this system. A solution of polyuridylic acid and polyadenylic acid (which forms triple-stranded helices) had no activity whatsoever, suggesting that single-strandedness is a necessary requisite for activity. Experiment 2 in Table 6 demonstrates that UMP, UDP, or UTP were unable to stimulate phenylalanine incorporation.

The data of Table 7 demonstrate that both ribosomes and 100,000 × *g* supernatant solution, as well as ATP and an ATP-generating system, were required for the polyuridylic acid–dependent incorporation of phenylalanine. Incorporation was inhibited by puromycin, chloramphenicol, and RNAase. The incorpora-

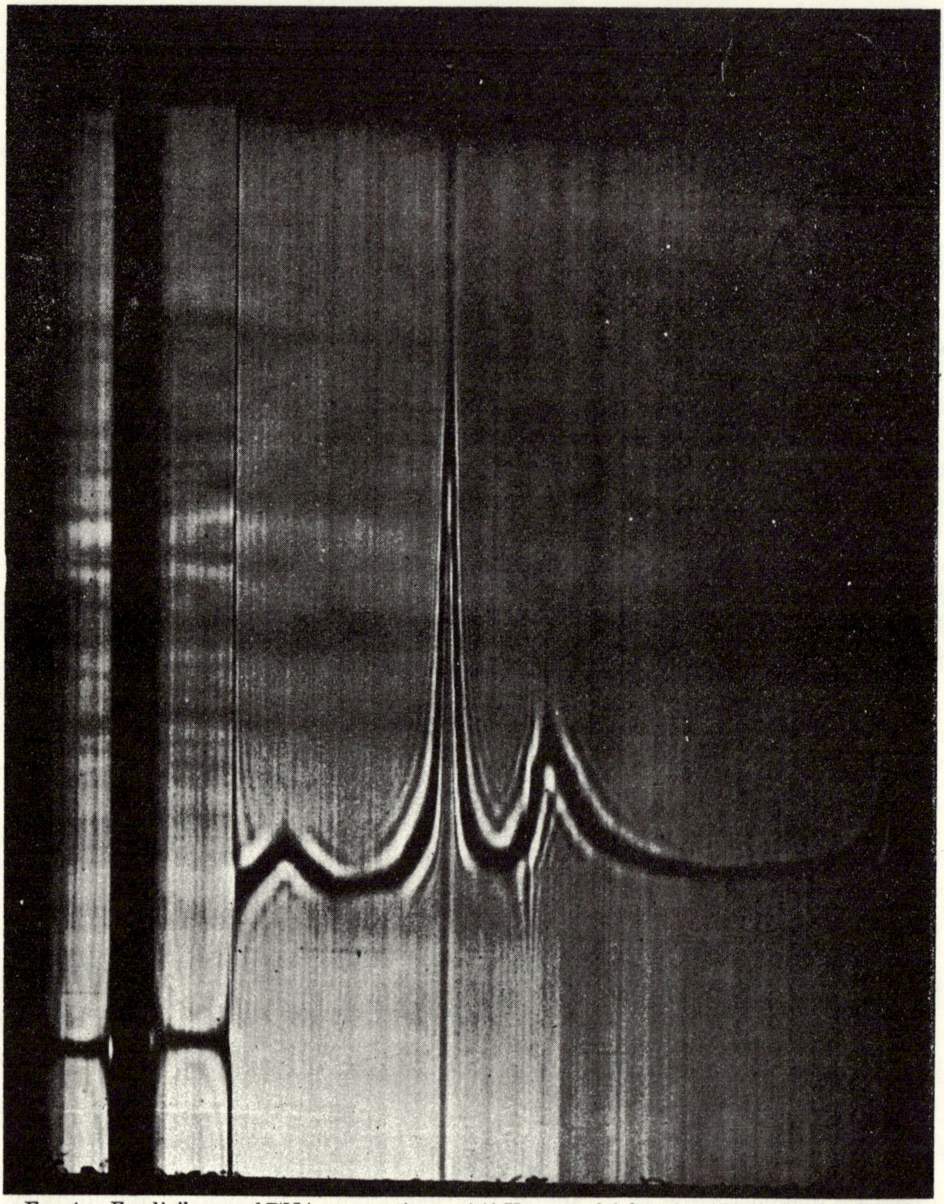

FIG. 4.—*E. coli* ribosomal RNA preparations. (*A*) Untreated (above). (*B*) digested with RNAase; (*C*) digested with trypsin. Preparation and digestion of samples presented under *Methods and Materials*. 9.8 and 10.5 mg/ml RNA were present in *A* and *C*. 11.5 mg/ml RNA was present in *B*
(*Continued on facing page*)

tion was not inhibited by addition of DNAase. Omitting a mixture of 19 L-amino acids did not inhibit phenylalanine incorporation, suggesting that polyuridylic acid stimulated the incorporation of L-phenylalanine alone. This conclusion was substantiated by the data presented in Table 8. Polyuridylic acid had little effect in stimulating the incorporation of 17 other radioactive amino acids. Each labeled amino acid was tested individually, and these data, corroborating the results given in Table 8, will be presented in a subsequent publication.

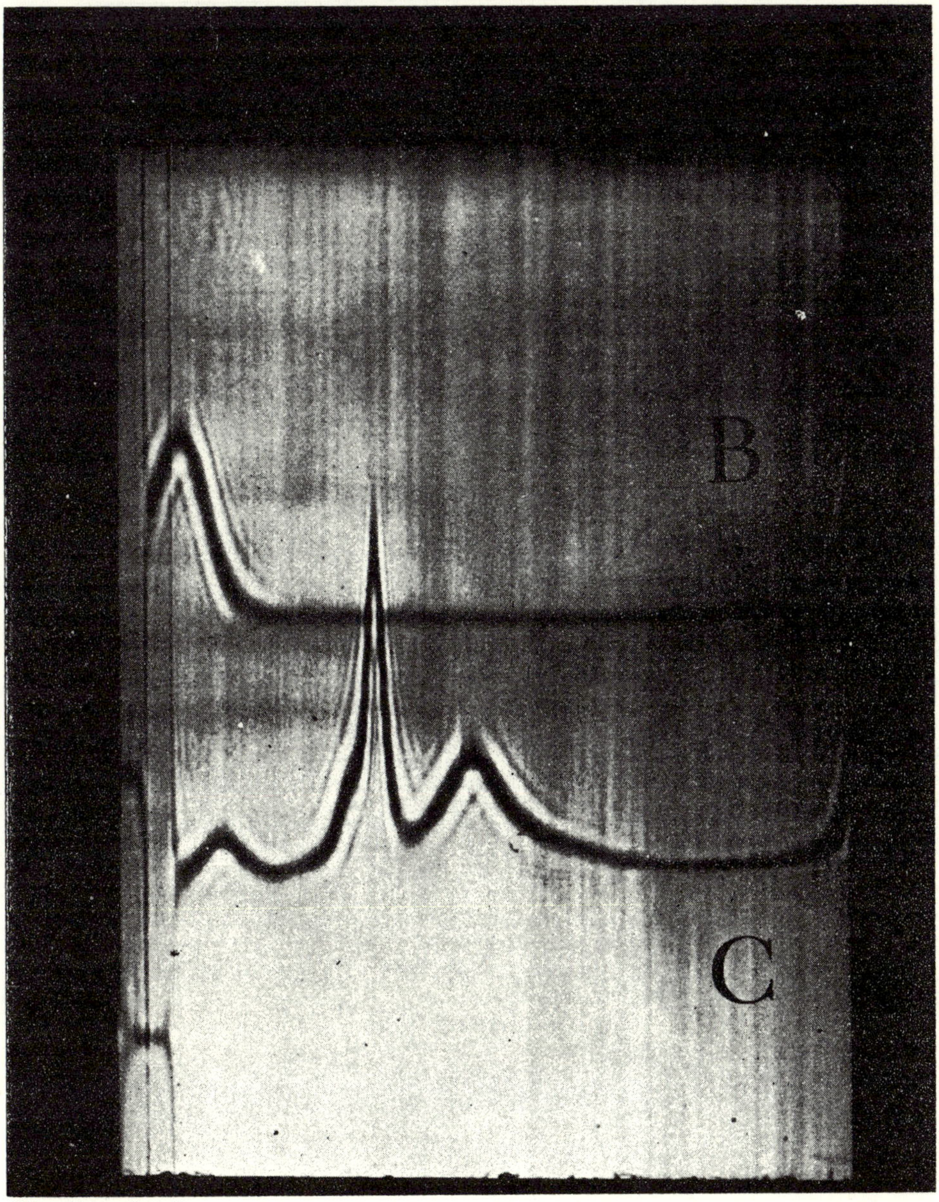

(*Fig. 4—continued*)
before digestion. Photographs were taken in a model E Spinco ultracentrifuge equipped with schlieren optics.

The product of the reaction was partially characterized and the results are presented in Table 9. The physical characteristics of the product of the reaction resembled those of authentic poly-L-phenylalanine, for, unlike many other polypeptides and proteins, both the product of the reaction and the polymer were resistant to hydrolysis by 6*N* HCl at 100° for 8 hr but were completely hydrolyzed by 12*N* HCl at 120–130° for 48 hr.

Poly-L-phenylalanine is insoluble in most solvents[25] but is soluble in 33 per cent

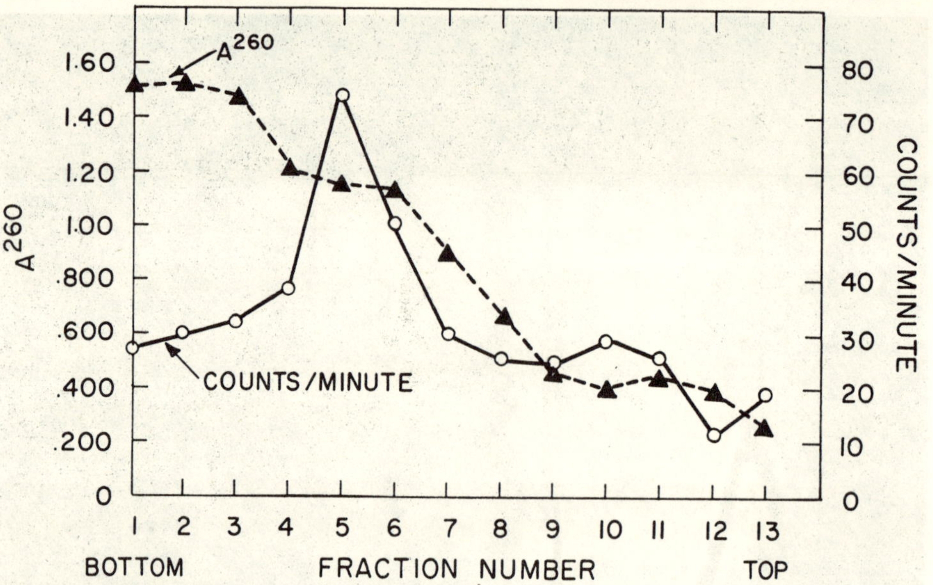

FIG. 5.—Sucrose density-gradient centrifugation of ribosomal RNA. A linear gradient of sucrose concentration ranging from 20 per cent at the bottom to 5 per cent at the top of the tube was prepared.[23] The sucrose solutions (4.4 ml total volume) contained 0.01 *M* Tris, pH 7.8, 0.01 *M* Mg acetate and 0.06 *M* KCl. 0.4 ml of ribosomal RNA (4.6 mg) was layered on top of each tube which was centrifuged at 38,000 × *g* for 4.5 hours at 3° in a swinging bucket rotor, Spinco type SW-39, using a Spinco Model L ultracentrifuge. 0.30 ml fractions were collected after piercing the bottom of the tube.[24]

0.025 ml aliquots diluted to 0.3 ml with H_2O were used for A^{260} measurements. 0.25 ml aliquots were used for amino acid incorporation assays. Reaction mixtures contained the components presented in Table 1. 0.7 mg of *E. coli* soluble RNA and 2.2 mg Incubated-S-30 protein were added. Control assays plus 0.25 ml 12.5 per cent sucrose in place of fractions gave 79 counts/min. This figure was subtracted from each value. Total volume was 0.7 ml. Samples were incubated at 35° for 20 min.

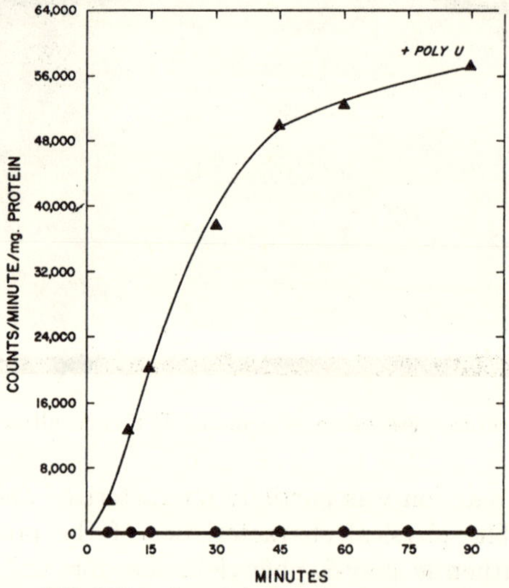

FIG. 6.—Stimulation of U-C^{14}-L-phenylalanine incorporation by polyuridylic acid. ● without polyuridylic acid; ▲ 10 μg polyuridylic acid added. The components of the reaction mixtures and the incubation conditions are given in Table 1. 0.024 μmole U-C^{14}-L-phenylalanine (~500,000 counts/min) and 2.3 mg Incubated-S-30 protein were added/ml of reaction mixture.

TABLE 7

CHARACTERISTICS OF POLYURIDYLIC ACID–DEPENDENT PHENYLALANINE INCORPORATION

Additions	Counts/min/mg protein
Minus polyuridylic acid	70
None	29,500
Minus 100,000 × *g* supernatant solution	106
Minus ribosomes	52
Minus ATP, PEP, and PEP kinase	83
+ 0.02 μmoles puromycin	7,100
+ 0.31 μmoles chloramphenicol	12,550
+ 6 μg RNAase	120
+ 6 μg DNAase	27,600
Minus amino acid mixture	31,700
Deproteinized at zero time	30

The components of the reaction mixtures are presented in Table 1. 10 μg of polyuridylic acid were added to all samples except the specified one. 2.3 mg of Incubated-S-30 protein were added to each reaction mixture except those in which ribosomes alone and 100,000 × *g* supernatant solution alone were tested. 0.7 mg W-Rib protein and 1.3 mg S-100 protein were used respectively. 0.02 μmoles U-C^{14}-L-phenylalanine, Sp. Act. = 10.3 mC/mM (~125,000 counts/minute) were added to each reaction mixture. Samples were incubated at 35° for 60 min.

TABLE 8

SPECIFICITY OF AMINO ACID INCORPORATION STIMULATED BY POLYURIDYLIC ACID

Experiment no.	C^{14}-amino acids present	Additions	Counts/min/mg protein
1	Phenylalanine	Deproteinized at zero time	25
		None	68
		+ 10 μg polyuridylic acid	38,300
2	Glycine, alanine, serine, aspartic acid, glutamic acid	Deproteinized at zero time	17
		None	20
		+ 10 μg polyuridylic acid	33
3	Leucine, isoleucine, threonine, methionine, arginine, histidine, lysine, tyrosine, tryptophan, proline, valine	Deproteinized at zero time	73
		None	276
		+ 10 μg polyuridylic acid	899
4	S^{35}-cysteine	Deproteinized at zero time	6
		None	95
		+ 10 μg polyuridylic acid	113

Components of the reaction mixtures are presented in Table 1. The unlabeled amino acid mixture was omitted. 0.015 μM of each labeled amino acid was used. The specific activities of the labeled amino acids are present in the *Methods and Materials* section. 2.3 mg of protein of preincubated S-30 enzyme fraction were added to each reaction mixture. All samples were incubated at 35° for 30 min.

TABLE 9

COMPARISON OF CHARACTERISTICS OF PRODUCT OF REACTION AND POLY-L-PHENYLALANINE

Treatment	Product of reaction	Poly-L-phenylalanine
6 *N* HCl for 8 hours at 100°	Partially hydrolyzed	Partially hydrolyzed
12 *N* HCl for 48 hours at 120–130°	Completely hydrolyzed	Completely hydrolyzed
Extraction with 33% HBr in glacial acetic acid	Soluble	Soluble
Extraction* with the following solvents: H_2O, benzene, nitrobenzene, chloroform, N,N-dimethylformamide, ethanol, petroleum ether, concentrated phosphoric acid, glacial acetic acid, dioxane, phenol, acetone, ethyl acetate, pyridine, acetophenone, formic acid	Insoluble	Insoluble

* The product was said to be insoluble if <0.002 gm of product was soluble in 100 ml of solvent at 24°. Extractions were performed by adding 0.5 mg of authentic poly-L-phenylalanine and the C^{14}-product of a reaction mixture (1800 counts/min) to 5.0 ml of solvent. The suspensions were vigorously shaken for 30 min at 24° and were centrifuged. The precipitates were plated and their radioactivity was determined.

HBr in glacial acetic acid.§ The product of the reaction had the same apparent solubility as authentic poly-L-phenylalanine. The product of the reaction was purified by means of its unusual solubility behavior. Reaction mixtures were deproteinized after incubation, and precipitated proteins were washed in the usual manner according to the method of Siekevitz.[22] Dried protein pellets containing added carrier poly-L-phenylalanine were then extracted with 33 per cent HBr in glacial acetic acid, and the large amount of insoluble material was discarded. Polyphenylalanine was then precipitated from solution by the addition of H_2O and was washed several times with H_2O. Seventy per cent of the total amount of C^{14}-L-phenylalanine incorporated into protein due to the addition of polyuridylic acid could be recovered by this procedure. Complete hydrolysis of the purified reaction product with 12*N* HCl followed by paper electrophoresis** demonstrated that the reaction product contained C^{14}-phenylalanine. No other radioactive spots were found.

Discussion.—In this investigation, we have demonstrated that template RNA is a requirement for cell-free amino acid incorporation. Addition of soluble RNA could not replace template RNA in this system. In addition, the density-gradient centrifugation experiments showed that the active fractions in the ribosomal RNA preparations sedimented much faster than soluble RNA. It should be noted that ribosomal RNA is qualitatively different from soluble RNA, since bases such as pseudouracil, methylated guanines, etc., found in soluble RNA, are not present in ribosomal RNA.[5]

The bulk of the RNA in our ribosomal RNA fractions may be inactive as templates, for tobacco mosaic virus RNA was 20 times as active in stimulating amino acid incorporation as equivalent amounts of *E. coli* ribosomal RNA. In addition, preliminary fractionation of ribosomal RNA indicated that only a portion of the total RNA was active.

It should be emphasized that ribosomal RNA could not substitute for ribosomes, indicating that ribosomes were not assembled from the added RNA *in toto*. The function of ribosomal RNA remains an enigma, although at least part of the total RNA is thought to serve as templates for protein synthesis and has been termed "messenger" RNA.[12–14] Alternatively, a part of the RNA may be essential for the synthesis of active ribosomes from smaller ribosomal particles.[15–21]

Ribosomal RNA may be an aggregate of subunits which can dissociate after proper treatment.[6–8] Phenol extraction of *E. coli* ribosomes yields two types of RNA molecules with S_w^{20} of 23 and 16 (Fig. 4), equivalent to molecular weights of 1,000,000 and 560,000, respectively.[9, 10] These RNA species can be degraded by boiling to products having sedimentation coefficients of 13.1, 8.8, and 4.4, corresponding to molecular weights of 288,000, 144,000, and 29,000. Although the sedimentation distributions of the latter preparations suggest a high degree of homogeneity among the molecules of each class, these observations do not eliminate the possibility that the subunits are linked to one another *via* covalent bonds.[8] Preliminary evidence indicates that the subunits may be active in our system, since the supernatant solution obtained after boiling *E. coli* ribosomal RNA for 10 min and centrifugation at $105{,}000 \times g$ for 60 min was active. Examination of boiled ribosomal RNA with the Spinco Model E ultracentrifuge showed a dispersed peak with a sedimentation coefficient of 4–8. This may be the same material found in the sucrose density-gradient experiment (using non-boiled RNA preparations), where a small peak of activity somewhat heavier than soluble RNA was usually noted (Fig. 5).

In our system, at low concentrations of ribosomal RNA, amino acid incorporation into protein was proportional to the amount of ribosomal RNA added, suggesting a stoichiometric rather than a catalytic action of ribosomal RNA. In contrast, soluble RNA has been shown to act in a catalytic fashion.[11]

The results indicate that polyuridylic acid contains the information for the synthesis of a protein having many of the characteristics of poly-L-phenylalanine. This synthesis was very similar to the cell-free protein synthesis obtained when naturally-occurring template RNA was added, i.e., both ribosomes and 100,000 $\times$ g supernatant solutions were required, and the incorporation was inhibited by puromycin or chloramphenicol. One or more uridylic acid residues therefore appear to be the code for phenylalanine. Whether the code is of the singlet, triplet, etc., type has not yet been determined. Polyuridylic acid seemingly functions as a synthetic template or messenger RNA, and this stable, cell-free *E. coli* system may well synthesize any protein corresponding to meaningful information contained in added RNA.

Summary.—A stable, cell-free system has been obtained from *E. coli* in which the amount of incorporation of amino acids into protein was dependent upon the addition of heat-stable template RNA preparations. Soluble RNA could not replace template RNA fractions. In addition, the amino acid incorporation required both ribosomes and 105,000 $\times$ g supernatant solution. The correlation between the amount of incorporation and the amount of added RNA suggested stoichiometric rather than catalytic activity of the template RNA. The template RNA–dependent amino acid incorporation also required ATP and an ATP-generating system, was stimulated by a complete mixture of L-amino acids, and was markedly inhibited by puromycin, chloramphenicol, and RNAase. Addition of a synthetic polynucleotide, polyuridylic acid, specifically resulted in the incorporation of L-phenylalanine into a protein resembling poly-L-phenylalanine. Polyuridylic acid appears to function as a synthetic template or messenger RNA. The implications of these findings are briefly discussed.

Note added in proof.—The ratio between uridylic acid units of the polymer required and molecules of L-phenylalanine incorporated, in recent experiments, has approached the value of 1:1. Direct evidence for the number of uridylic acid residues forming the code for phenylalanine as well as for the eventual stoichiometric action of the template is not yet established. As polyuridylic acid codes the incorporation of L-phenylalanine, polycytidylic acid‡ specifically mediates the incorporation of L-proline into a TCA-precipitable product. Complete data on these findings will be included in a subsequent publication.

* Supported by a NATO Postdoctoral Research Fellowship.

† Dr. Frankel-Conrat, personal communication.

‡ We thank Drs. Leon A. Heppel and Maxine F. Singer for samples of these polyribonucleotides, and Dr. George Rushizky for TMV-RNA.

§ We thank Dr. Michael Sela for this information.

** We thank Drs. William Dreyer and Elwood Bynum for performing the high-voltage electrophoretic analyses.

[1] Matthaei, J. H., and M. W. Nirenberg, these PROCEEDINGS, **47,** 1580 (1961).

[2] Matthaei, J. H., and M. W. Nirenberg, *Biochem. & Biophys. Res. Comm.*, **4,** 404 (1961).

[3] Matthaei, J. H., and M. W. Nirenberg, *Fed. Proc.*, **20,** 391 (1961).

[4] Crestfield, A. M., K. C. Smith, and F. W. Allen, *J. Biol. Chem.*, **216,** 185 (1955).

[5] Davis, F. F., A. F. Carlucci, and I. F. Roubein, *ibid.*, **234,** 1525 (1959).

[6] Hall, B. D., and P. Doty, *J. Mol. Biol.*, **1,** 111 (1959).

[7] Osawa, S., *Biochim. Biophys. Acta,* **43,** 110 (1960).

[8] Aronson, A. I., and B. J. McCarthy, *Biophys. J.*, **1,** 215 (1961).

[9] Kurland, C. G., *J. Mol. Biol.*, **2**, 83 (1960).
[10] Littauer, U. Z., H. Eisenberg, *Biochim. Biophys. Acta*, **32**, 320 (1959).
[11] Hoagland, M. B., and L. T. Comly, these PROCEEDINGS, **46**, 1554 (1960).
[12] Volkin, E., L. Astrachan, and J. L. Countryman, *Virology*, **6**, 545 (1958).
[13] Nomura, M., B. D. Hall, and S. Spiegelman, *J. Mol. Biol.*, **2**, 306 (1960).
[14] Hall, B. D., and S Spiegelman, these PROCEEDINGS, **47**, 137 (1961).
[15] Bolton, E. T., B. H. Hoyen, and D. B. Ritter, in *Microsomal Particles and Protein Synthesis*, ed. R. B. Roberts (New York: Pergamon Press, 1958), p. 18.
[16] Tissières, A., J. D. Watson, D. Schlessinger, and B. R. Hollingworth, *J. Mol. Biol.*, **1**, 221 (1959).
[17] Tissières, A., D. Schlessinger, and F. Gros, these PROCEEDINGS, **46**, 1450 (1960).
[18] McCarthy, B. J., and A. I. Aronson, *Biophys. J.*, **1**, 227 (1961).
[19] Hershey, A. D., *J. Gen. Physiol.*, **38**, 145 (1954).
[20] Siminovitch, L., and A. F. Graham, *Canad. J. Microbiol.*, **2**, 585 (1956).
[21] Davern, C. I., and M. Meselson, *J. Mol. Biol.*, **2**, 153 (1960).
[22] Siekevitz, P., *J. Biol. Chem.*, **195**, 549 (1952).
[23] Britten, R. J., and R. B. Roberts, *Science*, **131**, 32 (1960).
[24] Martin, R., and B. Ames, *J. Biol. Chem.*, **236**, 1372 (1961).
[25] Bamford, C. H., A. Elliott, and W. E. Hanby, *Synthetic Polypeptides* (New York: Academic Press, 1956), p. 322.

Reprinted from PROCEEDINGS OF THE NATIONAL ACADEMY OF SCIENCES, Vol. 47, 1961

*SYNTHETIC POLYNUCLEOTIDES AND THE AMINO ACID CODE**

BY PETER LENGYEL, JOSEPH F. SPEYER, AND SEVERO OCHOA

DEPARTMENT OF BIOCHEMISTRY, NEW YORK UNIVERSITY SCHOOL OF MEDICINE

Communicated October 25, 1961

The problem of coding in protein biosynthesis, i.e., of how a certain sequence of four different nucleotides in an RNA[1] chain can specify a given sequence of 20 different amino acids in a polypeptide chain, has been considered by several investigators in the last decade. Until now these studies have been either theoretical[2, 3] or statistical[4–6] in nature.

Increased knowledge of the mechanism of protein biosynthesis, the concept of messenger RNA as the actual template,[7] and especially the observation of Brenner *et al.*[8] (cf. also Gros *et al.*[9]) that after infection of *Escherichia coli* with T_2 phage the newly formed messenger RNA is added to preexisting ribosomes (on which most of the protein synthesis of the infected cell occurs) suggested the use of synthetic polyribonucleotides, with known nucleotide sequences, as messengers and a possible method of experimental approach to the nucleotide code. The simplest possible sequence is that in homopolynucleotides which, if active, should prescribe the formation of homopolypeptide chains. These homopolynucleotides as well as copolymers containing two or more different nucleotide species in predetermined ratios can be synthesized by polynucleotide phosphorylase.[10]

We wish to report that various kinds of synthetic polynucleotides are active as messengers in an *E. coli* system and indeed determine the incorporation of different amino acids into an acid-insoluble product. That is the case for phenylalanine, serine, and tyrosine with polymers containing uridylic acid, uridylic acid and cytidylic acid, and uridylic acid and adenylic acid respectively.

Recently Nirenberg and Matthaei[11] succeeded in obtaining a poly U-dependent synthesis of polyphenylalanine in a system of *E. coli* supernatant and ribosomes.

Preparations.—*E. coli* ribosomes were prepared from log phase cells grown on enriched media (0.5 per cent glucose, 0.5 per cent trypticase, 0.5 per cent yeast extract). The cells were washed twice with ice-cold 0.2 *M* KCl and once with 0.02 *M* Tris-HCl buffer, pH 7.9, containing 0.01 *M* $MgCl_2$ and 0.005 *M* mercaptoethylamine. They were then resuspended in 2 volumes of the same medium and disrupted for 3 min at maximum power in a 10 kc Raytheon oscillator at 5°. The suspension was centrifuged twice for 30 min at 3–4° and 30,000 *g* to remove debris and intact cells. The extract was centrifuged at low temperature for 2 hr in the preparative Spinco centrifuge at 105,000 *g*. The clear supernatant fluid was collected, the sediment containing the ribosomes was rinsed with Tris-HCl buffer, pH 7.9, containing 0.01 *M* $MgCl_2$, 0.02 *M* KCl, 0.2 *M* sucrose, and 0.5 per cent Lubrol (Imperial Chemical Industries) and suspended in this medium, with use of a glass homogenizer, to give a concentration of 20 mg of protein/ml. This suspension was layered over a solution containing 0.034 *M* Tris-HCl buffer, pH 7.9, 0.01 *M* $MgCl_2$, 0.6 *M* KCl, and 0.3 *M* sucrose, and again centrifuged for 3 hr at 105,000 *g*. The particles were then suspended in 0.1 *M* Tris-HCl buffer, pH 7.9, containing 0.01 *M* $MgCl_2$ and 0.02 *M* KCl, to give a concentration of 10 mg of protein/ml, frozen in dry ice, and kept in a freezer at −18°. The ribosomes are stable under these conditions. They can also be kept after lyophilization. Rat liver ribosomes were prepared by the method of Rendi and Hultin.[12] Rat liver supernatant was prepared by the method of Zamecnik and Keller.[13] For some experiments the precipitate obtained by bringing the *E. coli* supernatant to pH 5.0 was dissolved in 0.05 *M* Tris-HCl buffer, pH 7.9, containing 0.005 *M* mercaptoethylamine, and used in place of the supernatant. This treatment aimed at reducing the level of endogenous amino acids in the system. Transfer RNA (S-RNA) was prepared from *E. coli* supernatant as described by Hoagland *et al.*[14]

Polyribonucleotides were prepared with *Azotobacter vinelandii* polynucleotide phosphorylase (specific activity, 62) as previously described.[10] After precipitation with ethanol, the polymers were dissolved in 0.04 *M* NaCl–0.005 *M* sodium citrate and the solution was extracted with phenol according to Gierer and Schramm.[15] The aqueous solution was dialyzed against distilled water for 5 hr at 3° and the polymer recovered by lyophilization. Poly UC was prepared from a mixture of UDP and CDP in molar ratio 5:1, poly UA from a mixture of UDP and ADP in a molar ratio 5:1, and poly CU from UDP and CDP in molar ratio 1:5. The preparation of polythio U has been previously described.[16] 5-Fluorouridine 5′-diphosphate, used in the preparation of polyfluoro U,[17] was a synthetic product kindly supplied by Dr. Charles Heidelberger, University of Wisconsin, Madison, Wisconsin. The sedimentation coefficients of the polynucleotides used in this work are listed in Table 1. Nucleoside diphosphates were obtained from the Schwarz Laboratories, Mount Vernon, N. Y., and the Pabst Laboratories, Milwaukee, Wisconsin. C^{14}-

labeled amino acids were obtained from the California Biochemical Corporation under allocation from the Atomic Energy Commission.

Methods.—In order to decrease the "blank" incorporation of labeled amino acids in the absence of added polynucleotides, *E. coli* ribosomes and supernatant were preincubated for 15 min at 37° in a mixture containing in 1.0 ml the following components (in μmoles unless otherwise specified): Tris-HCl buffer, pH 7.9, 37; Lubrol, 1.8 mg; KCl, 55; $MgCl_2$, 13; mercaptoethylamine, 11; ATP, 0.9; GTP, 0.22; creatine phosphate, 12; creatine kinase, 44 μg; each of 20 nonlabeled amino acids, 0.09; ribosomes with 4 mg of protein; and supernatant with 4 mg of protein. For amino acid incorporation each sample contained 0.1 ml of the above preincubated mixture and 0.15 ml of a basic reaction mixture of the following composition (in μmoles unless otherwise specified): Tris-HCl buffer, pH 7.9, 10.5; Lubrol, 0.5 mg; KCl, 15.3; $MgCl_2$, 2.5; mercaptoethylamine, 3.1; ATP, 0.25; GTP, 0.06; creatine phosphate, 3.3; and creatine kinase, 12.5 μg. Other additions, as noted in the tables, included C^{14}-labeled amino acid (specific radioactivity 1 to 10 μc/μmole), 0.025 μc; transfer RNA, 0.5–1.0 mg; and polynucleotide, 0.04 μmole as mononucleotide. The latter amount was found to be optimal for poly U; higher concentrations were inhibitory. After incubation for 1 hr at 37° the reaction was stopped by the addition of 5 ml of 10 per cent trichloroacetic acid and the samples were worked up according to the procedure of Zamecnik *et al.*[18] The radioactivity of the acid-insoluble residue was measured with a windowless gas-flow counter. Amino acid incorporation is expressed throughout in mμmoles/mg ribosomal protein.

TABLE 1

SEDIMENTATION COEFFICIENTS OF SYNTHETIC POLYNUCLEOTIDES

Polynucleotide	$S_{20,w}$
Poly A	17.7
Poly U (sample 1)	10.3
Poly U (sample 2)	4.5
Poly C	8.6
Poly UC	9.5
Poly UA	9.1
Poly thio U	19.0
Poly fluoro U	9.4

In the experiment in which rat liver and *E. coli* ribosomes were compared, the preincubation described above was omitted and no transfer RNA was added. Protein was determined by the method of Lowry *et al.*[19]

Results.—*Poly U-dependent incorporation of phenylalanine in E. coli system:* Investigation of the incorporation of different C^{14}-labeled amino acids into an acid-insoluble product by an *E. coli* supernatant plus ribosomes system, with and without the addition of poly U, showed that out of 19 amino acids (C^{14}-labeled asparagine was not available to us) tested individually only the incorporation of phenylalanine was markedly stimulated by poly U. A small stimulation of the incorporation of leucine and isoleucine was also observed. The effect of poly U on the incorporation of phenylalanine is shown in Table 2. It may also be seen that addition of *E. coli* transfer RNA brought about a further pronounced increase of phenylalanine incorporation. Since poly U did not affect the loading of transfer RNA with phenylalanine, catalyzed by the amino acid activating enzymes in the supernatant, it appears that poly U affects the transfer of activated phenylalanine residues from the specific phenylalanine-transfer RNA to the ribosomes; this indicates that poly U acts as messenger RNA in this system. As shown by Nirenberg and Matthaei,[11] polyphenylalanine, which is exceedingly insoluble, is formed under these conditions. With the highest concentration of transfer RNA used (Table 2) one mole of phenylalanine was incorporated for every 3.25 uridylic acid

TABLE 2

EFFECT OF *E. coli* TRANSFER RNA ON THE POLY U-DEPENDENT INCORPORATION OF PHENYLALANINE*

Additions to basal system: Poly U	Transfer RNA (mg/ml)	Phenylalanine incorporation†
−	0	0.1
−	9	0.3
+	0	3.6
+	2	15.2
+	6	23.5
+	9	24.6

* Incubation 30 min at 37°.
† mμmoles/mg of ribosomal protein.

residues in the poly U added. The incorporation of phenylalanine was inhibited by puromycin and, to a lesser extent, by chloramphenicol.

Experiments with other homopolymers: The effect of poly A, poly thio U, and poly fluoro U was investigated in experiments similar to those of Table 2 but *without the addition of transfer RNA*. Poly A did not stimulate the incorporation of any of 19 amino acids tested. Addition of poly A to a system containing poly U completely inhibited the effect of the latter. This is undoubtedly due to formation of the double-stranded, helical poly A + U complex. Poly thio U (which appears to be multi-stranded[20, 21]) had no effect,[22] and poly fluoro U (which like poly U is single-stranded[20]) had but a small effect on phenylalanine incorporation. Poly C had a small but consistent effect on the incorporation of proline (cf. footnote 11) but had no influence on that of any other amino acid. It may be of interest that a short-chain poly U (poly U sample 2, Table 1) was about 40 per cent as effective as an equimolar amount of the longer-chain poly U used throughout this work (poly U sample 1, Table 1) in promoting phenylalanine incorporation.

Experiments with rat liver ribosomes: Poly U had no effect on the incorporation of phenylalanine by a system of rat liver supernatant plus ribosomes (Table 3).

TABLE 3

POLY U-DEPENDENT INCORPORATION OF PHENYLALANINE BY *E. coli* AND RAT LIVER RIBOSOMES

Supernatant	Ribosomes	Phenylalanine incorporation*: (a) No poly U	(b) With poly U	Ratio b/a
Liver	Liver	0.40	0.46	1.2
E. coli	*E. coli*	0.07	3.29	47.0
Liver	*E. coli*	0.06	0.17	2.8
E. coli	Liver	0.16	1.81	11.3

* mμmoles/mg of ribosomal protein.

However, no transfer RNA was added in these experiments and, considering the pronounced effect of transfer RNA on the *E. coli* system, this negative result cannot be considered conclusive. With the unsupplemented rat liver system there was no effect on either poly A, poly U, or poly C on the incorporation of any of 19 amino acids tried singly. However, as seen in Table 3, poly U had a fairly marked effect on the incorporation of phenylalanine by a system of *E. coli* supernatant plus rat liver ribosomes. This was not true for the opposite combination, i.e., liver supernatant plus *E. coli* ribosomes. Thus, the ineffectiveness of poly U with the rat liver system is related to the soluble components of this system. It is too early to speculate on the significance of this observation, but the finding that rat liver ribosomes can be substituted for *E. coli* ribosomes for poly U-dependent incorporation of phenylalanine is not without interest.

Effect of synthetic homo- and copolymers on amino acid incorporation in E. coli system: The results of these experiments are summarized in Table 4. The incorporation of phenylalanine was stimulated to a decreasing extent by poly U, poly UC, and poly UA. Most significant is the fact that some of the copolymers stimulated the incorporation of amino acids other than phenylalanine. Thus, whereas poly UC, and only this polymer, stimulated the incorporation of serine, only poly UA stimulated the incorporation of tyrosine. Less clearcut results were obtained with leucine and isoleucine, the incorporation of both of which was significantly stimulated by poly U, poly UC, and poly UA. Highest incorporation of leucine was promoted by poly UC whereas poly UA brought about the highest incorporation of isoleucine. Ambiguous results were also obtained with proline, the incorporation of which was stimulated by poly C, poly CU, and poly UC in order of increasing effectiveness.

TABLE 4

AMINO ACID INCORPORATION IN *E. coli* SYSTEM WITH VARIOUS POLYNUCLEOTIDES*

Amino acid	Polynucleotide: None	Poly U	Poly C	Poly UC	Poly UA	Poly CU
Phenylalanine	0.03	13		7	3	0.02
Serine	0.02	0.02	0.01	1.6	0.01	
Tyrosine	0.02			0.02	0.75	
Leucine	0.02	0.3		1.5	0.46	0.03
Isoleucine	0.01	0.09		0.32	0.62	0.007
Proline	0.02	0.02	0.06	0.6	0.03	0.14

* mμmoles/mg of ribosomal protein. 19 amino acids were tested individually in all cases, but the ones giving negative results have been omitted from the table. All values (except those for poly CU) are averages of at least two separate experiments.

Discussion.—In contrast to poly UC (U:C = 5:1) poly CU (U:C = 1:5) was almost ineffective in promoting amino acid incorporation (Table 4). The reason for this pronounced discrepancy may be sought in the fact that, contrary to natural messenger RNA (or to poly U), the synthetic copolymers do not provide an unbroken sequence of code units. Hence, as far as amino acid incorporation is concerned, there must be many gaps along their chains. This situation might result in the formation of short-chain, acid-soluble polypeptides which would escape detection by the procedure used here. Copolymers such as poly UC and poly UA containing relatively long unbroken U sequences together with some other code units might give rise to short, but acid-insoluble, polyphenylalanine chains with occasional serine, leucine, tyrosine, or isoleucine residues. Viewed in this way, a high proportion of U in a synthetic copolymer would provide a convenient "handle" for incorporation of amino acids other than phenylalanine into short, acid-insoluble polyphenylalanine chains.

If poly U codes for polyphenylalanine, a short sequence of three or more U residues would be the code letter for phenylalanine. If for the sake of simplicity we assume a triplet code, the ratio of UUU to UUC (or UCU, or CUU) triplets in a random copolymer of the composition of poly UC would be the same as the U:C ratio of the polymer, i.e., 5:1. The ratio of UUU to UCC (or CUC, or CCU) triplets would be 25:1. From Table 4 the ratio of phenylalanine to serine incorporation with poly UC was 4.4:1. On the basis of this result the triplet code letter for serine would be either UUC, UCU, or CUU. The ratio of phenylalanine to tyrosine incorporated with poly UA (U:A = 5:1) was 4.0. This would make either UUA, UAU, or AUU as the likely triplet code letter for tyrosine.

The overlapping results obtained with leucine and isoleucine are difficult to explain, but, if maximal stimulation is taken as the meaningful result, the triplet code letters for these two amino acids would contain 2U and 1C and 2U and 1A, respectively. Difficulties are also encountered with proline (Table 4). Using the same criterion as for leucine and isoleucine, one would tentatively assign a triplet code letter containing 1U and 2C to this amino acid. On the other hand, the greater effectiveness of poly UC as compared to poly C might be due to the provision of an insoluble polyphenylalanine "handle" for the proline residues. Thus it is possible that CCC is really the triplet code letter for proline. It is of interest in this connection that, in a nitrous acid mutant of tobacco mosaic virus described by Tsugita and Fraenkel-Conrat,[23] a proline residue was replaced by leucine. As the nitrous acid effect is due to deamination of C to U, replacement of C by U would be in line with the code letters suggested for these amino acids. Our results would also be compatible with the fact that proline:phenylalanine ratio of wild cucumber and tobacco mosaic virus protein varies in the same direction as the C:U ratio of their respective nucleic acids.[24]

Experiments now in progress with other homo- and copolymers prepared with polynucleotide phosphorylase and deductions from amino acid replacement data as used by Woese[5] may be expected to be of further help in deciphering the nucleotide code.

Summary. Polyribonucleotides synthesized with polynucleotide phosphorylase stimulated the incorporation of certain amino acids into an acid-insoluble product by a system of consisting of *E. coli* high-speed supernatant and ribosomes. Addition of *E. coli* transfer RNA brought about a further pronounced increase of the incorporation of phenylalanine in the presence of poly U. This indicates that the polymers affect the transfer of activated amino acid residues from transfer RNA to ribosomes and act as messenger or template RNA in this system. In one experiment with poly U and phenylalanine, without addition of transfer RNA, rat liver ribosomes could be substituted for their *E. coli* counterparts.

While poly U promoted phenylalanine incorporation, poly UC promoted the incorporation of phenylalanine and serine, and poly UA stimulated the incorporation of phenylalanine and tyrosine. These and other results reported in this paper would appear to open up an experimental approach to the study of the coding problem in protein biosynthesis.

We are indebted to Carlos Basilio for the preparation of poly CU, to R. C. Warner and Yo-Yun Chen for the sedimentation coefficients of the synthetic polynucleotides, and Albert Lenny and Horace Lozina for skillful technical assistance.

Note added in proof.—The results of further experiments to be reported in these PROCEEDINGS, with use of other copolymers (UG, UAC, where G stands for guanylic acid residues), extend to eleven the list of amino acids incorporated into an acid-insoluble product by the *E. coli* system with different polymers. It now includes cysteine, histidine, isoleucine, leucine, lysine, phenylalanine, proline, serine, threonine, tyrosine, and valine.

* Aided by grants from the National Institute of Arthritis and Metabolic Diseases (Grant A-1845) of the U.S. Public Health Service, and the Rockefeller Foundation.

[1] Abbreviations: RNA, ribonucleic acid; the capital letters A, U, and C are used for the nucleotides adenylic, uridylic, and cytidylic acid, respectively, or their corresponding residues in polynucleotide chains; ADP, UDP, and CDP, the 5'-diphosphates of adenosine, uridine, and cytidine; ATP and GTP, the 5'-triphosphates of adenosine and guanosine; Tris, tris(hydroxymethyl)aminomethane.

[2] Gamow, G., *Nature,* **173,** 318 (1954).

[3] Crick, F. H. C., J. S. Griffith, and L. E. Orgel, these Proceedings, **43,** 416 (1957).

[4] Yčas, M., *Nature,* **188,** 209 (1960).

[5] Woese, C. R., *Biochem. Biophys. Res. Comm.,* **5,** 88 (1961)

[6] Sueoka, N., these Proceedings, **47,** 1141 (1961).

[7] Jacob, F., and J. Monod, *J. Mol. Biol.,* **3,** 318 (1961).

[8] Brenner, S., F. Jacob, and M. Meselson, *Nature,* **190,** 576 (1961).

[9] Gros, F., H. Hiatt, W. Gilbert, C. G. Kurland, R. W. Risebrough, and J. D. Watson, *Nature,* **190,** 581 (1961).

[10] Grunberg-Manago, M., P. J. Ortiz, and S. Ochoa, *Biochim. Biophys. Acta,* **20,** 269 (1956).

[11] Nirenberg, M. W., and J. H. Matthaei, these Proceedings, **47,** 1558 (1961).

[12] Rendi, R., and T. Hultin, *Expl. Cell Research,* **19,** 253 (1960).

[13] Zamecnik, P. C., and E. B. Keller, *J. Biol. Chem.,* **209,** 337 (1954).

[14] Hoagland, M. B., M. L. Stephenson, J. F. Scott, L. I. Hecht, and P. C. Zamecnik, *J. Biol. Chem.,* **231,** 241 (1958).

[15] Gierer, A., and G. Schramm, *Nature,* **177,** 702 (1956).

[16] Lengyel, P., and R. W. Chambers, *J. Am. Chem. Soc.,* **82,** 752 (1960).

[17] Lengyel, P., and C. Basilio, to be published.

[18] Zamecnik, P. C., R. B. Lotfield, M. L. Stephenson, and J. M. Steele, *Cancer Research,* **11,** 592 (1951).

[19] Lowry, O. H., N. J. Rosebrough, A. L. Farr, and R. S. Randall, *J. Biol. Chem.,* **193,** 265 (1951).

[20] Lengyel, P., and R. C. Warner, unpublished results.

[21] A. Rich, personal communication.

[22] A. Rich (personal communication) also observed no effect of this polymer.

[23] Tsugita, A., and H. Fraenkel-Conrat, these Proceedings, **46,** 636 (1960).

[24] Yamazaki, H., and P. Kaesberg, *Nature,* **191,** 97 (1961).

Reprinted from PROCEEDINGS OF THE NATIONAL ACADEMY OF SCIENCES, Vol. 48, 1962

CHARACTERISTICS AND COMPOSITION OF RNA CODING UNITS*

BY J. HEINRICH MATTHAEI,† OLIVER W. JONES, ROBERT G. MARTIN, AND MARSHALL W. NIRENBERG

NATIONAL INSTITUTE OF ARTHRITIS AND METABOLIC DISEASES, BETHESDA

Communicated by Richard Roberts, February 27, 1962

The translation of a four-letter nucleotide code into a twenty-"word" amino acid dictionary has been the subject of much speculation. Although DNA and RNA polymers were inferentially involved in determining amino acid sequence, a cell-free protein synthesizing system dependent upon these polymers was not available. We have recently described such a system which is dependent upon the addition of template RNA.[1-3] This system affords a sensitive assay for both naturally occurring and synthetic template RNA.

Polyuridylic acid directed the synthesis of polyphenylalanine; thus, one or more uridylic acid residues in poly U appeared to be the coding unit corresponding to phenylalanine.[1, 2] Phenylalanine linked to soluble RNA is an intermediate in this process.[4] The finding that synthetic polyribonucleotides of known composition could be used to direct cell-free protein synthesis, suggested a reasonable experimental approach for establishing the characteristics of the genetic code, an approach which has been utilized by ourselves[5] and by others.[6, 7]

The purpose of this communication is to report further results concerning the influence of poly U in directing polyphenylalanine synthesis and the effects of randomly ordered copolymers upon the incorporation of other amino acids into protein. Some of these results have been reported in a preliminary communication.[5]

Methods and Materials.—The preparation of stable *E. coli* enzyme extracts (DNAase-treated, preincubated S-30 fractions) has been described.[2] Such extracts were dialyzed and stored under liquid nitrogen after preincubation. Reaction mixtures used in determining C^{14}-amino acid incorporation into protein contained the following components: 0.1 *M* tris(hydroxymethyl)-

aminomethane, pH 7.8; 0.01 *M* magnesium acetate; 0.05 *M* KCl; 6×10^{-3} *M* mercaptoethanol; 1×10^{-3} *M* ATP; 5×10^{-3} *M* potassium phosphoenolpyruvate; 20 μg/ml of crystalline phosphoenolpyruvate kinase (California Biochem. Corp.); $0.8 - 1.6 \times 10^{-4}$ *M* C^{14}-amino acids; 2×10^{-4} *M* each of 19 L-amino acids minus the C^{14}-amino acid; 3×10^{-5} *M* each of GTP, CTP, and UTP (except where stated); and *E. coli* extracts. Total volume was 0.50 ml except where specified. All assays were performed in duplicate. Techniques used in washing, plating, and counting protein precipitates have been reported.[2] Protein analyses were performed by a micro modification of the method of Lowry *et al.*[8] The lithium and sodium salts of nucleotide diphosphates were obtained from Schwarz Biochemical Corp. and Sigma Chemical Co., respectively. U-C^{14}-amino acids were obtained from Nuclear-Chicago Corp.

The purity of each C^{14}-amino acid was checked by high voltage electrophoresis followed by radioautography.[9] C^{14}-isoleucine was contaminated with C^{14}-leucine; therefore, C^{14}-isoleucine was purified electrophoretically before use. C^{14}-methionine was contaminated with C^{14}-methionine sulfoxide. All of the other C^{14}-amino acids were found to be free of C^{14}-contaminants. S^{35}-cysteine was reduced before use either electrolytically or with mercaptoethanol. The radioactive amino acids used, their source, and their respective specific radioactivities are as follows: U-C^{14}-glycine, U-C^{14}-L-isoleucine, U-C^{14}-L-tyrosine, U-C^{14}-L-leucine, U-C^{14}-L-proline, L-histidine-2(ring)-C^{14}, U-C^{14}-L-phenylalanine, U-C^{14}-L-threonine, L-methionine (methyl-C^{14}), U-C^{14}-L-arginine, and U-C^{14}-L-lysine obtained from Nuclear-Chicago Corporation, 5.8, 6.2, 5.95, 6.25, 10.5, 3.96, 10.3, 3.9, 6.5, 5.8, 8.3 mC/mM, respectively; C^{14}-L-aspartic acid, C^{14}-L-glutamic acid, C^{14}-L-alanine, obtained from Volk, 1.04, 1.18, 0.75 mC/mM, respectively; D-L-tryptophan-C^{14}, obtained from New England Nuclear Corporation, 2.5 mC/mM; S^{35}-L-cystine obtained from the Abbott Laboratories, 2.4 mC/mM; U-C^{14}-L-serine obtained from the Nuclear-Chicago Corporation, 0.2 mC/mM. The specific radioactivities of C^{14}-phenylalanine, valine, and leucine reported by Nuclear-Chicago Corp. were validated by assay. Amino acid concentration was determined by the ninhydrin method.[10] These assays agreed well with the commercially reported data. In a few instances the concentration of C^{14}-phenylalanine also was assayed by measurement of the absorption of the enol-borate complex of phenylpyruvic acid generated enzymatically from phenylalanine by L-amino acid oxidase.[11]

Polyribonucleotides were synthesized enzymatically with *Micrococcus lysodeikticus* polynucleotide phosphorylase purified by the method of Singer and Guss.[12] Reaction mixtures contained 0.15 *M* tris(hydroxymethyl)aminomethane pH 9.0; 0.01 *M* $MgCl_2$; 4×10^{-4} *M* ethylenediaminetetraacetate; 0.06 *M* nucleotide diphosphate; and polynucleotide phosphorylase purified through the acid ammonium sulfate fractionation step.[12] Approximately 1.4 units of polynucleotide phosphorylase were added per ml of reaction mixture. Reaction mixtures were incubated at 37° and the formation of polynucleotide was followed by determining phosphate liberation by the Fiske-SubbaRow method.[13] After the reaction had proceeded to 20–35% completion, the polynucleotides were precipitated by the addition of 3 volumes of cold absolute ethanol plus a few drops of *M* NaCl and the precipitates were collected by centrifugation. The polynucleotides were redissolved in H_2O and were reprecipitated as before. The pellet was dissolved in H_2O and the solution was deproteinized three times by the method of Sevag.[14] The aqueous phases obtained after deproteinization were combined and solid KCl was added to a final concentration of 0.025 *M*. The solution was dialyzed against 0.025 *M* KCl for 24 hr and against H_2O for an additional 48 hr. The polynucleotide solutions were then lyophilized and were stored at −15°. Some polynucleotides were the generous gifts of Drs. Leon Heppel, Maxine Singer, Daniel Bradley, David Davies, and Robert Steiner.

The base-ratio of each polynucleotide was determined by acid hydrolysis.[15] The bases were separated by descending chromatography on Whatman No. 1 paper for 18 hr using isopropanol-concentrated HCl-H_2O (130:33:37).

The base-ratio analyses obtained by acid hydrolysis were verified in some cases by comparison with base-ratio data obtained by alkaline hydrolysis[16] and by determination of phosphate.[17] Good agreement was found between base-ratio analyses obtained by the three methods.

Results.—Stoichiometry: In Table 1 the mμmoles of C^{14}-phenylalanine incorporated into protein are compared with the mμmoles of uridylic acid residues (pU) in poly U added to a reaction mixture. Limiting concentrations of poly U were

TABLE 1

RELATIONSHIP BETWEEN C^{14}-L-PHENYLALANINE INCORPORATED AND POLYURIDYLIC ACID PRESENT

Experiment no.	mμmole pU (in polyuridylic acid)	mμmole C^{14}-L-phenylalanine	mμmole pU / mμmole C^{14}-phenylalanine
1	11.35	9.12	1.24
2	17.7	10.25	1.73
3	25.7	16.9	1.52

The components of the reaction mixtures are presented under *Methods and Materials*. In addition, each reaction mixture contained 80 mμmole of C^{14}-L-phenylalanine with a specific activity of 1 mC per mM; 2.6 mg preincubated S-30 protein and the amounts of polyuridylic acid specified. In each experiment the amount of polyuridylic acid present was limiting. Final volume was 0.5 ml. Reaction mixtures were incubated for 60 minutes at 37°.

added and incubations were continued until phenylalanine incorporation had ceased. Approximately 1 mμmole of C^{14}-phenylalanine was incorporated into protein per 1.5 mμmole of uridylic acid residue in poly U. These data represent the lowest ratios obtained. The results of many similar experiments demonstrated that widely different stoichiometry ratios could be obtained depending upon the enzyme and poly U preparations used. In some experiments more than 50 mμmoles of uridylic acid residues in poly U were required to direct the incorporation of 1 mμmole of C^{14}-phenylalanine into protein. It should be noted that the presence of trace amounts of RNAase in the system, the molecular weight of poly U, etc., are variables which affect the incorporation of phenylalanine. For these reasons, it is likely that smaller ratios than those presented in Table 1 will be obtained. Since it is not known whether one molecule of poly U directs the synthesis of one or many molecules of polyphenylalanine, these data alone cannot be used to determine the number of uridylic acid residues in one phenylalanine coding unit (coding ratio).

Effect of Molecular Weight upon Template Activity of Poly U.—Poly U was separated into fractions of different molecular weights by sucrose density-gradient centrifugation.[18] Sixteen fractions were collected and both the absorbancy at 260 millimicrons of each fraction and its ability to stimulate C^{14}-phenylalanine incorpora-

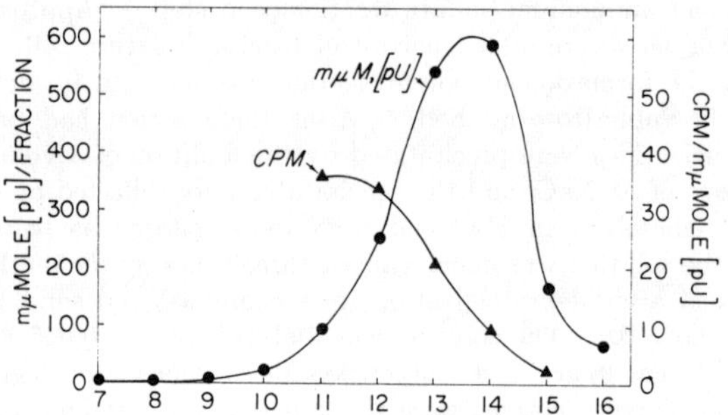

FIG. 1.—The relationship between the molecular weight of polyuridylic acid and its activity in stimulating C^{14}-phenylalanine into protein. Polyuridylic acid was separated into fractions of different sizes by sucrose density-gradient centrifugation. A linear gradient of sucrose concentration ranging from 16% at the bottom to 5% at the top of the tube was prepared. The sucrose solutions (4.4 ml total volume) contained 5×10^{-3} *M* imidazole, pH 6.8 and 0.1 *M* NaCl. 0.5 mg polyuridylic acid in 0.2 ml was layered on top of the tube which was centrifuged at 39,000 × *g* for 4.0 hours at 3°C in a swinging bucket rotor, Spinco type SW-39, using a Spinco Model L ultracentrifuge. Sixteen fractions containing 0.29 ml each were collected after piercing the bottom of the tube. Aliquots containing 25 mμmoles of uridylic acid residue in polyuridylic acid were used for C^{14}-phenylalanine incorporation assays. This amount of polyuridylic acid was limiting under the conditions of the assay. The components of each reaction mixture is presented under *Methods and Materials*. Total volume of each reaction mixture was 0.5 ml and incubations continued for 60 min at 37°.

tion into protein were determined (Fig. 1). Equal amounts of poly U were added to each reaction mixture and incubations were continued until the reactions had stopped. The concentration of poly U in reaction mixtures limited the rate of the reactions. The total activity of each fraction of poly U thus was determined. Poly U molecules of higher molecular weight were distributed toward the bottom of the tube; the lower molecular weight molecules were nearer the top of the tube. It can be seen that fractions of poly U of higher molecular weight were more active in directing polyphenylalanine into protein than lower molecular weight fractions. Although the biologic activity of poly U is related to its molecular weight, the

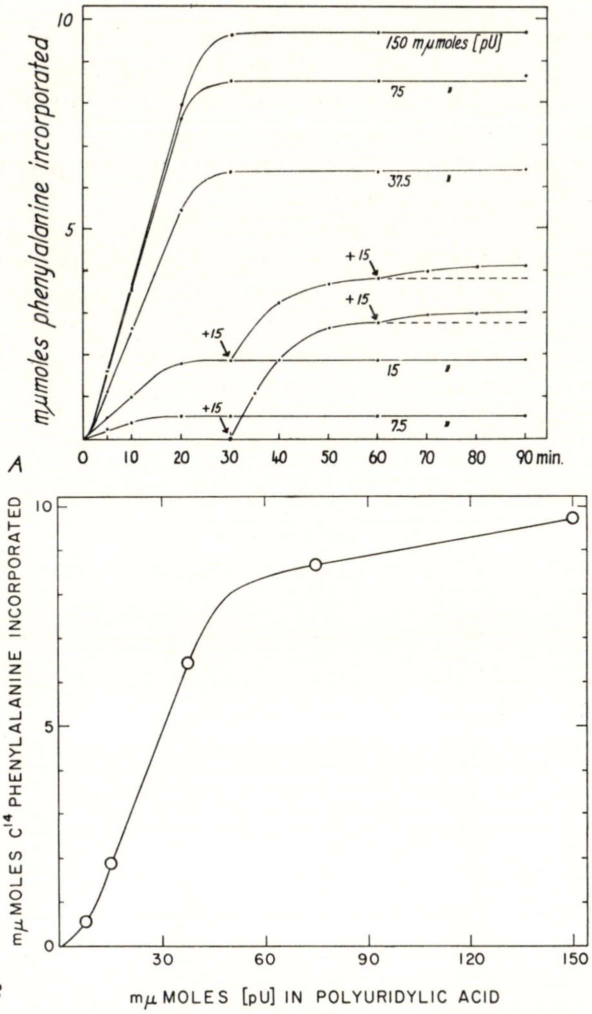

FIG. 2.—Kinetics of C^{14}-L-phenylalanine incorporation. (*A*) C^{14}-phenylalanine incorporation plotted as a function of time. The components of reaction mixtures are described under *Methods and Materials*. In (*A*) the mμmoles of uridylic acid residues (pU) in polyuridylic acid added to 0.5 ml of reaction mixture is shown. Each point represents a 0.5 ml reaction mixture containing 2.6 mg of preincubated S-30 protein. In (*B*) total C^{14}-phenylalanine incorporation (each final 90-minute analysis shown in (*A*)) is plotted as a function of polyuridylic acid concentration.

minimum molecular weight of poly U active as informational RNA is not known and experiments designed to answer this question are in progress.

Kinetics of Phenylalanine Incorporation.—In Figure 2 the rate of phenylalanine incorporation into protein is shown. From 7.5 to 150 mμmoles uridylic acid residue (pU) in poly U were added per 0.5 ml reaction mixture. C^{14}-phenylalanine incorporation ceased within 30 minutes after addition of poly U. Further addition of poly U after incorporation had stopped resulted in an additional stimulation of incorporation. After the reactions had again stopped, addition of more poly U did not appreciably stimulate C^{14}-phenylalanine incorporation. When the mixtures were incubated for 30 minutes in the absence of poly U, and at that time 15 mμmoles of pU were added, a somewhat greater incorporation of C^{14}-phenylalanine was observed compared with incorporation resulting from the addition of poly U at the onset of incubation. These experiments demonstrate that even though C^{14}-phenylalanine incorporation stops after 20–30 minutes of incubation, the enzyme extracts are fully active. The fact that C^{14}-phenylalanine incorporation stops after 30 minutes even though a large excess of poly U is present (150 mμmoles pU) shows that poly U is being inactivated. Such inactivation of excess poly U does not appear to be dependent upon protein synthesis.

The data of Figure 2 *B* show that phenylalanine incorporation is proportional to poly U concentration in the range of 10–50 mμmoles of pU in poly U per 0.5 ml reaction mixture. In the linear part of the curve about 6 mμmoles of pU were required to direct the incorporation of 1 mμmole of C^{14}-phenylalanine.

Stimulation of Amino Acid Incorporation by Polynucleotides Containing One Base.—The effect of "homopolynucleotides" upon amino acid incorporation into protein is presented in Table 2. As has been reported previously[12] poly U specifically stimulated phenylalanine incorporation into protein. Poly U also stimulated the incorporation of small amounts of C^{14}-leucine and C^{14}-valine into protein.

TABLE 2

SPECIFICITY OF AMINO ACID INCORPORATION STIMULATED BY HOMOPOLYNUCLEOTIDES AND YEAST RNA

C^{14}-L-amino acid	Minus polynucleotide	poly A	poly C	poly U	Yeast RNA
Alanine	27	18	21	26	82
Arginine	19	19	18	25	204
Aspartic acid	23	23	18	28	177
Cysteine	109	111	114	113	133
Glutamic acid	53	53	50	53	204
Glycine	32	31	28	31	284
Histidine	5	4	5	3	312
Isoleucine	50	22	55	74	229
Leucine	77	71	43	263	178
Lysine	14	22	11	26	129
Methionine	29	35	31	40	83
Phenylalanine	24	24	22	5424	128
Proline	22	20	336	27	118
Serine	250	250	250	281	687
Threonine	26	23	28	28	110
Tryptophan	245	239	238	262	514
Tyrosine	38	38	37	42	83
Valine	11	11	11	47	67

The figures represent the incorporation of C^{14}-amino acids in μμmoles. The components of reaction mixtures are presented under *Methods and Materials*. GTP, CTP, and UTP were omitted. Each 0.6 ml reaction mixture contained 80 mμmole of the appropriate C^{14}-amino acid (in addition to 19 C^{12}-amino acids); 2.6 mg preincubated S-30 protein; and either 100, 20, 10, or 500 μg of poly A, poly C, poly U or yeast RNA (prepared by the method of Crestfield *et al.*[14]) respectively. Samples were incubated for 60 minutes at 37°.

Base-ratio analyses (see *Methods and Materials*) showed that this polymer contained 2.5% guanylic acid present as an impurity. Thus, the small stimulation of leucine and valine incorporation by poly U probably is due to the presence of guanylic acid (see next section).

Poly C stimulated the incorporation of proline.[2] However, the effectiveness of poly C in stimulating proline incorporation varied with different preparations of poly C; for example, many preparations stimulated 5- to 10-fold and three preparations stimulated 75- to 100-fold. Since the effectiveness of poly C in stimulating proline incorporation was not fully reproducible, the purity of several preparations was examined (see *Methods and Materials* section). These preparations contained 4–14 per cent uridylic acid. Commerical preparations of CDP were analyzed and were found to contain no UDP. As will be seen in the next section, randomly-mixed poly UC markedly stimulated proline incorporation, therefore, the observed activity of poly C may result from the presence of small amounts of U in the polynucleotide preparations.

Poly A did not stimulate the incorporation of any amino acid into protein. The addition of smaller concentrations of poly A also gave similar results. The molecular weight of the poly A was approximately 30,000. In contrast, poly U of similar molecular weight markedly stimulated phenylalanine incorporation. The ineffectiveness of poly A in coding for any amino acid could be due to either its double-stranded structure in solution or to the possibility that a sequence of A does not specify any amino acid.

Previously, it was found that yeast ribosomal RNA prepared by the method of Crestfield *et al.*,[19] stimulated amino acid incorporation in this system.[2] Yeast RNA was used as a control to show that the system was active with respect to every amino acid. These experiments also demonstrate that the addition of naturally occurring template RNA stimulated the incorporation of every amino acid in contrast to the specificity displayed by poly U.

Since poly G is difficult to prepare enzymatically, it was not available to us in sufficient quantities to test with each individual amino acid. However, when poly G was tested with a C^{14}-algal protein hydrolysate which contained 16 C^{14}-amino acids, it did not stimulate the incorporation of any C^{14}-amino acid. An oligonucleotide primer (tetraadenylic acid) was required for the enzymic synthesis of poly G and primarily low molecular weight polynucleotides having an average chain length of 15 nucleotides were obtained. Therefore, the fact that poly G was unable to stimulate amino acid incorporation should be interpreted with caution.

The Effects of Randomly Ordered Polynucleotides upon Amino Acid Incorporation.— The data of Table 3 demonstrate that randomly ordered polyribonucleotides direct the incorporation of amino acids into protein in a highly specific manner. Since each preparation of polynucleotide will differ in molecular weight, it is difficult to compare directly the incorporation of activity of one polymer with another. Therefore, the figures in Table 3 represent the per cent of any amino acid incorporated compared to phenylalanine incorporation stimulated by the same polynucleotide. The counts/minute of C^{14}-L-phenylalanine incorporated due to the addition of each polynucleotide is presented in the legend accompanying Table 3; therefore, the percentage figures in the table readily may be converted into counts/minute. The base-ratio of each polynucleotide, determined experimentally, is given also. If a polynucleotide containing two bases stimulated incorporation of an amino acid, inclusion of a third base in the polynucleotide did not prevent this stimulation.

Poly UG and poly UGC previously were found to direct the incorporation of small amounts of methionine and glutamic acid respectively.[5] Further experiments have shown that poly UGA codes for these amino acids more effectively; thus, the coding units for methionine and glutamic acid contain U, G and A. Lysine incorporation was stimulated by the addition of poly UA ($^1/_4$ ratio) but not by randomly mixed poly UA containing lower proportions of A. The coding unit for lysine therefore probably contains UAA.... The stimulation of phenylalanine incorporation by poly UA ($^1/_4$ ratio) was negligible, so these data have not been expressed in Table 3 as the ratio of lysine to phenylalanine incorporation.

A series of polynucleotides each containing a different ratio of U to A were prepared. Similar series of polynucleotides containing either U and C or U and G were synthesized also. It was found that stimulation of incorporation of a given amino acid by a polynycleotide varied with the base-ratio of the polymer.[22] Thus the relative amounts of two nucleotides in a coding unit could be estimated. These data are too extensive to be reported here and will be the subject of a future communication.

The coding ratio is not known definitively. However, assuming a triplet code, the probability of a triplet occurring in a polynucleotide, relative to UUU, may be calculated from the base-ratio data presented in Table 3. For example, if poly UG had a base-ratio of 3U/1G, the probability of obtaining the sequence UUU would be $^3/_4 \times {}^3/_4 \times {}^3/_4 = {}^{27}/_{64}$. The probability of obtaining the sequence UUG would be $^3/_4 \times {}^3/_4 \times {}^1/_4 = {}^9/_{64}$. Thus, 3 UUU would occur for 1 UUG and, assuming the frequency of UUU to be 100 per cent, the frequency of UUG would be 33 per cent. The theoretical frequency of each possible triplet, relative to UUU is presented in Table 3. Amino acid incorporation agreed with predictions based upon probability theory in most cases. The nucleotide composition of coding units corresponding to each amino acid may be derived from these data and a summary is presented in the last column of Table 3. Dots after each "word" are used to indicate the possible presence of additional uridylic acid residues.

Discussion.—Since the coding units of arginine, alanine, and glutamic acid each contain three different nucleotides, the minimum coding ratio would appear to be three. Poly GC does not code for arginine or alanine and poly AG does not code for glutamic acid.[22] Possibly the coding ratio is larger, but these data rule out the possibility of singlet and doublet codes. In this preliminary study we will assume that every amino acid has the same coding ratio[20] and that in this system, as in cellular systems,[21] an overlapping triplet code is improbable.

It is important to determine whether one molecule of poly U directs the synthesis of either one or several molecules of polyphenylalanine, i.e., whether poly U functions stoichiometrically or catalytically. If, for example, each molecule of poly U directed the synthesis of only one molecule of polyphenylalanine and a nonoverlapping sextuplet code were operative, six uridylic acid residues in poly U would be necessary to mediate the incorporation of one phenylalanine. Instead, the data of Table 1 show that almost one mμmole of phenylalanine was incorporated into

TABLE 3

Amino Acid Incorporation into Protein Stimulated by Randomly Mixed Polynucleotides

Polynucleotide	UA	UC	UG	UAC	UGC	UGA	Composition of coding units*
Base-ratio	U = 0.87 A = 0.13	U = 0.39 C = 0.61	U = 0.76 G = 0.24	U = 0.834 A = 0.050 C = 0.116	U = 0.341 G = 0.152 C = 0.502	U = 0.675 G = 0.291 A = 0.034	
Probability of triplet relative to phenylalanine (UUU) = 100%	UUU—100 UUA— 13 UAA— 2.2 AAA— 0.3	UUU—100 UUC—157 UCC—244 CCC—382	UUU—100 UUG— 32 UGG— 10.6 GGG— 3.4	UUU—100 UUA— 6.0 UAA— 0.4 AAA— 0.02 UUC— 13.9 UCC— 1.9 CCC— 0.3 UAC— 0.8 AAC— 0.05 ACC— 0.12	UUU—100 UUG— 46.2 UGG— 21.0 GGG— 1.0 UUC—147 UCC—218 CCC—322 UGC— 68.1 GGC— 31.7 GCC—101	UUU—100 UUG— 43 UGG— 19 GGG— 8.1 UUA— 5.1 UAA— 0.26 AAA— 0.01 UGA— 2.2 GGA— 0.1 GAA— 0.01	
Amino Acid							
Phenylalanine	100	100	100	100	100	100	UUU...
Arginine	0	0	1.1	0	49.3	2.9	UCG...
Alanine	1.9	0	0	1.0	40.4	0.9	UCG...
Serine	0.4	160	3.2	3.6	170	2.3	UUC.. + UCG...
Proline	0	285	0	0	188	0	UCC...
Tyrosine	13	0	0	8	1.0	8.6	UUA...
Isoleucine	12	1.0	1.0	4.8	5.4	8.4	UUA...
Valine	0.6	0	37	0.4	29.8	75	UUG...
Leucine	4.9	79	36	5.1	157	44	UUC.. + UUG...
Cysteine	4.9	0	35	0	5.4	46	UUG... or UGG
Tryptophan	1.1	0	14	0	1.6	23	UGG...
Glycine	4.7	0	12	0.5	9.7	15	UGG...
Methionine	0.6	0	0	0.6	1.5	8	UGA...
Glutamic acid	1.5	0	0	1.2	0.44	6.2	UGA...
Lysine							UAA... (?)

The figures in the main part of the table represent the incorporation of any amino acid compared to phenylalanine incorporation expressed as percentages (mμmoles amino acid incorporated/mμmoles phenylalanine incorporated × 100). Underlined figures refer to the polynucleotide containing the nucleotides necessary to stimulate the incorporation of a given amino acid. The components of the reaction mixtures are presented under *Methods and Materials*. 0.1 μmole each of 19 L-amino acids minus the appropriate C^{14}-amino acid, 0.075 μmole of the C^{14}-amino acid, and approximately 25 μg of each polynucleotide were added to each 0.5 ml reaction mixture. Samples were incubated at 37° for 15 minutes. Incorporation of C^{14}-phenylalanine in counts per minute due to the addition of polynucleotides UA, UC, UG, UAC, UGC, and UGA were 731, 2900, 714, 804, 2144, and 2744, respectively. The reproducibility of the above percentage figures was ±3.

* Sequence of nucleotides in a coding unit is not specified.

protein per mμmole of uridylic acid residue in poly U. Therefore, each phenylalanine coding unit appears to function catalytically for a limited time.

A degenerate code is one in which two or more different coding units can direct the incorporation of the same amino acid into protein. In a completely degenerate code, every permutation of nucleotides would code for an amino acid. In a partially degenerate code, certain coding units would direct amino acid incorporation, whereas others would not. Coding units which would not direct amino acids into protein will be called "nonsense words." Up to now it has not been possible to determine directly whether the code contains nonsense units.

The data of Table 2 demonstrate that poly A does not direct the incorporation of any amino acid into protein. Other polynucleotides such as poly AG also do not direct incorporation of amino acids.[22] Such experiments indicate that nonsense words exist and thus would preclude the possibility of a completely degenerate code. Nonsense regions in template RNA may be functionally important, for it is possible that these nucleotide sequences serve as periods, i.e., may specify C- or N-terminal groups in proteins.

The data of Table 3 demonstrate that a coding unit corresponding to leucine can contain either U and C or U and G. Since two words containing different nucleotides correspond to leucine, the code is partially degenerate when synthetic polynucleotides are used to direct amino acid incorporation.

Whereas the U content of RNA viruses is not excessive, a surprisingly high proportion of U has been found in coding units thus far (Table 3). This dichotomy cannot be explained at the present time. However, it is probable that additional degenerate code words will be found which will not contain U. Trichloracetic acid insoluble peptides only have been measured in this study. Since polyphenylalanine becomes insoluble when four or five phenylalanine residues are linked together, such an insoluble "handle" may ensure the precipitation of a polypeptide which otherwise would be soluble. If a code with much degeneracy is assumed, it is possible that coding units containing U have been selected for by the assay method. However, other explanations are possible and are being considered.

The foregoing data demonstrate that the code is partially degenerate, the minimum coding ratio is three and that nonsense coding units exist. Crick *et al.*, on the basis of ingenious genetic experiments employing the rII region in T_4 phage, have come to similar conclusions.[23] It is important to note that such information, obtained by means of both biochemical and genetic approaches, is in complete accord with one another.

The treatment of tobacco mosaic virus (TMV) RNA with nitrous acid results in the deamination of nucleotides[24] and the formation of mutant strains of TMV with altered plaque morphology.[25] Cytidine is converted to uridine, adenosine to hypoxanthine and guanosine to xanthosine. The recent striking work of Wittmann[28] at Tübingen and Tsugita and Fraenkel-Conrat[26, 27] at Berkeley has shown that the protein isolated from such mutant strains is, in many cases, different from the protein produced by wild type TMV. The amino acid sequences of both wild type and mutant TMV proteins were determined and certain amino acids were found to be replaced by others rather frequently. A comparison has been made in Table 4 between nitrous acid-induced amino acid substitutions and the nucleotide compositions of corresponding coding units. Only amino acid substitutions which have

TABLE 4

COMPARISON OF NITROUS ACID INDUCED REPLACEMENTS IN TOBACCO MOSAIC VIRUS PROTEIN WITH THE NUCLEOTIDE COMPOSITION OF RNA CODING UNITS

No. of mutant strains A	B*	Amino acid replacement	Nucleotide composition of corresponding coding units	Possible nucleotide changes
1	2	Ser → Phe	UUC... → UUU...	C → U
1	2	Glu → Gly	UAG... → UGG...	A → G
1	2	Prol → Leu	UCC... → UUC...	C → U
	2	Isoleu → Val	UUA... → UUG...	A → G
5		Arg → Gly	UCG... → UGG...	C → G

* The mutant amino acid replacement data cited and obtained either by (*A*) Tsugita and Fraenkel-Conrat[26, 27] or (*B*) Wittmann.[28]

been found more than once have been cited. Thus, the chance of a substitution occurring spontaneously rather than by deamination is lessened. Only those amino acid substitutions are included which correspond to RNA coding units with experimentally determined nucleotide compositions. We are indebted to Drs. Tsugita and Fraenkel-Conrat for allowing us to cite some of their unpublished replacement data.

Replacement of serine by phenylalanine, glutamic acid by glycine, proline by leucine, glutamic acid by valine, and arginine by glycine have been observed. Most of the nucleotide conversions corresponding to the amino acid replacements thus would be the conversion of either cytidine to uridine or of adenine to guanine.

As has been predicted by Freese,[29] a fairly high proportion of nitrous acid induced amino acid substitutions should be the result of conversion of cytidine to uridine in a coding unit. Wittmann has also shown that nitrous acid will convert adenine to guanine.[28] Deamination of guanosine has no mutagenic effect. Comparison of amino acid substitution data with the proposed nucleotide compositions of coding units shown in Table 4 strikingly confirm these predictions.

Previously we have shown that sRNA is an intermediate in polyphenylalanine synthesis.[4] It is likely that amino acids lose their identity after being linked to sRNA and that aminoacyl-sRNA molecules recognize template RNA coding units by base-pairing. During the course of evolution, mutations affecting the nucleotide sequence of sRNA or the specificity of amino acid activating enzymes might occur and might result in a series of phylogenetically related codes rather than one code universal to all species.

From an evolutionary point of view, a mutation resulting in a changed code wherein one amino acid substitutes for another in all proteins probably would be lethal to a highly organized cellular system. The use of synthetic polynucleotides affords a unique opportunity to determine whether the code is universal. The data of Table 4 demonstrate that both TMV coding units directing protein synthesis in tobacco plants, and coding units functioning in *E. coli* have similar nucleotide compositions. It is not improbable that minor changes in the code will be found in different species; however, these data strongly suggest that at least part of the code may be universal.

Summary.—The activity of polyuridylic acid in directing polyphenylalanine synthesis was found to vary with the molecular weight of polyuridylic acid; longer polynucleotide chains were more active than shorter ones. Evidence was presented which suggested that one molecule of polyuridylic acid may direct the synthesis of a number of molecules of polyphenylalanine. Randomly mixed polynucleotides as well as "homopolynucleotides" were used to direct cell-free amino acid incorporation. Using this technique the nucleotide compositions of RNA coding units corresponding to 15 amino acids were determined. Characteristics of the code such as degeneracy, the existence of nonsense words, the minimum coding ratio, and the universality of the code were investigated. Two coding units corresponding to leucine were found; thus, part of the code was shown to be degenerate. Certain nucleotide sequences in a polynucleotide did not code for any amino acid: therefore, the presence of nonsense coding units was suggested. The *minimum* number of nucleotides per coding units appeared to be three. Comparison between the composition of RNA coding units in *E. coli* and amino acid replacement data in tobacco mosaic virus suggested that at least part of the code may be universal.

The authors wish to express their appreciation to Mrs. Linda Greenhouse for valuable technical assistance. It is also a pleasure to express our gratitude to Drs. Leon Heppel and Maxine Singer for invaluable advice and helpful discussions.

* The following abbreviations were used: Poly U, polyuridylic acid; poly A, polyadenylic acid; poly C, polycytidylic acid; poly G, polyguanylic acid; poly UA, polyuridylic-adenylic acid; poly UC, polyuridylic-cytidylic acid; poly UG, polyuridylic-guanylic acid; poly UAC, polyuridylic-adenylic-cytidylic acid; poly UCG, polyuridylic-cytidylic-guanylic acid; poly UGA, polyuridylic-guanylic-adenylic acid; ADP, adenosine diphosphate; UDP, uridine diphosphate; GDP; guanosine diphosphate; CDP, cytidine diphosphate, pU, uridylic acid residue in poly U; TMV, tobacco mosaic virus.

† NATO Postdoctoral Research Fellow.

[1] Nirenberg, M. W., and J. H. Matthaei, *V International Congress of Biochemistry*, Moscow (August, 1961).

[2] Nirenberg, M. W., and J. H. Matthaei, these PROCEEDINGS, **47,** 1588 (1961).

[3] Matthaei, J. H., and M. W. Nirenberg, these PROCEEDINGS, **47,** 1580 (1961).

[4] Nirenberg, M. W., J. H. Matthaei, and O. W. Jones, these PROCEEDINGS, **48,** 104 (1962).

[5] Martin, R. G., J. H. Matthaei, O. W. Jones, and M. W. Nirenberg, *Biochem. Biophys., Research Commun.*, **6,** 410 (1962).

[6] Lengyel, P., J. F. Speyer, and S. Ochoa, these PROCEEDINGS, **47,** 1936 (1961).

[7] Speyer, J. F., P. Lengyel, C. Basilio, and S. Ochoa, these PROCEEDINGS, **48,** 63 (1962).

[8] Lowry, O. H., N. J. Rosebrough, A. L. Farr, and R. J. Randall, *J. Biol. Chem.*, **193,** 265 (1951).

[9] Dreyer, W. J., and E. Bynum, in preparation.

[10] Spies, Joseph R., in *Methods in Enzymology*, ed. S. P. Colowick and N. O. Kaplan (New York: Academic Press, 1957), vol. 3, p. 467.

[11] La Du, B., and P. J. Michael, *J. Lab. and Clin. Med.*, **55,** 491 (1960).

[12] Singer, M. F., and J. K. Guss, *J. Biol. Chem.*, **237,** 182 (1962).

[13] Fiske, C. H., and Y. SubbaRow, *ibid.*, **66,** 375 (1925).

[14] Sevag, M. G., D. B. Lackman, and J. Smolens, *ibid.*, **124,** 425 (1938).

[15] Smith, J. D., and R. Markham, *Biochem. J.*, **46,** 509 (1950).

[16] Davidson, J. H., and R. M. S. Smellie, *Biochem. J.*, **52,** 594 (1952).

[17] Ames, B., and D. T. Dubin, *J. Biol. Chem.*, **235,** 769 (1960).

[18] Britten, R. J., and R. B. Roberts, *Science,* **131,** 32 (1960).
[19] Crestfield, A. M., K. C. Smith, and F. W. Allen, *J. Biol. Chem.,* **216,** 185 (1955).
[20] Gamow, G., *Nature,* **173,** 318 (1954).
[21] Brenner, S., these PROCEEDINGS, **43,** 687 (1957).
[22] Nirenberg, M. W., J. H. Matthaei, and O. W. Jones, unpublished data.
[23] Crick, F. H. C., L. Barnett, S. Brenner, and R. J. Watts-Tobin, *Nature,* **192,** 1227 (1961).
[24] Levene, P. A., and R. S. Tipson, *J. Biol. Chem.,* **111,** 313 (1935).
[25] Gierer, A., and K. W. Mundry, *Nature,* **182,** 1457 (1958).
[26] Tsugita, A., and H. Fraenkel-Conrat, *J. Mol. Biol.,* in press.
[27] Tsugita, A., and H. Fraenkel-Conrat, personal communication.
[28] Wittmann, H. G., *Naturwissenschaften,* **48,** 729 (1961).
[29] Freese, E., in *Structure and Function of Genetic Elements,* Brookhaven Symposia in Biology, no. 12 (1959), p. 63.

Reprinted from SCIENCE, Vol. 145, 1964

RNA Codewords and Protein Synthesis

The Effect of Trinucleotides upon the Binding of sRNA to Ribosomes

Marshall Nirenberg and Philip Leder

Although many properties of the RNA code and protein synthesis have been clarified with the use of synthetic polynucleotides containing randomly ordered bases, a more comprehensive understanding of certain aspects of the code clearly requires investigation with nucleic acid templates of demonstrated structure. Since oligonucleotides of known base sequence are readily prepared and characterized, we have tried, in many ways, to use defined oligonucleotide fractions for studies relating to the RNA code. In this article we describe a simple, direct method which should provide a general method for determining the genetic function of triplets of known sequence. The system is based upon interactions between ribosomes, aminoacyl sRNA (*1*), and mRNA which occur during the process of codeword recognition, prior to peptide-bond formation.

The binding of sRNA to ribosomes has been observed in many studies (*2, 3*); however, this interaction is not fully understood. An exchangeable binding of sRNA to ribosomes was reported by Cannon, Krug, and Gilbert (*4*). However, the addition of polyU induced, with specificity, Phe-sRNA binding to ribosomes, as demonstrated in the laboratories of Schweet (*5–7*) and Lipmann (*8*), by Kaji and Kaji (*9, 10*), and by Spyrides (*11*). Binding was reported to be dependent upon GTP (*6, 7*) and the first transfer enzyme (*5–7*), but not upon peptide-bond synthesis. However, the mechanism of binding and the possibility of a prior, nonenzymatic binding of amino-acyl sRNA induced by mRNA have not been clarified. The second transfer enzyme was shown to be required for peptide bond formation (*6, 7*).

To determine the minimum chain length of mRNA required for codeword recognition and to test the ability of chemically defined oligonucleotides to induce C^{14}-aminoacyl-sRNA binding to ribosomes, we have devised a rapid method of detecting this interaction and have found that trinucleotides are active as templates.

The authors are affiliated with the Section of Biochemical Genetics of the National Heart Institute, National Institutes of Health, Bethesda, Maryland. These data were presented at the VIth International Congress of Biochemistry, 26 July–1 August 1964, New York City.

Methods

Preparation, purification, and characterization of oligonucleotides. To obtain oligonucleotides with different terminal groups, polyA, polyU, and polyC (*12*) were digested as follows: (i) *Oligonucleotides with 5′-terminal phosphate*; 100 mg of polynucleotide were incubated at 37°C for 18 hours in a 28-ml reaction mixture containing 29m*M* tris, *p*H 7.2; 0.18m*M* $MgCl_2$; 0.23m*M* 2-mercaptoethanol; 8.0 mg crystalline bovine albumin and 0.5 mg pork liver nuclease (*13*). (ii) *Oligonucleotides with 3′(2′)-terminal phosphate*; 100 mg of polynucleotide were incubated at 37°C for 24 hours in 20 ml of 7.0*M* NH_4OH. (iii) *Oligonucleotides without terminal phosphate*; Oligonucleotides with terminal phosphate were treated with *Escherichia coli* alkaline phosphatase (*14*) free of diesterase activity as described by Heppel *et al.* (*15*).

Oligonucleotide fractions were separated on Whatman 3 MM paper by chromatography with solvent A (H_2O:*n*-propanol:NH_3, 35:55:10, by volume) for 36 hours (fractions with terminal phosphate) or for 18 hours (fractions without terminal phosphate). This procedure fractionates oligonucleotides containing less than eight nucleotide residues according to chain length. Oligonucleotides were eluted with H_2O and further purified on Whatman 3 MM paper by electrophoresis at *p*H 2.7 (0.05*M* ammonium formate, 80 v/cm for 15 to 30 minutes).

After elution the purity of each fraction was estimated by subjecting 2.5 A^{260} units of each to paper chromatography (Whatman 54 paper) both with solvent A and with solvent B (40 g ammonium sulfate dissolved in 100 ml 0.1*M* sodium phosphate, *p*H 7.0). In addition 3.0 A^{260} units of each oligonucleotide were subjected to chromatography on Whatman DE 81 (DEAE) paper with solvent C (0.1*M* ammonium formate), and 3.0 A^{260} units with solvent D (0.3*M* ammonium formate). The four chromatographic systems described separate homologous series of oligonucleotides according to chain length. Contaminating oligonucleotides present in amounts greater than 2 percent could be detected. Several preparations of each oligonucleotide were used during the course of this study. In almost all preparations, no contaminants were detected. The following preparations, specified in legends of figures or tables when used, contained contaminants in the proportions indicated: No. 591, $(Ap)_4$ [$(Ap)_3$, 11 percent]; No. 599, $(pA)_6$ [$(pA)_5$, 37 percent]; No. 610, $(pU)_2$ [$(pU)_3$, 10 percent], No. 613 $(pU)_5$ [$(pU)_6$, 14 percent]; No. 617, $(Up)_4$ [$(Up)_3$, 12 percent].

Base composition and position of terminal phosphate were determined by digesting 2.0 A^{260} units of each oligonucleotide preparation with 3.5 × 10^3 units of T_2 ribonuclease (*16*) in 0.1*M* NH_4HCO_3 at 37°C for 2.5 hours. The nucleotide and nucleoside products were separated by electrophoresis at *p*H 2.7 and identified by their mobilities and ultraviolet spectra at *p*H 2.0. Oligonucleotides with 5′-terminal phosphate yielded the appropriate 5′-3′(2′)-nucleoside diphosphate, 3′(2′)-nucleoside monophosphate, and nucleoside. From the ratio of these compounds, the average chain length of the parent oligonucleotide was calculated. Since oligonucleotides with 3′(2′)-terminal phosphate yielded only the appropriate 3′(2′)-nucleoside monophosphate (confirming its structure), terminal and total inorganic phosphate was determined (*15, 17*) in order to estimate the average chain length of each.

Oligo-d(pT) and oligo-d(pA) fractions were prepared and characterized by B. F. C. Clark as described previously (*18*). The UpUpUp with 3′-

Table 1. Characteristics of the system. Complete reactions contained the components described in the text, 15 mμmole uridylic acid residues in polyU, and 20.6 μμmole C^{14}-Phe-sRNA (2050 count/min, 0.714 A^{260} units). Incubation was at 0°C for 60 minutes. Deacylated sRNA was added either at zero time or after 50 minutes of incubation, as indicated.

Modifications	C^{14}-Phe-sRNA bound to ribosomes (μμ mole)
Complete	5.99
−PolyU	0.12
−Ribosomes	0
$-Mg^{++}$	0.09
+sRNA (deacylated) at 50 min	
0.500 A^{260} units	5.69
2.500 A^{260} units	5.36
+sRNA (deacylated) at zero time	
0.500 A^{260} units	4.49
2.500 A^{260} units	2.08

terminal phosphate only was prepared and characterized by M. Bernfield (*19*).

Assay of ribosomal bound C^{14}-aminoacyl-sRNA. Each 50-μl reaction mixture contained: 0.1*M* tris-acetate, *p*H 7.2; 0.02*M* magnesium acetate; 0.05*M* KCl; 2.0 A^{260} units of ribosomes (washed three times) and, as indicated for each experiment, oligo- or polynucleotide, and C^{14}-aminoacyl sRNA. Tubes were kept at 0°C and C^{14}-aminoacyl sRNA was added last to initiate binding (less binding was obtained if polynucleotide was added last). Incubation for 20 minutes at 24°C was often convenient for studies requiring maximum binding, and incubation for 3 minutes at 24°C or 30 minutes at 0°C for rate studies.

After incubation, tubes were placed in ice and each reaction was immediately diluted with 3 ml of buffer containing 0.10*M* tris-acetate, *p*H 7.2; 0.02*M* magnesium acetate; and 0.05*M* KCl, at 0° to 3°C. A cellulose nitrate filter (HA Millipore filter, 25 mm diameter, 0.45μ pore size) in a stainless steel holder was washed under gentle suction with 5 ml of buffer at 0° to 3°C. The diluted reaction mixture was immediately poured on the filter under suction and washed to remove unbound C^{14}-aminoacyl sRNA with three, 3-ml portions of buffer at 0° to 3°C. Ribosomes and bound sRNA remained on the filter. Since reaction mixtures are not deproteinized, it is important to dilute and wash the ribosomes immediately after incubation, to use cold buffer, and to allow relatively little air to be pulled through the filter during the washing procedure. The filter was removed from the holder, glued with rubber cement to a disposable planchette, and dried. Radioactivity was determined in a thin-window, gas-flow counter (*20*) with a C^{14}-counting efficiency of 23 percent. In some experiments, radioactivity was determined in a liquid scintillation counter with a C^{14}-counting efficiency of 65 percent (*21*) and dried filters (not glued) were placed in vials containing 10 ml of toluene-PPO-POPOP phosphor solution (*22*).

Table 2. Polynucleotide specificity. Reaction mixtures containing C^{14}-Phe- and C^{14}-Lys-sRNA were incubated for 60 minutes at 0°C; mixtures containing C^{14}-Pro-sRNA were incubated for 20 minutes at 24°C. In addition to the components described in the text, reaction mixtures contained, in a final volume of 50 μl, the specified polynucleotide and C^{14}-amino-acyl-sRNA (14.7 μμmole C^{14}-Phe-sRNA, 2015 count/min, 0.960 A^{260} units; 16.5 μμmole C^{14}-Lys-sRNA, 1845 count/min, 0.530 A^{260} units; 30 μμmole C^{14}-Pro-sRNA, 2750 count/min, 1.570 A^{260} units).

Polynucleotide (mμmole base residues)	C^{14}-Aminoacyl-sRNA bound to ribosomes (μμmole)		
	C^{14}-Phe-sRNA	C^{14}-Lys-sRNA	C^{14}-Pro-sRNA
None	0.19	0.99	0.25
PolyU, 25	6.00	.67	.15
PolyA, 16	0.22	4.35	.17
PolyC, 19	.21	0.72	.80

Preparation of sRNA. Except where noted *E. coli* B sRNA (*23*), was used. Uniformly labeled C^{14}-L-phenylalanine, C^{14}-L-lysine, and C^{14}-L-leucine with specific radioactivities of 250, 305, and 160 μc/μmole, respectively, were obtained commercially (*24*). The *E. coli* W 3100 sRNA was prepared as described by Zubay (*25*) from cells grown to late log phase in 0.9 percent Difco nutrient broth, containing 1 percent glucose. The C^{14}-amino acyl-sRNA was prepared by modifications of meth-

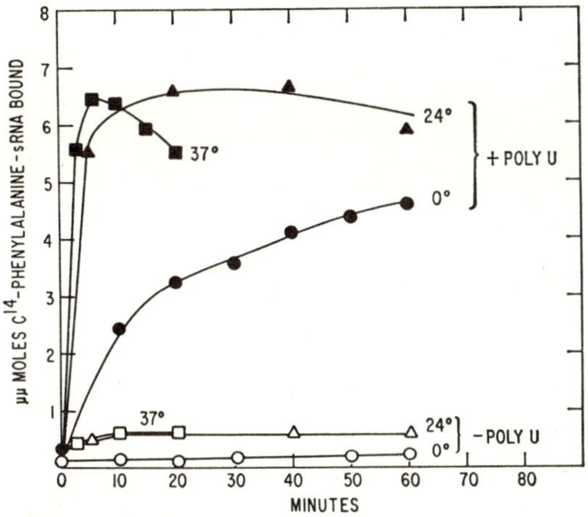

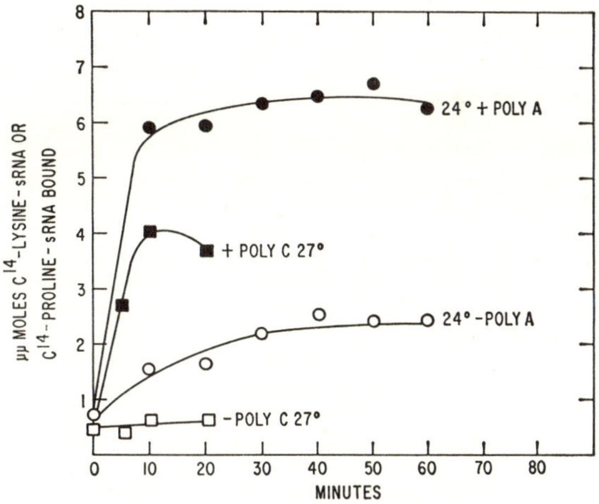

Fig. 1 (left). Effect of polyU upon the rate of C^{14}-Phe-sRNA binding to ribosomes at 0°, 24°, and 37°C. Each point represents a 50 μl reaction mixture incubated for the time and at the temperature indicated. Reaction mixtures contain the components described under *Methods*; 9.65 μμmole of C^{14}-Phe-sRNA (1180 count/min, 0.380 A^{260}); and polyU, 25 mμmole base residues, where specified. Fig. 2 (right). The effect of polyA and polyC upon the rates of C^{14}-Lys- and C^{14}-Pro-sRNA binding to ribosomes, respectively. Each point represents a 50-μl reaction mixture as described under *Methods*. The reactions specified contained 27.5 μμmole C^{14}-Lys-sRNA (3080 count/min, 0.880 A^{260} units) and polyA, 25 mμmole base residues, or 11.8 μμmole C^{14}-Pro-sRNA (2660 count/min, 0.905 A^{260} units) and polyC, 25 mμmole base residues. The temperature and the time of incubation are shown in the figure.

ods described previously (*26*). Unless otherwise specified, sRNA was acylated with one C^{14}-amino acid plus 19 C^{12}-amino acids. The formation of C^{14}-aminoacyl sRNA was catalyzed by the supernatant solution obtained by centrifugation of *E. coli* (W-3100) extracts at 100,000*g*.

Elution and characterization of C^{14}-phenylalanine product bound to ribosomes. Reaction mixtures (0.5 ml) incubated at 24°C for 10 minutes with C^{14}-Phe-sRNA and polyU were washed on cellulose nitrate filters in the usual manner. The ribosomal bound C^{14}-product was eluted from filters by washing with 0.01*M* tris-acetate, *p*H 7.2; $10^{-5}M$ magnesium acetate; and 0.05M KCl at 0°C.

The C^{14}-product eluted from ribosomes was precipitated in 10 percent TCA at 3°C in the presence of 200 μg of bovine serum albumin. Specified samples were heated in 10 percent TCA at 90° to 95°C for 20 minutes and then were chilled. Precipitates were washed on filters with 5 percent TCA at 3°C.

Aminoacyl sRNA was deacylated in 0.1*M* ammonium carbonate solution adjusted with NH_4OH to *p*H 10.2 to 10.5 by incubation at 37°C for 60 minutes.

Digestions with ribonuclease were performed by incubating 0.4-ml portions (each containing 1500 count/min precipitable by 10 percent TCA at 3°C) with and without 10 μg of purified pancreatic ribonuclease A (purified chromatographically) (*27*) at 37°C for 15 minutes.

Results and Discussion

Assay of ribosomal bound C^{14}-aminoacyl-sRNA. The assay is based upon the retention of ribosomes and C^{14}-aminoacyl sRNA bound to ribosomes by cellulose nitrate filters. After unbound C^{14}-aminoacyl sRNA is removed by washing with buffered salts solution, as already described, the radioactivity remaining on the filter is determined. Thirty reaction mixtures can be washed per hour easily. The sensitivity of the assay is limited primarily by the specific radioactivity of the aminoacyl-sRNA used. With sRNA which has accepted a C^{14}-amino acid of specific radioactivity 100 to 300 μc/μmole, the binding to ribosomes of 0.2 μμmoles of C^{14}-aminoacyl sRNA readily can be detected. A filter 25 mm in diameter with a pore size of 0.45 μ retains up to 1200 μg of *E. coli* ribosomes. The use of larger filters or columns packed with cellulose nitrate may be useful for preparative procedures.

The retention of ribosomes by cellulose nitrate filters may be the result of absorption rather than of filtration, for filters with pores 100 times larger than *E. coli* 70*S* ribosomes can be used. The rapidity of this assay, compared to others which depend upon the centrifugation of ribosomes, has greatly simplified this study.

The data of Table 1 show that little C^{14}-Phe-sRNA was retained on filters after incubation with ribosomes in the absence of polyU. Incubation in the presence of both polyU and ribosomes resulted in marked retention of C^{14}-Phe-sRNA by the filter. Ribosomes, polyU, and Mg^{++} were required for retention of C^{14}-Phe-sRNA. Spyrides (*28*) and Conway (*29*) have reported that polyU-directed binding of Phe-sRNA to ribosomes is dependent upon K^+ or NH_4^+.

The addition of deacylated sRNA to reactions shortly before incubation was terminated, after C^{14}-Phe-sRNA binding had ceased, had little effect upon ribosomal bound C^{14}-Phe-sRNA. The bound C^{14}-Phe-sRNA fraction apparently is not readily exchangeable.

In contrast, the addition of deacylated sRNA at the start of incubation inhibited C^{14}-Phe-sRNA binding. In other experiments, the extent of inhibition was found to be affected by the ratio of deacylated to acylated sRNA.

Deacylated sRNA added near the end of incubation often reduces background binding without polynucleotide and may afford a way of differentiating between exchangeable and nonexchangeable binding. It should be noted that the presence of a polynucleotide which is not recognized by a C^{14}-aminoacyl sRNA (for example, polyA and C^{14}-Phe-sRNA) also reduces background sRNA binding, perhaps by saturating ribosomal sites with specified nonexchangeable sRNA.

Characteristics of binding. The assay was validated further by demonstrating that the binding of sRNA to ribosomes was directed with specificity by different polynucleotides. As shown in Table 2, polyU, polyA and polyC specifically directed the binding to ribosomes of C^{14}-Phe-, C^{14}-Lys-, and C^{14}-Pro-sRNA, respectively. These data agree well with specificity data obtained with a sucrose-gradient centrifugation assay, reported by Nakamoto *et al.* (*8*) and Kaji and Kaji (*9, 10*) (also compare *5–7, 11*) and with data on their specificity for directing amino acid incorporation into protein (*30, 31*).

The rate of binding of Phe-sRNA to ribosomes at 0°, 24° and 37°C, in the presence and absence of polyU, is shown in Fig. 1. During incubation at each temperature polyU markedly stimulated C^{14}-Phe-sRNA binding; how-

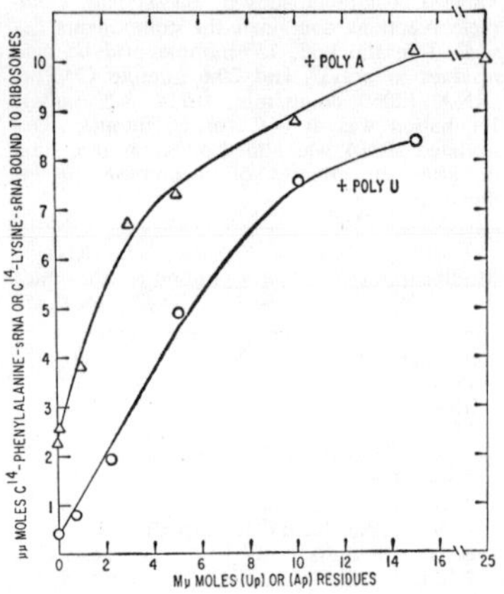

Fig. 3. Relation between polynucleotide concentration and C^{14}-aminoacyl-sRNA binding to ribosomes. Each point represents a 50-μl reaction mixture with the components described under *Methods*; △, polyA as specified and 27.5 μμmole C^{14}-Lys-sRNA (3080 count/min, 0.880 A^{260} units); ○, polyU as specified and 15 μμmole of C^{14}-Phe-sRNA (2050 count/min, 0.714 A^{260} units). Incubation was at 24°C for 20 minutes.

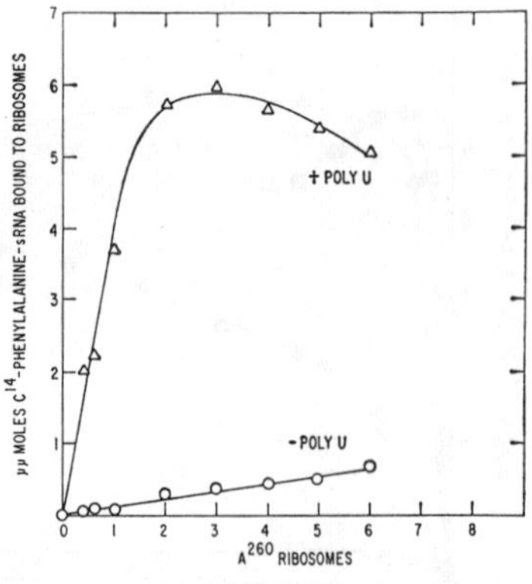

Fig. 4. Relation between ribosome concentration and C^{14}-Phe-sRNA binding. Each point represents a 50-μl reaction mixture containing the amount of ribosomes indicated; ○ No additions, (-poly U); △ + poly U, 25 mμmole base residues. In addition, reaction mixtures contain the components described under *Methods* and 9.65 μμmole of C^{14}-Phe-sRNA (1180 count/min, 0.380 A^{260}). Incubations were for 10 minutes at 24°C.

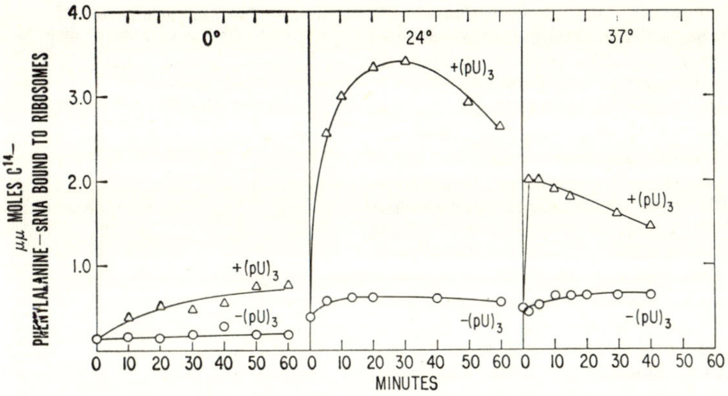

Fig. 5. Effect of pUpUpU upon the rate of C^{14}-Phe-sRNA binding to ribosomes at 0°, 24°, and 37°C. ○, No addition; △, addition of 3.67 mμmole pUpUpU. Each point represents a 50-μl reaction mixture incubated for the time and at the temperature specified. Reaction mixtures contain the components described under *Methods*; 9.65 μμmole of C^{14}-Phe-sRNA (1180 count/min, 0.380 A^{260}); and oligoU as specified.

ever, the rate of binding increased as the temperature of incubation was raised. Although polyU induced C^{14}-Phe-sRNA binding at 0°C, maximum binding was not observed after 60 minutes of incubation. Maximum binding concentrations of C^{14}-Phe-sRNA, 50 minutes of incubation at 24°C and after 6 minutes at 37°C. In this experiment equal amounts of C^{14}-Phe-sRNA were bound at 24° and 37°C. In other experiments, with limiting concentrations of C^{14}-Phe-sRNA, 50 to 98 percent of the C^{14}-Phe-sRNA was induced to bind to ribosomes by polyU. Kaji and Kaji have suggested the possible utility of this system for the purification of sRNA species (*9*).

In the absence of polyU, relatively little C^{14}-Phe-sRNA associated with ribosomes. Such binding may be due to endogenous mRNA on ribosomes or in sRNA preparations. Alternatively, this binding may be nonspecific, possibly similar to that described by Cannon, Krug, and Gilbert (*4*).

The effect of polyA and polyC upon the rates of C^{14}-Lys- and C^{14}-Pro-sRNA binding to ribosomes is shown in Fig. 2. Maximum stimulation of C^{14}-Lys-sRNA binding by polyA, and of C^{14}-Pro-sRNA binding by polyC, occurred after 10 minutes of incubation at 24° and 27°C, respectively. In this experiment, C^{14}-Lys-sRNA binding observed in the absence of poly*A* was higher than that found in experiments with other sRNA preparations.

The relation between polyU or polyA concentration and the amount of C^{14}-Phe- or C^{14}-Lys-sRNA bound to ribosomes is shown in Fig. 3. Binding of sRNA was proportional to polynucleotide concentration in both cases.

In experiments not presented here, the effect of *p*H upon sRNA binding was studied. PolyU directed C^{14}-Phe-

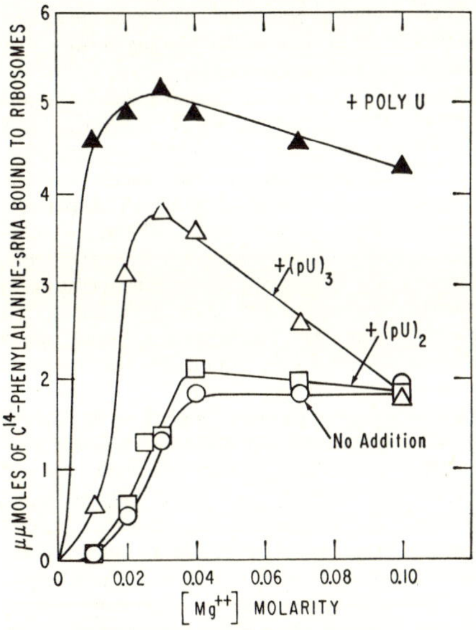

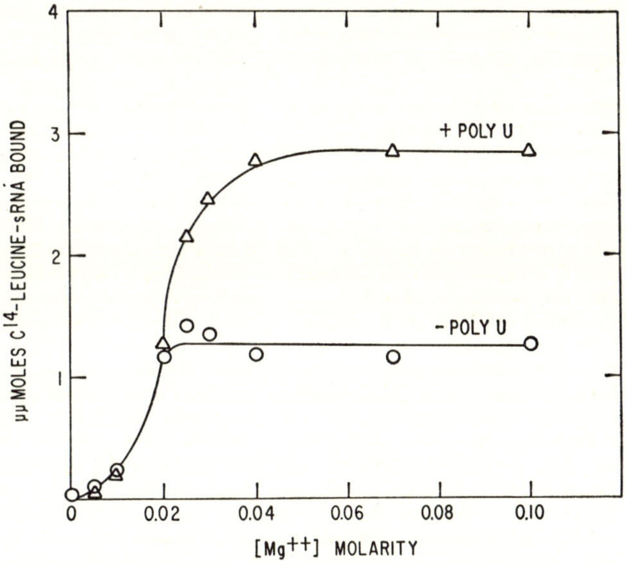

Fig. 6 (left). Effect of (Mg^{++}) concentration upon C^{14}-Phe-sRNA binding to ribosomes. Each symbol represents a 50-μl reaction mixture containing: ○, polymer (no addition); □, pUpU, 10 mμmole base residues; △, pUpUpU, 10 mμmole base residues; ▲, polyU, 25 mμmole base residues; and the components described under *Methods;* the magnesium acetate concentrations of reaction mixtures and the washing buffer are shown in the figure; 3.5 A^{260} units of ribosomes; and 8.63 μμmole C^{14}-Phe-sRNA (1025 count/min, 0.357 A^{260} units. Incubation was at 24°C for 20 minutes. Fig. 7 (right). The polyU-leucine ambiguity. The effect of polyU upon C^{14}-Leu-sRNA binding to ribosomes at different Mg^{++} concentrations. Each 50-μl reaction contained the components described under *Methods*; the magnesium acetate concentration specified in the figure 23.7 μμmole C^{14}-Leu-sRNA (1710 count/min, 0.424 A^{260} units of *E. coli* W 3100 sRNA, charged only with C^{14}-leucine); and polyU, 25 mμmole base residues where specified. Incubation was at 24°C for 20 minutes. Each reaction mixture was washed with a buffer containing Mg^{++} at the concentration present during incubation.

Table 3. Oligonucleotide specificity. Reaction mixtures containing either C^{14}-Phe-, C^{14}-Lys-, or C^{14}-Pro-sRNA, and oligonucleotide were incubated at 24°C for 20 minutes. Components of reaction mixtures are described in the legend accompanying Table 2. The numbers in parentheses are millimicromoles base residues.

Oligonucleotide (mμmoles base residues)	C^{14}-Aminoacyl-sRNA bound to ribosomes (μμmole)		
	C^{14}-Phe-sRNA	C^{14}-Lys-sRNA	C^{14}-Pro-sRNA
None	0.34	0.80	0.24
pUpUpU (10)	1.56	0.56	0.20
pApApA (7)	0.20	6.13	0.18
pCpCpC (8)	0.30	0.60	0.73

sRNA binding to ribosomes throughout the *p*H range tested, from 5.5 to 7.8. Binding was maximum in reactions buffered either with 0.1*M* tris, *p*H 7.2, or with 0.05*M* cacodylate, *p*H 6.5.

The addition of GTP, PEP, and PEP kinase to reactions, with or without polyU, did not stimulate C^{14}-Phe-sRNA binding; however, further study is necessary to determine whether binding in this system is dependent upon enzymatic catalysis or GTP or both. Polyphenylalanine synthesis (that is, radioactivity in the fraction precipitable with hot TCA) was not detected in reaction mixtures incubated under optimum conditions for C^{14}-Phe-sRNA binding in the presence of polyU. In additional experiments, C^{14}-Phe-sRNA induced by polyU to bind to ribosomes was eluted with $10^{-5}M$ Mg^{++} as already described. Although 50 to 98 percent of the ribosomal-bound C^{14}-product was eluted (in different experiments), few ribosomes were released from filters. The C^{14}-product was insoluble in TCA at 3°C, but was converted quantitatively to a soluble product by (i) incubation at *p*H 10.2 for 60 minutes at 37°C, (ii) heating in TCA, or (iii), incubation with pancreatic ribonuclease. Further characterization of the bound C^{14}-product is necessary; however, these preliminary results are in accord with the demonstration by Schweet and his co-workers that the binding of Phe-sRNA to ribosomes induced by polyU is not dependent upon peptide-bond formation (*6*, *7*).

The relation between ribosome concentration and the amount of bound C^{14}-Phe-sRNA is illustrated in Fig. 4. C^{14}-Phe-sRNA binding was stimulated markedly by polyU and was proportional to the ribosome concentration, within the range 0 to 1.0 A^{260} units of ribosomes. The number of 70*S* *E. coli* ribosomes and ribosomal-bound C^{14}-Phe-sRNA molecules can be estimated from such data; however, various factors, such as the inhibitory effects of deacylated sRNA (see above) and mRNA terminal phosphate (described below) undoubtedly reduce the accuracy of such calculations. However, in the presence of polyU approximately 4.0 μμmole of C^{14}-Phe-sRNA became bound to 23.2 μμmole of 70*S* ribosomes (*4*); therefore, the C^{14}-Phe-sRNA ribosome ratio was 1 : 5.8. Arlinghaus *et al.* (*7*) and Warner and Rich (*32*) recently reported two binding sites for each ribosome for sRNA. One site is thought to hold the nascent polypeptide chain to the ribosome; the other, to bind the next aminoacyl-sRNA molecule specified by mRNA.

Effect of oligonucleotides on the binding of C^{14}-aminoacyl sRNA to ribosomes. OligoU preparations of different chain length were prepared, and their effect upon C^{14}-Phe-sRNA binding to ribosomes was determined. The effect of the trinucleotide, pUpUpU, upon C^{14}-Phe-sRNA binding to ribosomes, at 0°, 24°, and 37°C, is shown in Fig. 5. The C^{14}-Phe-sRNA binding was stimulated by pUpUpU at each temperature; however, binding was maximum in reactions incubated at 24°C for 20 to 30 minutes. These results demonstrate that a trinucleotide can direct C^{14}-aminoacyl-sRNA binding to ribosomes and suggest a general method of great simplicity for determining the genetic function of other trinucleotide sequences.

The binding of C^{14}-Phe-, C^{14}-Lys-, and C^{14}-Pro-sRNA was induced with apparent specificity by pUpUpU, pApApA, and pCpCpC, respectively (Table 3). In additional experiments, each trinucleotide had no discernible effect upon the binding to ribosomes of 15 aminoacyl-sRNA preparations, each charged with a different C^{14}-amino acid (C^{14}-asparagine and C^{14}-glutamine-sRNA were not tested). Therefore, the specificity of each trinucleotide for inducing sRNA binding to ribosomes was high and clearly paralleled that of the corresponding polynucleotide.

The T_m of the interaction between pApApA and polyU is 17°C (*33*). Therefore, hydrogen-bonding between a triplet codeword in mRNA and a complementary "anticodeword" in sRNA would not by itself appear sufficient to account for the stability of the interaction between C^{14}-Phe-sRNA, polyU, and ribosomes. An interaction between the -CpCpA end of sRNA and ribosomes is possible, for several laboratories (*3*, *4*, *34*) have reported that removal of the terminal adenosine of sRNA greatly reduces its ability to bind to ribosomes. The participation of an enzyme in the binding process also must be considered.

Table 4. Template activity of oligodeoxynucleotides. The components of each 50 μl reaction mixture are presented in the text. In addition, each reaction mixture in Expt. 1 contained 9.65 μμmoles of C^{14}-Phe-sRNA (1370 count/min, 0.380 A^{260} units); and in Expt. 2, 16.5 μμmoles C^{14}-Lys-sRNA (1845 count/min, 0.530 A^{260} units) and the oligonucleotides specified. Mixtures were incubated at 24°C for 10 minutes.

Oligonucleotide (mμmole)	C^{14}-Aminoacyl-sRNA bound to ribosomes (μμmole)
Experiment 1: C^{14}-Phe-sRNA	
None	0.29
1.00 pUpUpU	1.29
3.33 pUpUpU	2.40
6.67 pUpUpU	2.90
3.33 oligo $d(pT)_3$	0.35
6.67 oligo $d(pT)_3$	0.39
10.00 oligo $d(pT)_3$	0.40
1.67 oligo $d(pT)_{12}$	0.31
2.50 oligo $d(pT)_{12}$	0.44
Experiment 2: C^{14}-Lys-sRNA	
None	1.27
0.17 pApApA	3.48
1.37 pApApA	5.99
3.60 pApApA	7.40
0.65 oligo $d(pA)_8$	1.45
1.30 oligo $d(pA)_8$	1.68
2.00 oligo $d(pA)_8$	1.61

The data of Table 4 indicate that the 2′-hydroxyl of RNA codewords may be necessary for codeword recognition. Oligodeoxynucleotides such as oligo-$d(pT)_{3-12}$ and oligo-$d(pA)_8$ apparently were inactive as templates. In additional experiments not presented here, the effects of time and temperature of incubation (0°, 24°, 37°, and 43°C), template concentration, and chain length were studied. No template activity was found.

The template activities of pUpU, pUpUpU and polyU at different concentrations of Mg^{++} are shown in Fig. 6. Both tri- and polyU induced maximal binding at approximately 0.03*M* Mg^{++} (0.02 to 0.03*M* in other experiments). Although the dinucleotide, pUpU, stimulated binding slightly, it is not known whether the activity of pUpU indicates partial recognition of a triplet codeword as previously suggested (*35*). At 0.02*M* Mg^{++}, the concentration used throughout our study, little binding of C^{14}-Phe-sRNA was found in the absence of polyU. However, at higher Mg^{++} concentrations, Phe-sRNA binding in the absence of template RNA increased;

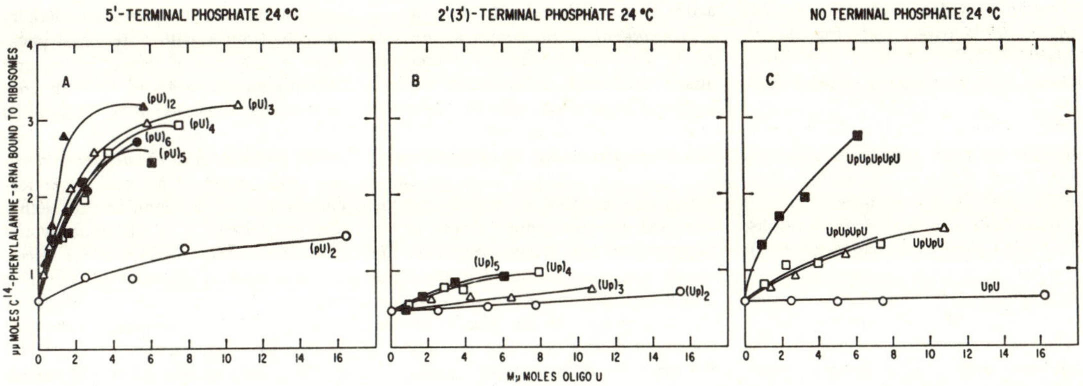

Fig. 8. *A*, *B*, and *C*, Relation between template activity and oligoU chain-length, concentration and end-group. The activities of oligoU with 5′-terminal phosphate are shown in Fig. 8*A*; with 2′(3′)-terminal phosphate in Fig. 8*B*; and without terminal phosphate in Fig. 8*C*. Symbols represent oligoU chain-lengths as follows: ○ 2; △ 3; □ 4; ■ 5; ● 6; ▲ 12. Each reaction mixture contained, in a volume of 50 μl, the components specified in the *Methods* section; oligoU preparations and concentrations specified; and 10.8 μμmole C^{14}-Phe-sRNA (1880 count/min, 0.714 A^{260}). Oligonucleotide preparations $(pU)_2$, No. 610; $(pU)_5$, No. 613; $(Up)_4$, No. 617; with the contaminants specified under *Methods* were used in these experiments. Incubations were at 24°C for 30 minutes.

whereas binding induced by tri- and polyU, decreased. These data suggest that certain ribosomal binding sites become saturated with sRNA at Mg^{++} concentrations greater than 0.03*M*, but are not saturated at lower concentrations.

Leucine-polyU ambiguity. Since polyU is known to direct some leucine incorporation into protein in cell-free systems (*36*), especially at high Mg^{++} concentrations (*37*), the effect of polyU upon the binding of C^{14}-Leu-sRNA was determined (Fig. 7). In the absence of polyU, background binding saturated at 0.02*M* Mg^{++}. The addition of polyU clearly stimulated the binding of C^{14}-Leu-sRNA. It is possible that the Mg^{++}-dependent leucine-polyU ambiguity occurs before peptide-bond synthesis.

As shown in Table 5, C^{14}-Leu-sRNA binding was stimulated with specificity by polyU, but not by the trinucleotide, pUpUpU, polyA, or polyC at 0.07*M* Mg^{++}. In additional experiments, pUpUpU had no effect upon the binding of C^{14}-Leu-sRNA at other Mg^{++} concentrations. These data suggest that pUpUpU may be recognized by aminoacyl sRNA with greater specificity than polyU is recognized.

Effect of oligonucleotide chain length, concentration, and terminal phosphate. In Fig. 8 the template activity of oligoU preparations differing in chain length and position of terminal phosphate are compared at different oligoU concentrations. The activity of oligoU with 5′-terminal phosphate is shown in Fig. 8*A*; preparations with 2′(3′)-terminal phosphate in Fig 8*B*; and preparations without terminal phosphate, in Fig. 8*C*.

As shown in Fig. 8*A*, the dinucleotide with 5′-terminal phosphate, pUpU, had little template activity, whereas the trinucleotide, pUpUpU, markedly stimulated C^{14}-Phe-sRNA binding. This ob-

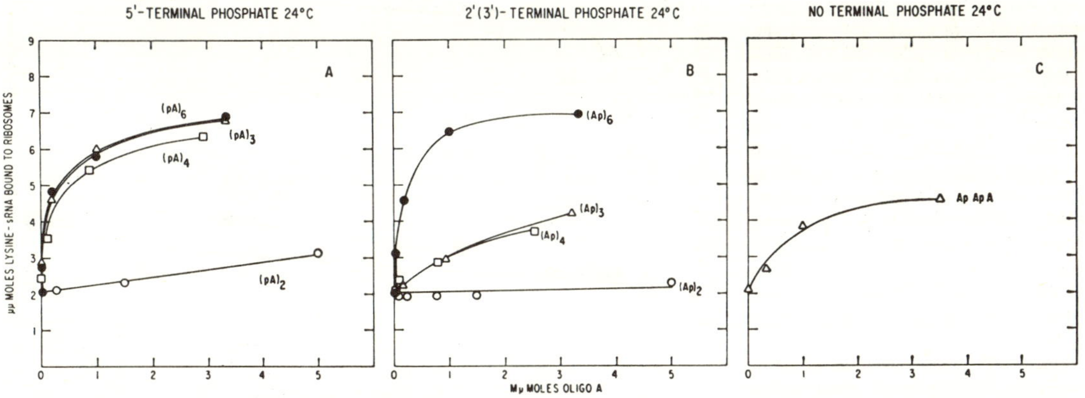

Fig. 9. *A*, *B*, and *C*, Relation between template activity and oligo*A* concentration, chain-length, and end group. The activity of oligoA with 5′-terminal phosphate is shown in Fig. 9*A*; preparation without terminal phosphate in Fig. 9*C*. The symbols indicate the chain-length of oligo*A* as follows: ○ 2; △ 3; □ 4; ● 6. Each 50 μl reaction mixture contained the components specified under methods; 15.9 μμmole of C^{14}-Lys-sRNA (1780 count/min, 0.562 A^{260} units); and the oligoA preparation specified. Incubation was at 24°C for 20 minutes. Oligonucleotide preparations $(pA)_6$, No. 599 and $(Ap)_4$, No. 591, containing the contaminants specified under *Methods*, were used in this experiment.

servation provides direct experimental support for a triplet code for phenylalanine and is in full accord with earlier genetic and biochemical studies (*30, 38*).

In addition, the data demonstrate that the template activity of pUpUpU equals that of the corresponding tetra-, penta-, and hexanucleotides.

In striking contrast to these results, the trinucleotide, UpUpUp, with 2′(3′)-terminal phosphate, induced little or no binding of C^{14}-Phe-sRNA to ribosomes (Fig. 8*B*). The template activity of the tetra- and pentauridylic acid fractions with 2′(3′)-terminal phosphate also were markedly reduced when compared to similar fractions with 5′-terminal phosphate.

As shown in Fig. 8*C*, UpUpU and UpUpUpU, without terminal phosphate, induced C^{14}-Phe-sRNA binding, but less actively than pUpUpU. The template activity of the pentamer, UpUpUpUpU, was almost equal to that of pUpUpU.

Since the sensitivity of an oligonucleotide to digestion by a nuclease often is influenced by terminal phosphate, the relative stability of pUpUpU, UpUpUp, and UpUpU incubated with ribosomes at 37°C for 60 minutes was estimated by recovering mono- and oligonucleotides from reaction mixtures and separating them by paper chromatography. In each case, the expected trinucleotide was the only component found after incubation. Hydrolysis of oligoU was not observed.

The template activities of oligoA fractions with different end groups are shown in Fig. 9, *A*, *B*, and *C*. The results were similar to those obtained with oligoU; however, pApApA induced maximum C^{14}-Lys-sRNA binding at one-fifth the oligonucleotide concentration required previously (compare with Fig. 8*A*). The hexamer with 3′-terminal phosphate induced as much C^{14}-Lys-sRNA binding to ribosomes as the trimer with 5′-terminal phosphate, pApApA. When reaction mixtures were incubated at 0°C (Fig. 10) the difference between the template activities of ApApAp and pApApA was more marked than when incubations were at 24°C.

Since each oligonucleotide preparation with 2′(3′)-terminal phosphate is a mixture of molecules, some chains terminating with 2′-phosphate and others with 3′-phosphate, a trinucleotide, UpUpUp, with 3′-terminal phosphate only, was prepared and found to be inactive as a template for C^{14}-Phe-sRNA.

Attachment of ribosomes to mRNA. It is not known whether ribosomes attach to 5′-ends, 3′-ends, or internal positions of mRNA. The template activity of trinucleotides indicates that (i) ribosomes can attach to terminal codewords of mRNA, (ii) terminal codewords are capable of specifying the first and the last amino acids to be incorporated into protein, and (iii) the attachment of a ribosome to *only* the terminal triplet of mRNA may provide the minimum stability necessary for codeword recognition, and possibly for the initiation of protein synthesis.

The demonstration that terminal and internal codeword phosphates strongly influence the codeword recognition process indicates that phosphate may take part in the binding of codewords to ribosomes. Watson has suggested interaction between phosphate of mRNA and amino groups of ribosomes, because 30*S* ribosomes treated with formaldehyde were found by Moore and Asano to bind less polyU than did ribosomes not so treated (*39*).

Although terminal codeword phosphate is not required for the recognition of a codeword on a ribosome, the observation that the template activity of trinucleotides with 5′-terminal phosphate equals that of tetra-, penta-, and hexanucleotides, even at limiting concentrations, suggests that 5′-terminal codewords may *attach to sites on ribosomes where codewords are recognized, in correct phase to be read*. A preferential, phased recognition of either terminal codeword by ribosomes would provide a simple mechanism for selecting the polarity of reading, the first word to be read, and the phase of reading. Since 5′-terminal codewords of mRNA most actively induce sRNA binding, such codewords would appear to serve these functions best. Although the polarity with which mRNA is read may be from the 5′- towards the 3′-terminal codeword, further work is necessary to clarify this point. The opposite polarity has been suggested (*45*)

We have reported (*40*) that trinucleotides can be used in this system to determine the base sequence of codewords and have shown that the sequence of the valine RNA codeword is GpUpU. Codewords are recognized with polarity in this system, for GpUpU induced C^{14}-Val-sRNA binding to ribosomes, whereas UpUpG did not.

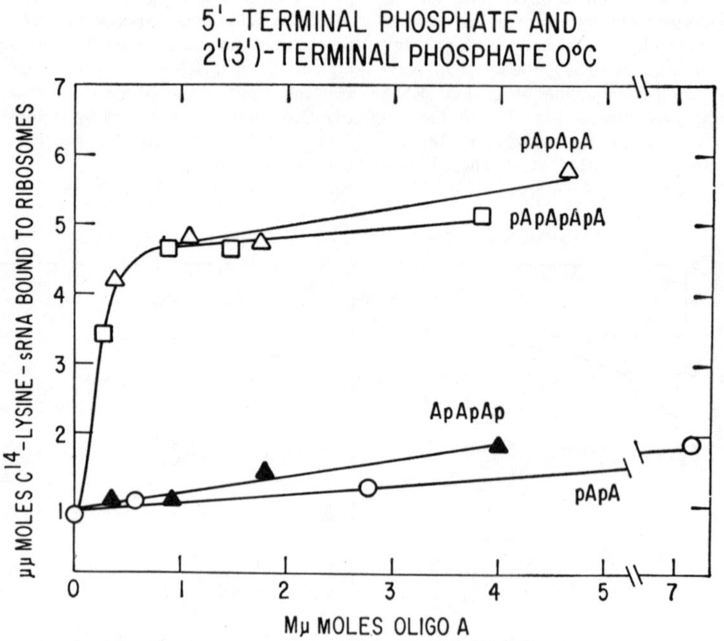

Fig. 10. Template activity of oligoA with 5′- or 2′(3′)-terminal phosphate in reaction mixtures incubated at 0°C for 60 minutes. OligoA concentrations, chain-lengths, and end-groups are shown in the figure. Each 50-μl reaction mixture contained the components specified under *Methods*; 11.8 μμmole C^{14}-Lys-sRNA (2840 count/min, 0.767 A^{260} units); and the oligoA preparation specified.

Regulatory Codewords

Terminal codewords. We suggest that 5′-terminal, 3′-terminal and internal codewords of RNA and DNA constitute separate classes of codewords, for each differs in chemical structure as shown diagrammatically in Fig. 11. The 5′-terminal codeword contains a 5′-terminal hydroxyl, the internal codeword is attached to adjacent codewords on both sides by way of (3′-5′)-phosphodiester bonds, and the 3′-terminal codeword contains 2′- and 3′-terminal hydroxyl groups. It should be noted that phosphate, linked to a terminal hydroxyl is a monoester, whereas, an internal phosphate is a diester. Therefore, the 5′-terminal, 3′-terminal, and internal codewords differ in chemical structure. To avoid confusion between the three codeword classes, the codeword with free or substituted 5′-hydroxyl will be designated the 5′-terminal codeword, and the codeword with a free or substituted 3′-hydroxyl, will be designated the 3′-terminal codeword. Internal codewords will not be designated by position. These differences raise the possibility that codewords may occur in three chemically distinct forms and that RNA and DNA may contain either one, two, or three forms of any triplet codeword, depending upon the position of the codeword in the molecule.

Sense-missense-nonsense codewords. Genetic evidence suggests that certain mutations result in the conversion of readable into nonreadable codewords, that is, sense-nonsense interchanges (*41*). The addition of terminal phosphate to a 3′-terminal codeword similarly changes a readable into a nonreadable codeword and resembles a sense-nonsense interchange. Two additional mechanisms for converting readable into nonreadable words have been found in cell-free systems; that is, an increase in secondary structure (*37*, *42*) and specific base methylations (*37*). It should be noted that each type of sense-nonsense interchange involves a modification of codewords rather than modification of a component required for codeword recognition.

It is possible that the synthesis of certain proteins may be regulated in vivo by sense-nonsense interchanges involving either modification of a codeword or, as proposed by Ames and Hartman (*43*), modification of sRNA, that is, codeword recognition. It seems probable that terminal codewords may have special functions in addition to directing amino acids into protein. For example, in mRNA they may specify (i) attachment and detachment of ribosomes, (ii) the first codeword to be read, (iii) the phase of reading, and (iv) the sensitivity of the message to degradation by exonucleases. Similarly, terminal DNA codewords may influence the rate with which DNA is copied by DNA or RNA polymerase. Experimental observations support this possibility, for DNA without terminal phosphate has been found to serve as a template for DNA polymerase, whereas DNA with 3′-terminal phosphate has no template activity (*44*).

Table 5. Specificity of polyU-leucine-sRNA bound to ribosomes. Reaction mixtures contained 0.07*M* magnesium acetate; other components as described in the text; 3.1 $\mu\mu$moles of C^{14}-Leu-sRNA (2280 count/min, 0.565 A^{260} units of *E. coli* W 3100 sRNA acylated only with C^{14}-Leucine); and oligo- or polynucleotides as specified. Incubations were at 24°C for 20 minutes. The magnesium acetate concentration of the solution used to wash ribosomes on filters after incubation was 0.07*M*.

Addition (base residues) (mμmole)	C^{14}-Leu-sRNA bound to ribosomes ($\mu\mu$mole)
None	1.00
25 PolyU	2.02
15 pUpUpU	0.70
25 PolyA	.99
25 PolyC	.85

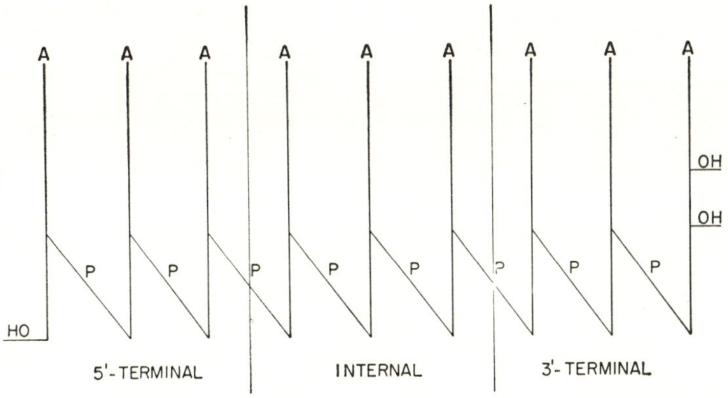

Fig. 11. The structures of three classes of codewords, 5′-terminal, internal, and 3′-terminal, are illustrated diagramatically. The figure represents oligoA of chain-length 9, ApApApApApApApApA. Although the three triplet codewords have identical base-sequences, AAA, each differs in structure. The 5′-terminal codeword contains a 5′-terminal hydroxyl; the internal codeword is linked to adjacent codewords on either side by way of 3′, 5′-phosphodiester bonds; and the 3′-terminal codeword contains 2′, 3′-terminal hydroxyl groups (3′-hydroxyl only if deoxyribose). Phosphate, when present at 5′-, 3′-, or 2′-terminal hydroxyl positions, may be a phosphomonoester, and therefore may differ from an internal phosphate which is a phosphodiester.

Terminal words with 3′-phosphate may be members of a larger class of DNA and mRNA nonsense words, with substituted 5′-, 3′-, or 2′-hydroxyl groups. Many enzymes have been described which catalyze the transfer of nucleotides, amino acids, methyl groups, carbohydrates, and other molecules, to or from mononucleotides or terminal ribose or deoxyribose of nucleic acids. Such enzymes may recognize terminal bases or conformations of nucleic acids and catalyze group transfer reactions with great specificity.

The data of Figs. 8 to 10, and the chemical and biological considerations described, suggest that *codeword modification* may serve a regulatory or operator function. Modification of codewords, at both terminal and internal positions, may regulate the reading of DNA or RNA by converting a readable word into one read incorrectly or not read. It should be noted that both 3′- and 5′-terminal codewords could serve, in different ways, as operator words.

The capacity of trinucleotides to direct the binding of sRNA to ribosomes and the ease with which the process can be assayed should provide a general method of great simplicity for studying the base sequence and genetic functions of each triplet codeword. In addition, this method should permit the detailed study of interactions between codewords, sRNA, and ribosomes during the codeword recognition process.

Summary

A rapid, sensitive method is described for measuring C^{14}-aminoacyl-sRNA interactions with ribosomes which are specifically induced by the appropriate RNA codewords prior to peptide-bond formation. Properties of the codeword recognition process and the minimum oligonucleotide chain length required to induce such interactions are presented. The trinucleotides, pUpUpU, pApApA, and pCpCpC, but not dinucleotides, specifically direct the binding to ribosomes of phenylalanine-, lysine-, and proline-sRNA, respectively.

Since 5′-terminal, 3′-terminal, and internal codewords differ in chemical structure, three corresponding classes of codewords are proposed. The recognition of each class in this system is described. The template efficiency of trinucleotide codewords is modified greatly by terminal phosphate. Triplets with 5′-terminal phosphate are more active as templates than triplets without terminal phosphate. Triplets with 3′- or 3′ (2′)-terminal phosphate are markedly less active as templates. These findings are discussed in relation to the probable functions of terminal codewords. The modification of RNA and DNA codewords, converting sense into missense or nonsense codewords, is suggested as a possible regulatory mechanism in protein synthesis.

References and Notes

1. The following abbreviations and symbols are used: Phe-, phenylalanine; Leu-, leucine; Lys-, lysine; Pro-, proline; Val-, valine; polyU, polyuridylic acid; polyC, polycytidylic acid; polyA, polyadenylic acid; TCA, trichloracetic acid; d(pT), deoxythymidylic acid; d(pA), deoxyadenylic acid; sRNA, transfer RNA; mRNA, messenger RNA; DEAE, diethylaminoethyl cellulose: PPO. 2,5-diphenyloxazole; POPOP, 1,4-bis-2′-(5′-phenyloxazolyl) benzene; PEP, phosphoenolpyruvate; T_m, melting temperature; A^{260}, at 260 mμ.
   For mono- and oligonucleotides of specific structure, the p to the left of a terminal nucleoside initial indicates a 5′-terminal phosphate; the p to the right, a 2′ (3′)-terminal phosphate. Internal phosphates of oligonucleotides are (3′-5′)-linkages.
2. A Von der Decken and T. Hultin, *Exptl. Cell. Res.* **15**, 254, (1958); L. Bosch, H. Bloemendal, M. Sluyser, *Biochim. Biophys. Acta* **34**, 272 (1959); M. B. Hoagland and L. T. Comly, *Proc. Natl. Acad. Sci. U.S.* **46**, 1554 (1960).
3. M. Takanami, *Biochim. Biophys. Acta* **55**, 132 (1962); L. Bosch, F. Huizinga, H. Bloemendal, *ibid.* **61**, 220 (1962).
4. M. Cannon, R. Krug, W. Gilbert, *J. Mol. Biol.* **7**, 360 (1963).
5. R. Arlinghaus, G. Favelukes, R. Schweet, *Biochem. Biophys. Res. Commun.* **11**, 92 (1963).
6. B. Hardesty, R. Arlinghaus, J. Schaeffer, R. Schweet, in *Cold Spring Harbor Symp. Quant. Biol.* **28**, 215 (1963).
7. R. Arlinghaus, J. Shaeffer, R. Schweet, *Proc. Natl. Acad. Sci. U.S.* **51**, 1291 (1964).
8. T. Nakamoto, T. Conway, J. Allende, G. Spyrides, F. Lipmann, in *Cold Spring Harbor Symp. Quant. Biol.* **28**, 227 (1963).
9. A. Kaji and H. Kaji, *Biochem. Biophys. Res. Commun.* **13**, 186 (1963).
10. ———, *Federation Proc.* **23**, 478 (1964).
11. G. J. Spyrides, Ph.D. dissertation, Rockefeller Institute (1963); *Federation. Proc.* **23**, 318 318 (1964).
12. Miles Chemical Company.
13. L. A. Heppel, personal communication.
14. Worthington Biochemical Corporation.
15. L. A. Heppel, D. Harkness, R. Hilmoe, *J. Biol. Chem.* **237**, 841 (1962).
16. G. W. Rushizky and H. A. Sober, *ibid.* **238**, 371 (1963).
17. P. S. Chen, Jr., T. Y. Toribara, H. Warner, *Anal. Chem.* **28**, 1756 (1956); B. N. Ames and D. T. Dubin, *J. Biol. Chem.* **235**, 769 (1960).
18. P. Leder, B. F. C. Clark, W. S. Sly, S. Pestka, M. W. Nirenberg, *Proc. Natl. Acad. Sci. U.S.* **50**, 1135 (1963).
19. M. Bernfield and M. W. Nirenberg, *Abstracts, 148th Meeting, American Chemical Society*, Chicago, Ill., August 1964.
20. Nuclear Chicago Corporation.
21. Packard Instrument Company.
22. J. R. Sherman and J. Adler, *J. Biol. Chem.* **238**, 873 (1963).
23. General Biochemicals Corporation.
24. New England Nuclear Corporation.
25. G. Zubay, *J. Mol. Biol.* **4**, 347 (1962).
26. M. W. Nirenberg, J. H. Matthaei, O. W. Jones, *Proc. Natl. Acad. Sci. U.S.* **48**, 104 (1962).
27. Sigma Chemical Company.
28. G. J. Spyrides, *Proc. Natl. Acad. Sci. U.S.* **51**, 1220 (1964).
29. T. W. Conway, *ibid.* **51**, 1216 (1964).
30. M. W. Nirenberg, O. W. Jones, P. Leder, B. F. C. Clark, W. S. Sly, S. Pestka, in *Cold Spring Harbor Symp. Quant. Biol. Vol.* **28**, 549 (1963).
31. J. S. Speyer, P. Lengyel, C. Basilio, A. J. Wahba, R. S. Gardner, S. Ochoa, *ibid.* **28**, 559 (1963).
32. J. R. Warner and A. Rich, *Proc. Natl. Acad. Sci. U.S.* **51**, 1134 (1964).
33. M. N. Lipsett, L. A. Heppel, D. F. Bradley, *J. Biol. Chem.* **236**, 857 (1961).
34. H. Bloemendal, F. Huizinga, M. de Vries, L. Bosch, *Biochim. Biophys. Acta* **61**, 209 (1962).
35. M. W. Nirenberg and O. W. Jones, in *Symposium on Informational Macromolecules*, H. Vogel, V. Bryson, J. Lempen, Eds. (Academic Press, New York, 1963); O. W. Jones and M. W. Nirenberg, *Proc. Natl. Acad. Sci. U.S.* **48**, 2115 (1962); R. S. Gardner, A. J. Wahba, C. Basilio, R. S. Miller, P. Lengyel, J. F. Speyer, *ibid.*, p. 2087.
36. J. H. Matthaei, O. W. Jones, R. G. Martin, M. W. Nirenberg, *Proc. Natl. Acad. Sci. U.S.* **48**, 666 (1962); M. S. Bretscher and M. Grunberg-Manago, *Nature* **195**, 283 (1962).
37. M. Grunberg-Manago and A. M. Michelson, *Biochim. Biophys. Acta* **80**, 431 (1964); W. Szer and S. Ochoa, *J. Mol. Biol.* **8**, 823 (1964).
38. F. H. C. Crick, L. Barnett, S. Brenner, R. J. Watts-Tobin, *Nature* **192**, 1227 (1961).
39. J. D. Watson, personal communication.
40. P. Leder and M. W. Nirenberg, *Proc. Natl. Acad. Sci. U.S.* **51** (1964), in press.
41. S. Benzer and S. P. Champe, *ibid.* **48**, 1114 (1962); A. Garen and O. Siddiqi, *ibid.*, p. 1121.
42. M. W. Nirenberg and J. H. Matthaei, *ibid.* **47**, 1588 (1961); O. W. Jones, R. G. Martin, S. H. Barondes, *Federation Proc. Symp.* **22**, 55 (1963); M. F. Singer, O. W. Jones, M. W. Nirenberg, *Proc. Natl. Acad. Sci. U.S.* **49**, 392 (1962); R. Haselkorn and V. A. Fried, *ibid.* **51**, 308 (1964).
43. B. M. Ames and P. E. Hartman, in *Cold Spring Harbor Symp. Quant. Biol.* **28**, 349 (1963).
44. C. C. Richardson, C. L. Schildkraut, H. V. Aposhian, A. Kornberg, W. Bodner, J. Lederberg, in *Informational Macromolecules*, H. Vogel, V. Bryson, J. Lampen, Eds. (Academic Press, New York, 1963), p. 13.
45. A. J. Wahla, C. Basilio, J. F. Speyer, P. Lengyel, R. S. Miller, S. Ochoa, *Proc. Natl. Acad. Sci. U.S.* **48**, 1683 (1962); presented by P. Doty, R. E. Thach, and T. A. Sundararajan at the VIth International Congress of Biochemistry (1964).
46. We thank Miss Norma Zabriskie and Mrs. Theresa Caryk for invaluable technical assistance; L. A. Heppel for his advice regarding preparation of oligonucleotides; and J. C. Keresztesy and D. Rogerson for providing *E. coli*.

Reprinted from PROCEEDINGS OF THE NATIONAL ACADEMY OF SCIENCES, Vol. 52, 1964

RNA CODEWORDS AND PROTEIN SYNTHESIS, II. NUCLEOTIDE SEQUENCE OF A VALINE RNA CODEWORD

BY PHILIP LEDER AND MARSHALL NIRENBERG

NATIONAL HEART INSTITUTE, NATIONAL INSTITUTES OF HEALTH, BETHESDA, MARYLAND

Communicated by R. B. Roberts, June 30, 1964

Recent studies demonstrate that synthetic polynucleotides, such as poly U, induce with specificity the binding of C^{14}-amino-acyl-sRNA to ribosomes.[1-4] Characteristics of binding in the absence of poly U also have been reported.[5] C^{14}-Phe-sRNA binding induced by poly U is thought to represent an early step in protein synthesis before peptide bond formation; however, the precise nature of the interaction has not been clarified.

To establish the sequence of nucleotides in RNA codewords, we devised a rapid, sensitive method for measuring C^{14}-amino-acyl-sRNA binding to ribosomes, and have investigated both characteristics of binding and the minimum oligonucleotide chain length required to direct such binding.[6] pUpUpU, pApApA, and pCpCpC directed the binding of Phe-, Lys-, and Pro-sRNA, respectively, whereas dinucleotides had no effect. Trinucleotides with 5′-terminal phosphate were more active than those with no terminal phosphate, whereas trinucleotides with 2′-(3′)-terminal phosphate were inactive.[6]

In this report, trinucleotides of known sequence were used to direct sRNA binding. GpUpU, but not its sequence isomers, UpGpU and UpUpG, was shown to induce Val-sRNA binding to ribosomes. These data indicate that the nucleotide sequence of an RNA codeword for valine is GpUpU.

Materials and Methods.—Analyses of poly- and oligonucleotides: (*a*) Paper electrophoresis was performed on Whatman 54 or 3 MM paper in 0.05 *M* NH_4COOH buffer, pH 2.7, at 80 v/cm for 0.5 hr with authentic reference markers. If a marker was not available, the expected mobility was calculated.[7] (*b*) Descending paper chromatography was performed at room temperature with Whatman 3 MM paper and the following solvents: (*A*) conc. NH_4OH-N-propanol-H_2O, 10/55/35, v/v; (*B*) 40 gm $(NH_4)_2SO_4$ dissolved in 100 ml 0.1 *M* sodium phosphate, pH 7.0.[8] Bands were visualized under UV light. (*c*) Ultraviolet absorption measurements were made in a Zeiss spectrophotometer, and spectra were obtained in a Cary recording spectrophotometer The absorbancy of eluates from paper was read against blank paper eluates. (*d*) The base ratio of the poly UG preparation (D-132) was 0.74/0.26 (U/G). The base ratio of UpUpG, UpGpU, and GpUpU preparations was determined by incubating 2.0 A^{260} units with 3.5×10^{-3} units of T_2 RNase[9] in 0.02 ml of 0.5 *M* ammonium acetate, pH 4.5, for 2.5 hr at 45°. The digestion products were separated by paper electrophoresis, eluted, and the absorbancy of each at appropriate λ_{max} was determined.

Preparation and separation of oligonucleotides: UpU was characterized as described.[6] UpUpG and UpG were obtained by incubating 3×10^3 A^{260} units of poly UG with 1.3×10^3 units of T_1-RNase (Sankyo Co., Ltd., Tokyo) in 4.7 ml of 0.1 *M* NH_4HCO_3 at 37° for 6 hr.[10] The reaction mixture was lyophilized, dissolved in 0.1 *M* $(NH_4)_2$ CO_3, and terminal phosphates were removed by incubation with *E. coli* alkaline phosphatase, free of diesterase, as described by Heppel *et al.*[11] Oligonucleotide fractions were separated by paper chromatography with solvent *A* for 18 hr. Ultraviolet-absorbing bands with mobilities of UpG and UpUpG were eluted with H_2O, lyophilyzed, and purified separately by electrophoresis.

GpUpU and GpU were obtained by digesting poly UG with purified pork liver nuclease,[6] removing terminal phosphates, and separating fractions by paper chromatography and electrophoresis as described above. This procedure yielded a fraction containing UpG and GpU and also one containing GpUpU, UpGpU, and UpUpG. Paper chromatography with solvent *B* for

36 hr resolved each dinucleotide. The trinucleotide fraction was separated into two bands by chromatography with solvent *B*. The band nearest the origin contained GpUpU of high purity. It was eluted with H_2O, absorbed on acid-washed Norite, and eluted with 45% ethanol 0.5 *M* NH_4OH. GpUpU was purified again by chromatography with solvent *B*, eluted, and desalted as described.

A derivative of pancreatic RNase A was used by Dr. Merton Bernfield to catalyze the transfer of uridine-2′,3′-cyclic phosphate to GpU.[12] UpGpU was purified from the reaction mixture by paper chromatography and electrophoresis as described for UpUpG.

Characterization of triplets: GpUpU, UpGpU, and UpUpG preparations were purified as described. Only one spot was observed after each was subjected to paper chromatography with solvents *A* or *B* and to paper electrophoresis. Base-ratio and sequence analyses established the purity, chain length, and base sequence of each triplet preparation. Base-ratio analyses were as follows: GpUpU, 1.2 U, 1.0 Up, 1.0 Gp; UpGpU, 1.0 U, 1.0 Up, .09 Gp; UpUpG, 2.1 Up, 1.0 G.

Base sequence was determined as follows: 2.0 A^{260} units of each trinucleotide was digested separately with T_1-RNase (as described earlier) and in other reactions, with 2.0 μg of chromatographically purified pancreatic RNase (Sigma) in 0.02 ml of 0.1 *M* $(NH_4)_2CO_3$, at 37° for 6 hr. Reaction products were separated by paper electrophoresis. Dinucleotide products were eluted, lyophilized, and digested with the RNase to which they were susceptible, and the products again were separated by electrophoresis. Nucleosides and nucleotides were identified by their electrophoretic mobility as well as by their spectra.

After digestion of GpUpU, UpGpU, and UpUpG with pancreatic RNase, only the following products were obtained: GpUp + U; Up + GpU; and Up + G, respectively. After digestion of GpUp and GpU with T_1-RNase, Gp + Up and Gp + U were obtained, respectively. T_1-RNase digestion of GpUpU, UpGpU, and UpUpG yielded Gp + UpU, UpGp + U, and UpUpG, respectively. Digestion of UpU, UpGp, and UpUpG with pancreatic RNase yielded Up + U, Up + Gp, and Up + G, respectively. No other UV-absorbing material was found. Since a 2% contaminant could have been detected, the purity of each trinucleotide was estimated to be >98%. GpUpU and UpGpU have not been purified previously.

Ribosomes, sRNA, and E. coli extracts: E. coli W 3100 sRNA was prepared from cells grown to late log phase in 0.5% nutrient broth, 1% glucose.[25] *E. coli* B sRNA was obtained from General Biochemicals, Inc. Each C^{14}-amino-acyl-sRNA was prepared in the presence of 19 C^{12}-amino acids as described elsewhere.[6] In additional experiments C^{14}-Leu-sRNA also was prepared in the absence of other amino acids. C^{14}-L-Val, C^{14}-L-Phe, and C^{14}-L-Leu, uniformly labeled with radioactivities of 293, 229, and 435 cpm/μμmole (Packard Corp. scintillation counter), respectively, were obtained from New England Nuclear Corp. Ribosomes for binding studies and *E. coli* extracts (DNase-treated, preincubated, S-30 fraction) for amino acid incorporation into protein studies were prepared as described elsewhere.[6, 13]

Assays: Each 50-μl reaction mixture contained 0.1 *M* tris-acetate, pH 7.2; 0.02 *M* magnesium acetate; 0.05 *M* KCl; and 1.0 A^{260} units of ribosomes unless otherwise specified. Incubations were at 24° for 20 min unless otherwise indicated. C^{14}-amino-acyl-sRNA binding to ribosomes was determined, as reported elsewhere,[6] by washing the ribosomes on Millipore filters. Ribosomes with bound sRNA remained on the filter.

C^{14}-amino acid incorporation into protein was determined as reported previously.[13]

Results.—Effect of GpUpU and Poly UG upon the binding of C^{14}-valine sRNA to ribosomes: The data of Figure 1 show that the binding of Val-sRNA to ribosomes is stimulated by the addition of GpUpU or poly UG and is dependent upon the concentration of either template. More Val-sRNA is bound to ribosomes in the presence of poly UG than in the presence of an equivalent amount of GpUpU. We have observed similar differences between the activity of pUpUpU and poly U.[6]

In Figure 2, the rates of binding at 0° and 24° are compared, and the specificity of GpUpU and poly UG for amino-acyl-sRNA is shown. At 0°, poly UG markedly stimulated Val- and Phe-sRNA binding to ribosomes, whereas GpUpU stimulated only Val-sRNA binding. At 24° the rates of binding directed by poly UG were higher than at 0°. Slightly higher binding in control reactions was observed

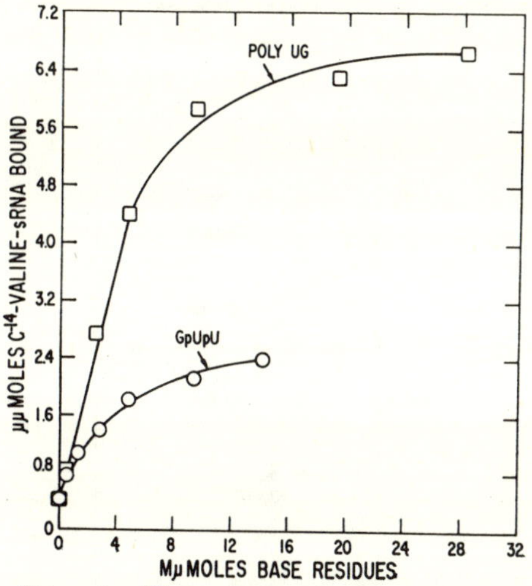

FIG. 1.—The relation between GpUpU (○) and poly UG (□) concentration and C^{14}-Val-sRNA binding to ribosomes. Each reaction mixture contained 2.0 A^{260} units of ribosomes and 18.9 $\mu\mu$moles of C^{14}-Val attached to 0.44 A^{260} units of sRNA. Poly UG and GpUpU were added as specified.

(especially of Leu-sRNA). Under the conditions employed, poly UG had little activity in directing Leu-sRNA binding to ribosomes, compared with its ability to direct amino acid incorporation into protein.

Codeword specificity and the effect of (Mg^{++}) *concentration:* The data of Figure 3 demonstrate that poly UG stimulates binding of Val- and Phe-sRNA to ribosomes optimally at 0.02–0.03 *M* (Mg^{++}), whereas GpUpU directs binding of Val-sRNA optimally at higher concentrations. At higher (Mg^{++}) concentrations, a small and apparently nonspecific increase in Val-, Leu-, and Phe-sRNA binding was observed in control reaction mixtures (no addition). Tri- and polynucleotides induce specific sRNA binding. GpUpU directed only Val-sRNA binding, whereas UpGpU and UpUpG had no effect upon either Val-, Leu-, or Phe-sRNA. The addition of tri- or polynucleotides did not appreciably increase Leu-sRNA binding over that of control reactions within the range of (Mg^{++}) concentrations tested.

The specificity of di-, tri-, and polynucleotides in directing amino-acyl-sRNA binding is shown in Table 1. Poly UG directed Val- and Phe-sRNA binding, whereas poly U directed the binding only of Phe-sRNA. UpU, GpU, and UpG did not stimulate Val-, Phe-, or Leu-sRNA binding.

The effect of GpUpU was tested upon the binding of 17 other sRNA preparations, each with a different C^{14}-amino accepted (C^{14}-asparagine and C^{14}-glutamine-sRNA not tested). GpUpU was found to direct the binding only of Val-sRNA.

Amino acid incorporation into protein: Since poly UG had little effect upon Leu-sRNA binding to ribosomes, we investigated the ability of poly UG to direct

TABLE 1

CODEWORD SPECIFICITY

Expt.	Addition, mμmoles base residues	C^{14}-Amino-Acyl-sRNA Bound to Ribosomes, $\mu\mu$moles: C^{14}-Val-sRNA	C^{14}-Phe-sRNA	C^{14}-Leu-sRNA
1	None	0.38	0.22	0.37
	4.7 Poly U	0.23	4.73	0.22
	4.7 Poly UG	2.65	1.93	0.24
2	None	0.40	0.22	0.62
	4.7 GpUpU	1.11	0.27	0.56
	4.7 UpGpU	0.40	0.25	0.55
	4.7 UpUpG	0.37	0.25	0.44
3	None	0.18	0.22	0.69
	4.7 GpU	0.20	0.22	0.69
	4.7 UpG	0.19	0.21	0.71
	4.7 UpU	0.18	0.20	0.67

Oligonucleotide specificity in directing sRNA binding to ribosomes. Where indicated, 9.10 $\mu\mu$moles C^{14}-Val attached to 0.3 A^{260} units of sRNA, 17.0 $\mu\mu$moles C^{14}-Phe attached to 0.7 A^{260} units of sRNA, or 21.0 $\mu\mu$moles C^{14}-Leu attached to 1.5 A^{260} units of sRNA were added to reaction mixtures.

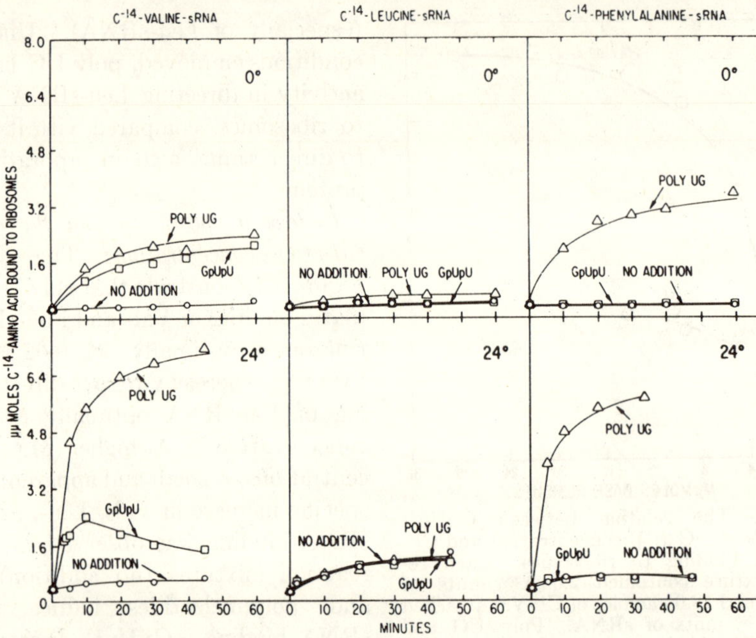

FIG. 2.—Effect of GpUpU and poly UG upon the rates of binding of Val-, Leu-, and Phe-sRNA to ribosomes at 0° and 24°. The symbols represent the addition of: (△) poly UG, 4.70 mμmoles of base residues; (□) GpUpU, 2.35 mμmoles of base residues; (○) no addition. Where indicated, 17 mμmoles of C^{14}-Val attached to 0.44 A^{260} units of sRNA, 8.5 μμmoles C^{14}-Phe attached to 1.5 A^{260} units sRNA, or 21.0 μμmoles of C^{14}-Leu attached to 1.5 A^{260} units of sRNA were added. Samples were incubated at the temperatures and for the times indicated.

amino acid incorporation into protein in *E. coli* extracts at 0° and 24° (also at 37° to make these data comparable to previously reported studies[14, 15]). As shown in Table 2, poly UG directed Phe, Val, and Leu into protein at 37° and 24°, and as reported previously,[14, 15] Leu incorporation almost equaled that of Val. Amino acid incorporation into protein was not detected at 0°.

Discussion.—Previously we have shown that UpUpU, ApApA, and pCpCpC specifically induce the binding to ribosomes of Phe-, Lys-, and Pro-sRNA, respectively. The present study demonstrates that the trinucleotide, GpUpU, can induce the binding of Val-sRNA to ribosomes, whereas sequence isomers such as UpGpU and UpUpG, and dinucleotides have no activity. When the specificity of poly UG was compared with that of GpUpU, the polynucleotide was found to direct both

TABLE 2

C^{14}-AMINO ACID INCORPORATION INTO PROTEIN

Temperature of incubation	Addition	Amino Acid Incorporated into Protein, mμmoles C^{14}-valine	C^{14}-phenylalanine	C^{14}-leucine
0°	None	<0.03	<0.03	<0.03
	Poly UG, 53 mμmoles	<0.03	<0.03	<0.03
24°	None	<0.03	0.04	0.04
	Poly UG, 53 mμmoles	0.70	1.57	0.67
37°	None	0.10	0.09	0.08
	Poly UG, 53 mμmoles	1.04	2.05	0.96

Each 0.1-ml reaction mixture contained 0.3 mg *E. coli* protein (DNase-treated, preincubated S-30 fraction[20]), 20 mμmoles of either C^{14}-Val, C^{14}-Phe, or C^{14}-Leu (4.8, 4.0, 4.0 μcuries/μmole, respectively), in addition to components previously described.[20] Reactions were incubated for 10, 20, or 60 min at 37°, 24°, or 0°, respectively.

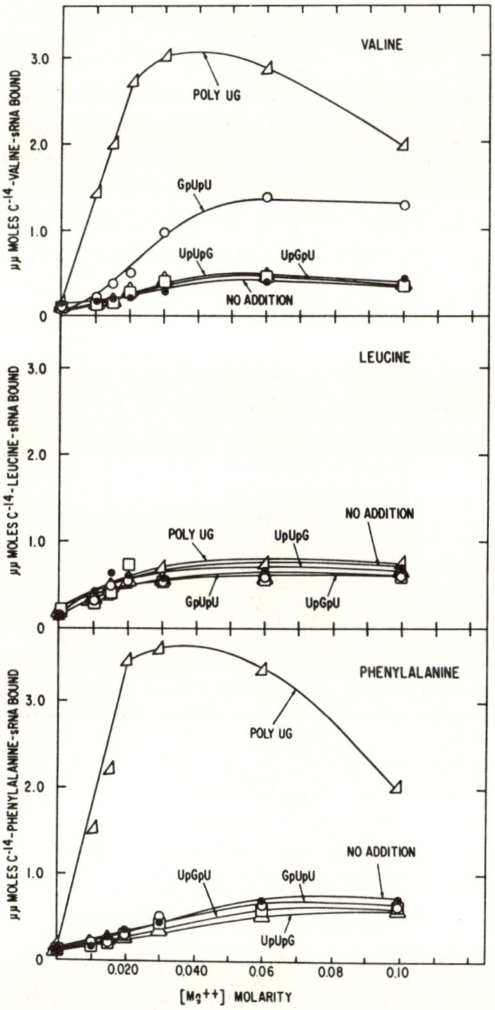

FIG. 3.—Relationship between magnesium acetate concentration and C^{14}-amino-acyl-sRNA binding to ribosomes. The symbols represent the addition of: (◿) poly UG, 4.70 mμmoles of base; (○) GpUpU, 2.35 mμmoles of base; (□) UpGpU, 2.35 mμmoles of base; (△) UpUpG, 2.35 mμmoles of base; (●) no addition. 9.10 μμmoles C^{14}-Val attached to 0.3 A^{260} units of sRNA, 25.0 μμmoles C^{14}-Leu attached to 1.5 A^{260} units of sRNA, or 10.0 μμmoles C^{14}-Phe attached to 0.7 A^{260} units of sRNA were added where indicated. The (Mg^{++}) concentration of each reaction mixture is shown on the abscissa. Samples were washed with buffer containing the appropriate concentration of (Mg^{++}).

Phe- and Val-sRNA binding, but GpUpU induced the binding of only Val-sRNA. We conclude that the nucleotides of a valine codeword in mRNA are arranged in the sequence GpUpU. Since UpUpG does not replace GpUpU, recognition occurs with polarity.

As discussed elsewhere,[6] the ability of tri- but not dinucleotides to induce sRNA binding demonstrates directly that triplets may code for amino acids, and agrees with conclusions derived from both genetic[19] and biochemical studies.[20]

Studies of the incorporation of amino acids stimulated by various polynucleotides together with the data on amino acid replacements in mutants provide a basis for assignment of codewords to amino acids.[16–18] Unfortunately, widespread degeneracy and the difficulty of assigning words which contain three different bases prevent unique assignments. Several self-consistent schemes are possible. The additional information provided by the definite knowledge of even a single codeword such as GpUpU provides a distinction among these different possibilities. The reassignments suggested to make GpUpU the codeword for valine consistent with the other data are listed in Table 3. For example, ApUpU was assigned to isoleucine on the basis of studies employing randomly ordered polynucleotides to direct amino acid incorporation into protein, and the HNO_2-induced replacement of isoleucine by valine in TMV coat protein.

TABLE 3

Predicted Nucleotide Sequences of RNA Codewords

Amino acid	Sequence	Reference
Valine	GpUpU	Derived experimentally
Isoleucine	ApUpU	17, 18
	Ap (UA)	
Tyrosine	Up (AU)	
Alanine	GpCpC	16
	Gp (CA)	
	Gp (CU)	
Glutamic acid	GpApA*	16
	Gp(AU)	
Glycine	Gp (GU)	16
	Gp (GA)	
	Gp (GC)	
	Note: UpGpG is ruled out	
Arginine	ApGpA	16
	CpGpC	
	(C)Gp(A)	
Threonine	Ap(CA)	17, 18
	(ACC)	

* The assignment of GpApA to glutamic acid is based upon limited amino acid incorporation into protein data and is, therefore, tentative.[23, 24]
Nucleotides within parentheses have not been arranged in sequence. Amino acid replacements used for these predictions were found in *E. coli* by Yanofsky and co-workers[16] or were induced by HNO_2 in TMV by Wittmann and Wittmann-Liebold[17] and also by Tsugita.[18]

It should be noted that the predicted sequence for isoleucine, ApUpU, and also the GpUpU sequence found for valine do not agree with reported sequence determinations[21] which, on the basis of experiments with uncharacterized polynucleotides, assign the sequence ApUpU to tyrosine, tentatively assign GpUpU to cysteine, and derive the polarity of mRNA for protein synthesis.

Although our results show only marginal stimulation of Leu-sRNA binding by poly UG, we have consistently induced Leu- and also Cys-sRNA binding with poly UG when reactions were incubated at 0° and at a (Mg^{++}) concentration of 0.03 *M*. Leu-sRNA binding was readily induced by poly UC which confirms results reported by Kaji and Kaji.[3] Further studies with Leu- and Cys-sRNA are in progress.

We have shown that specific binding of sRNA can be induced at 0° and have described preliminary evidence which indicates that intact amino-acyl-sRNA can be eluted from ribosomes after incubation.[6] Although these data suggest that formation of peptide bonds is not required for binding to ribosomes, a requirement for the first transfer enzyme and GTP may exist (cf. refs. 1–4, and 6).

Since synonym codewords corresponding to one amino acid often differ in base composition by only one nucleotide,[20] it is possible that bases in common may occupy identical positions within each triplet. A triplet code can be constructed wherein recognition between two out of three nucleotide pairs may, in some cases, suffice for coding; or, alternatively, a base at one position in the triplet may pair optionally with more than one base.[22-24] Ambiguous and synonymous codewords may involve such recognitions. The development of a rapid method for measuring sRNA binding to ribosomes and the use of trinucleotides of known sequence to direct such binding should provide a general method of great simplicity to test these possibilities and also to study the genetic function of other nucleotide sequences and interactions between specific codewords, sRNA, and ribosomes.

Summary.—The binding of C^{14}-valine-sRNA to ribosomes was directed both by the trinucleotide, GpUpU, and by poly UG, but not by UpGpU, UpUpG, or dinucleotides. GpUpU had no effect upon the binding to ribosomes of sRNA corresponding to 17 other amino acids. The nucleotide sequence GpUpU was proposed for a valine RNA codeword. The implications of these findings and predictions based on amino acid replacement data are discussed briefly.

We are grateful to Drs. Leon Heppel and George Rushizky for their advice on oligonucleotide separations. It is also a pleasure to thank Norma Zabriskie and Theresa Caryk for their extremely skillful technical assistance.

The following abbreviations are used: Val, valine; Phe, phenylalanine; Leu, leucine; p oly U, polyuridylic acid; poly C, polycytidylic acid; poly A, polyadenylic acid; poly UG, copolymer of uridylic and guanylic acids; TCA, trichloroacetic acid; sRNA, transfer RNA; mRNA, messenger RNA. For mono- and oligonucleotides of specific structure, the "p" to the left of a terminal nucleoside initial indicates a 5′-terminal phosphate; the "p" to the right, a 2′-(3′)-terminal phosphate. Internal phosphates of oligonucleotides are (3′,5′)-linkages.

[1] Arlinghaus, R., G. Favelukes, and R. Schweet, *Biochem. Biophys. Res. Commun.*, **11**, 92 (1963).

[2] Nakamoto, T., T. W. Conway, J. E. Allende, G. J. Spyrides, and F. Lipmann, in *Synthesis and Structure of Macromolecules*, Cold Spring Harbor Symposia on Quantitative Biology, vol. 28 (1963), p. 227.

[3] Kaji, A., and H. Kaji, *Biochem. Res. Commun.*, **13**, 186 (1963); *Federation Proc. (Abstr.)*, **23**, 478 (1964).

[4] Spyrides, G. J., *Federation Proc. (Abstr.)*, **23**, 318 (1964).

[5] Cannon, M., R. Krug, and W. Gilbert, *J. Mol. Biol.*, **7**, 360 (1963).

[6] Nirenberg, M. W., and P. Leder, *Science*, in press.

[7] Markham, R., and J. D. Smith, *Biochem. J.*, **52**, 558 (1952).

[8] Rushizky, G. W., and H. A. Sober, *J. Biol. Chem.*, **237**, 2883 (1962).

[9] *Ibid.*, **238**, 371 (1963).

[10] *Ibid..*, **237**, 834 (1962).

[11] Heppel, L. A., D. R. Harkness, and R. J. Hilmoe, *J. Biol. Chem.*, **237**, 841 (1962).

[12] Bernfield, M. R., and M. W. Nirenberg, *Abstracts*, 148th National Meeting, American Chemical Society, Chicago, Illinois, August 1964.

[13] Nirenberg, M. W., in *Methods in Enzymology*, ed. S. P. Colowick and N. O. Kaplan (New York: Academic Press, 1964), vol. 6, p. 17.

[14] Nirenberg, M. W., J. H. Matthaei, O. W. Jones, R. G. Martin, and S. H. Barondes, *Federation Proc.*, **22**, 55 (1963).

[15] Singer, M. F., O. W. Jones, and M. W. Nirenberg, these PROCEEDINGS, **49**, 392 (1963).

[16] Yanofsky, C., in *Synthesis and Structure of Macromolecules*, Cold Spring Harbor Symposia on Quantitative Biology, vol. 28 (1963), p. 581.

[17] Wittmann, H. G., and B. Wittmann-Liebold, in *Synthesis and Structure of Macromolecules*, Cold Spring Harbor Symposia on Quantitative Biology, vol. 28 (1963), p. 589.

[18] Tsugita, A., personal communication.

[19] Crick, F. H. C., L. Barnett, S. Brenner, and R. J. Watts-Tobin, *Nature*, **192**, 1227 (1961).

[20] Nirenberg, M., O. Jones, P. Leder, B. Clark, W. Sly, and S. Pestka, *Synthesis and Structure of Macromolecules*, Cold Spring Harbor Symposia on Quantitative Biology, vol. 28 (1963), p. 549.

[21] Wahba, A. J., C. Basilio, J. F. Speyer, P. Lengyel, R. S. Miller, and S. Ochoa, these PROCEEDINGS, **48,** 1683 (1962).

[22] Nirenberg, M. W., and O. W. Jones, in *Informational Macromolecules*, ed. H. Vogel *et al.*, (New York: Academic Press, 1963).

[23] Jones, O. W., and M. W. Nirenberg, these PROCEEDINGS, **48,** 2115 (1962).

[24] Gardner, R. S., A. J. Wahba, C. Basilio, R. S. Miller, P. Lengyel, and J. F. Speyer, these PROCEEDINGS, **48,** 2087 (1962).

[25] Zubay, G., *J. Mol. Biol.* **4,** 347 (1962).

Reprinted from JOURNAL OF MOLECULAR BIOLOGY, Vol. 13, 1965

Studies on Polynucleotides

XLVIII.† The *in vitro* Synthesis of a Co-polypeptide containing Two Amino Acids in Alternating Sequence dependent upon a DNA-like Polymer containing Two Nucleotides in Alternating Sequence

S. NISHIMURA,‡ D. S. JONES AND H. G. KHORANA

Institute for Enzyme Research, University of Wisconsin
Madison, Wisc., U.S.A.

(*Received 5 May 1965*)

The DNA-like polymers, poly d-TC:AG and poly d-TG:AC, which contain in each strand only two bases in alternating sequence, have been used as templates for DNA-dependent RNA polymerase. In this way, four single-stranded long-chain ribopolynucleotides, poly UC, poly AG, poly UG and poly AC containing in every case the two nucleotides in alternating sequence have been prepared. The messenger activity of poly UC in the *Escherichia coli* cell-free amino acid incorporation system has been studied. The ribopolynucleotide stimulates the incorporation of only two amino acids, serine and leucine, in equimolar amounts. The product of the reaction has been characterized as a co-polypeptide containing serine and leucine in alternating sequence.

1. Introduction

The chemical synthesis of short-chain deoxyribopolynucleotides containing repeating dinucleotide sequences has been reported from this laboratory (Jacob, Ohtsuka, Moon, Narang, & Khorana, 1964; Khorana, Jacob, Moon, Narang & Ohtsuka, 1965; Ohtsuka, Moon & Khorana, 1965). These synthetic deoxyribopolynucleotides have been shown to serve as templates for the DNA polymerase of *Escherichia coli* (Byrd, Ohtsuka, Moon & Khorana, 1965; Wells, Ohtsuka & Khorana, 1965) and the products of the enzymic reaction have been characterized as high molecular weight DNA-like polymers containing repeating dinucleotide sequences. Two such polymers have so far been described. One polymer, poly d-TC:AG,§ contains alternating thymidylate and deoxycytidylate units in one strand and alternating deoxyadenylate and deoxyguanylate units in the complementary strand (Byrd *et al.*, 1965). The second polymer, poly d-TG:AC, contains alternating thymidylate and deoxyguanylate units in one strand and alternating deoxyadenylate and deoxycytidylate units in the complementary strand (Wells *et al.*, 1965).

† Paper XLVII in this series is Nishimura, Jones, Ohtsuka, Hayatsu, Jacob & Khorana, 1965.

‡ Present address: National Cancer Center Research Institute, Tokyo, Japan.

§ Poly d-TC:AG refers to the DNA-like polymer which contains alternating thymidylate and deoxycytidylate units in one strand and alternating deoxyadenylate and deoxyguanylate units in the complementary strand. Similarly, poly d-TG:AC refers to the polymer which contains alternating thymidylate and deoxyguanylate units in one strand and deoxyadenylate and deoxycytidylate units in the complementary strand. Poly UC, poly UG, poly AG and poly AC are ribopolynucleotides containing in every case two ribonucleotide units, as indicated by the letters U, C, A or G, in alternating sequence.

In the present work the action of DNA-dependent RNA polymerase on the above DNA-like polymers has been studied. In particular, it has been found that by providing in the reaction mixture a set of two ribonucleoside triphosphates, which are appropriate for the copying of any given strand of the DNA-like polymers, single-stranded ribopolynucleotides containing two nucleotides in alternating sequence may be prepared. The formation and characterization of four such ribopolynucleotides, poly UC, poly AG, poly UG and poly AC, containing in every case the two nucleotides in alternating sequence, are described.

The messenger activity of one of the ribopolynucleotides, poly UC, in the *in vitro* amino acid incorporation system has been studied. The ribopolynucleotide stimulates the incorporation of two amino acids, serine and leucine, and the incorporation of these amino acids has been found to be interdependent. The product of the reaction has been characterized as a co-polypeptide containing serine and leucine in alternating sequence. A brief report of a part of this work has already appeared (Nishimura, Jones, Wells, Jacob & Khorana, 1965)

2. Materials and Methods

(a) *Materials*

Poly d-TC:AG was prepared as described previously (Byrd *et al.*, 1965). Poly d-TG:AC was prepared in collaboration with Dr R. D. Wells by a procedure which is being published separately (Wells *et al.*, 1965). [^{14}C]ATP, [^{14}C]GTP, [^{14}C]CTP and [^{14}C]UTP were purchased from Schwarz BioResearch, Inc. [α-^{32}P]ATP and [α-^{32}P]UTP were purchased from International Chemical and Nuclear Corp. Unlabeled ribonucleoside triphosphates were also commercial products. Uniformly labeled [^{14}C]L-amino acids were of the following specific activities: alanine, 123; arginine, 246; aspartic acid, 164; asparagine, 29·8; DL-cystine, 7·5, glutamic acid, 205; glutamine, 6·58; glycine, 66; histidine, 198; isoleucine, 246; leucine, 198; lysine, 198; methionine, 10·6; phenylalanine, 297; proline, 165; serine, 123; threonine, 164; tryptophan, 9·2; tyrosine, 297; valine, 185. Dinitrophenyl amino acids were commercial products.

Methods for the preparation of *E. coli* sRNA, [^{14}C]aminoacyl-sRNA, *E. coli* ribosomes and 100,000 *g* supernatant fraction are described in the previous paper (Nishimura, Jones, Ohtsuka, Hayatsu, Jacob & Khorana, 1965). In some cases, the ribosomes were further dialyzed against the standard buffer for 12 hr with one change of the buffer during the course of dialysis. RNA polymerase was prepared from *E. coli* B according to Chamberlin & Berg (1962). The preparation represented a purification of 100-fold over the crude extract and the specific activity corresponded to an incorporation of 2000 mμmoles of [^{14}C]ATP/hr/mg of fraction 4, using calf thymus DNA as the template under the assay conditions of Chamberlin & Berg (1962). Fraction 4 was precipitated by adding an equal volume of saturated ammonium sulfate, and the precipitate was dissolved in 0·05 M-tris–HCl (pH 8·0), containing 0·05 M-$(NH_4)_2SO_4$, 0·001 M-$MgCl_2$ and 0·005 M-glutathione and stored in liquid nitrogen (Chamberlin & Berg, 1964).

(b) *Methods*

(i) *Assay of ribopolynucleotide synthesis and large-scale preparation of poly UC*

RNA polymerase-catalyzed synthesis of ribopolynucleotides with DNA-like polymers as templates was carried out using the following general conditions: reaction mixture contained, per ml.: 40 μmoles of tris–HCl buffer (pH 7·9), 4 μmoles $MgCl_2$, 1 μmole $MnCl_2$, 12 μmoles β-mercaptoethanol, 0·16 to 0·32 μmole ribonucleoside triphosphates, 4·0 to 150 mμmoles poly d-TC:AG or poly d-TG:AC and 0·19 to 0·5 mg (350 to 920 units) RNA polymerase. The incorporation of the labeled nucleoside triphosphate into acid-insoluble material was carried out as described previously (Nishimura, Jacob & Khorana, 1964).

For a large-scale preparation of poly UC, the reaction mixture (6·25 ml.) contained 2·5 O.D. units† of poly d-TC:AG, 2·5 mg of RNA polymerase, 1 μmole of CTP and 1 μmole of UTP. After incubation for 50 min at 37°C, 805 mμmoles of poly UC had been synthesized. To the reaction mixture was then added 0·6 ml. of 0·1 M-EDTA (pH 7·0), and incubation was continued for 2 min. The clear solution which resulted was shaken with an equal volume of 90% phenol for 6 min. The phenol layer was re-extracted with an equal volume of the buffer containing 0·015 M-NaCl and 0·0015 M-potassium citrate (pH 7·0). The combined aqueous layer was dialyzed against 1 l. of the same buffer for 6 hr and then dialyzed against 2 l. of distilled water for 10 hr with one change of water during the course of dialysis. The volume of the solution was reduced to 1·1 ml. by evaporation under reduced pressure. The yield was 11 O.D. units. No attempt was made to remove poly d-TC:AG from this product.

(ii) *Paper chromatography*

Paper chromatography was performed by the descending technique using Whatman paper no. 1, 40 or 3 MM. The following solvent systems were used for the nearest-neighbor frequency analysis of ribopolynucleotide products and for the identification of serine–leucine polypeptide, seryl-leucine, the free amino acids and their dinitrophenyl derivatives: solvent I, isobutyric acid–concn NH_4OH–water (66:1:33), pH 3·7; solvent II, ethanol–1·0 M-ammonium acetate, pH 7·5 (7:3); solvent III, *n*-butanol–water–acetic acid (4:2:1, by vol.); solvent IV, *n*-butanol–pyridine–water–acetic acid (30:20:24:6, by vol.) (Waley & Watson, 1953); solvent V, *n*-butanol saturated with 2 N-ammonium hydroxide; solvent VI, phenol–water (8:2, v/v); solvent VII, 1·5 M-sodium phosphate (pH 6·0) (Levy, 1954); solvent VIII, *n*-butanol–water–acetic acid (4:5:1, by vol.).

(iii) *Nearest-neighbor frequency analysis of poly UC, poly AG, poly UG and poly AC*

These ribopolynucleotides were prepared by using either [α-^{32}P]ATP or [α-^{32}P]UTP as one of the two triphosphates and the acid-insoluble products were collected on filter paper disks as described previously (Nishimura *et al.*, 1964). The disks were well soaked in 0·3 N-sodium hydroxide (0·5 ml. for a total area of 1 cm^2) for 18 to 21 hr at 23 or 37°C. Cold mononucleotides (Ap, Cp, Up or Gp) were added as required for markers and the total supernatant fraction was collected by centrifugation. The filter paper was washed twice with 0·5 ml. of distilled water. The combined alkaline solution was neutralized by the addition of pyridinium Dowex-50 ion exchange resin. The resin was removed by centrifugation and washed twice with 0·5 ml. of 0·1 M-ammonium hydroxide. The combined supernatant fraction and wash was evaporated in a rotary evaporator at low temperature. The dry residue was dissolved in 0·05 ml. of water and appropriate portions used for analysis by paper chromatography in appropriate solvents.

(iv) *Determination of chain length of ribopolynucleotides*

The filter paper disks containing ribopolynucleotides prepared by using the two ^{14}C-labeled triphosphates of equal specific activity were treated with alkali and the hydrolysis products isolated as described above. Appropriate markers of nucleosides and nucleotides were added and portions were chromatographed on DEAE-cellulose paper strips using 0·3 M-ammonium formate (pH 5) for about 5 hr. The areas corresponding to the carrier nucleosides and nucleotides and the remainder of the chromatogram were scanned for radioactivity by counting in the liquid-scintillation counter.

(v) *Fractionation of sRNA*

E. coli B-sRNA was fractionated by counter-current distribution following the method of Holley and co-workers (1963) as described in the accompanying paper (see Fig. 2, Nishimura, Jones, Ohtsuka, Hayatsu, Jacob & Khorana, 1965). [^{14}C]Leucyl-sRNA used in the binding experiment was prepared by charging of leucine-specific sRNA present in fraction 114. This fraction contains material corresponding to peak I of Weisblum, Benzer & Holley (1962).

† O.D. units refers to absorbance measurements in 1 ml., 1-cm light path at 260 mμ and at neutral pH.

(vi) *Assay of aminoacyl-sRNA binding to ribosomes*

This was carried out as described by Nirenberg & Leder (1964). The reaction mixture (0·05 ml.) contained 5 μmoles tris–HCl (pH 7·5), 2·5 μmoles KCl, 2·0 O.D. units of ribosomes, 1 μmole of magnesium acetate, [^{14}C]aminoacyl-sRNA and poly UC as indicated. Incubation was carried out at 25°C for 20 min.

(vii) *Assay of polypeptide synthesis*

The two-step reaction procedure described by Wood & Berg (1962) was used to test for polypeptide synthesis as directed by ribopolynucleotides in the cell-free amino acid incorporation system. After the synthesis of the ribopolynucleotide poly UC, using poly d-TC:AG as template, RNA polymerase, UTP and CTP (step 1), the reaction mixture was supplemented with the components of the amino acid incorporation system except for the 100,000 **g** supernatant fraction.

This reaction mixture (step 2) contained, per ml.: 60 μmoles of tris–HCl (pH 7·8), 9 μmoles of magnesium acetate, 2 μmoles of $MgCl_2$, 12 μmoles of β-mercaptoethanol, 0·5 μmole of $MnCl_2$, 50 μmoles of KCl, 0·25 μmole of GTP, 2 μmoles of ATP, 5 μmoles of phosphoenolpyruvate, 20 μg of phosphoenolpyruvate kinase, 20 O.D. units of ribosomes, 16 O.D. units of sRNA, 0·5 to 2 μc of [^{14}C]amino acid as specified, 63 mμmoles of [^{12}C]amino acid where indicated and poly d-TC:AG, poly UC, excess nucleoside triphosphates and RNA polymerase as introduced from the first stage of the reaction. This reaction mixture was allowed to incubate at 0°C for 10 min, and to it was added 200 μg of 100,000 **g** supernatant fraction. Incubation was now carried out at 37°C. For routine measurement of amino acid incorporation, portions (0·01 to 0·02 ml.) of the reaction mixture (total volume, usually 0·1 to 0·2 ml.) were taken at different intervals and were applied to Whatman 3 MM filter paper disks (approximately 1 cm^2 in area). The filter paper disks were immersed in cold 5% trichloroacetic acid, washed once with a fresh portion of the acid solution, then heated for 30 min at 90°C in 5% trichloroacetic acid solution. The filter paper disks were washed once again with cold 5% trichloroacetic acid, then with ethyl alcohol–ether (1:1, v/v) mixture and finally with ether. The disks were counted for radioactivity in a Packard TriCarb scintillation counter as reported by Mans & Novelli (1961).

The incorporation of radioactive material from ^{14}C-labeled amino acids into polypeptidic material was also followed by applying portions of the reaction mixture, after an alkaline treatment, to paper chromatograms and developing in solvent III. The radioactive material incorporated stayed at the origin. The area was cut out and counted for radioactivity. The results thus obtained were identical with those obtained by the assay described above.

(viii) *Large-scale preparation of ^{14}C-labeled co-polypeptide*

The RNA polymerase reaction mixture (the first stage) contained 640 mμmoles each of UTP and CTP, 300 mμmoles of poly d-TC:AG and 0·8 mg of RNA polymerase, in addition to other components as described above in a final volume of 2·0 ml. The incubation was carried out at 37°C for 45 min. Under these conditions, 225 mμmoles of poly UC were synthesized, as judged from a parallel small-scale experiment in which the incorporation of [^{14}C]UTP was measured. After the first-stage reaction, 2 μc of each of [^{14}C]leucine and [^{14}C]serine were introduced into the reaction mixture in addition to other components as described above, the total volume being 4·0 ml. After incubation for 25 min at 37°C, [^{14}C]serine and [^{14}C]leucine giving 2,350,000 cts/min were incorporated. The reaction mixture was chilled in ice and 0·04 ml. of 10 N-NaOH was added. The solution was allowed to stand for 1 hr at room temperature to hydrolyze aminoacyl- and polypeptidyl-sRNA linkages. The total protein fraction including [^{14}C]serine–leucine polypeptide was precipitated by the addition of 0·2 ml. of 100% trichloroacetic acid. The precipitate was washed 3 times with 1-ml. portions of cold 5% trichloroacetic acid, twice with 1·5 ml. of ethyl alcohol–ether (1:1, v/v) and finally twice with 2·0 ml. of ether.

(ix) *Preparation of [^{14}C]seryl-leucine from [^{14}C]serine–leucine polypeptide by acidic hydrolysis*

[^{14}C]Serine–leucine polypeptide (600,000 cts/min) was dissolved in 1·5 ml. of 10 N-HCl, and the solution incubated at 30°C for 5 days as described by Desnuelle & Casal (1948).

After the incubation, HCl was removed by evaporation, using a rotary evaporator at room temperature. The dry residue was dissolved in 1·5 ml. of water, evaporated again and finally dissolved in 1·5 ml. of water.

(x) *N-Terminal amino acid determination of [^{14}C]serine–leucine dipeptide*

An aqueous solution of [^{14}C]serine–leucine dipeptide (0·75 ml., giving about 300,000 cts-min) as prepared above was evaporated to dryness. To the residue 0·2 ml. of 1% triethyl/amine (in water) and 0·4 ml. of 5% 1-fluoro-2,4-dinitrobenzene (in ethyl alcohol) were added, and the mixture incubated for 8 hr at 30°C with occasional shaking. During the incubation, 0·002 ml. of 100% triethylamine was added 3 times to adjust pH to alkaline region. A part (0·5 ml.) of the above reaction mixture was taken and evaporated to dryness. Water was added to the dried residue and the solution evaporated again. Without removal of the excess of fluorodinitrobenzene, the total material was subjected to acidic hydrolysis by adding 1·0 ml. of 6 N-HCl. The hydrolysis was carried out at 105°C for 10 hr in a sealed tube. The solution was then diluted to 6 ml. with water to obtain 1 N-HCl solution. After the addition of 50 mμmoles of cold leucine, 50 mμmoles of cold serine, 200 mμmoles of cold dinitrophenyl-leucine, and 200 mμmoles of dinitrophenyl-serine as carriers, the mixture was extracted four times with ether. The ether fraction was evaporated to dryness and the residue dissolved in 0·2 ml. of acetone. The aqueous layer was also evaporated by using a rotary evaporator, and the residue dissolved in 0·2 ml. of water.

(xi) *Edman degradation of [^{14}C]serine–leucine dipeptide*

The microscale procedure described by Gray & Hartley (1963) was followed. Thus, 0·2 ml. of a 5% solution of phenyl isothiocyanate (in pyridine) was added to an aqueous solution of [^{14}C]serine–leucine dipeptide (0·1 ml., 18,000 cts/min) and the resulting single-phase mixture incubated overnight at room temperature. After addition of 0·2 ml. of water, the pyridine and excess phenyl isothiocyanate were removed by 4 extractions with 3 ml. of benzene. A portion of the aqueous solution was applied to a paper chromatogram alongside a marker of [^{14}C]serine–leucine dipeptide and developed in solvent V. The remainder of the aqueous solution was evaporated and 0·2 ml. of acetic acid saturated with anhydrous hydrogen chloride was added. After 1 hr at room temperature, the acid was removed *in vacuo* and to the residue was added 1 ml. of water. After re-evaporation, 25 mμmoles of cold leucine and 25 mμmoles of leucyl-leucine were added as carriers and the solution made up to a total volume of 0·05 ml. Portions of this solution were added to paper chromatograms alongside leucine and leucyl-leucine markers and developed in solvent V or solvent VIII.

3. Results

(a) *Synthesis of ribopolynucleotides containing two nucleotides in alternating sequence using poly d-TC:AG and poly d-TG:AC as templates for RNA polymerase*

(i) *Kinetics of incorporation of individual nucleoside triphosphates*

Poly d-TC:AG and poly d-TG:AC contain only two bases in any one strand; therefore, two types of experiments are possible using RNA polymerase. In one type of experiment it might be expected that by providing in the reaction mixture the two triphosphates containing bases complementary to those present in one given strand of the above DNA-like polymers, only that strand should be copied. In the second type of experiment, all the four ribonucleoside triphosphates may be used to determine the rate and amount of the copying of both strands of the deoxypolynucleotide templates. Figure 1 and Table 1 show the kinetics of incorporation of the individual ribonucleoside triphosphates using poly d-TC:AG as the template, whereas Fig. 2 and Table 2 give the results of parallel experiments carried out using poly d-TG:AC as the template. It is seen that in every case two triphosphates, fulfilling the Watson–Crick base-pairing requirement for the copying of one of the deoxypolynucleotide

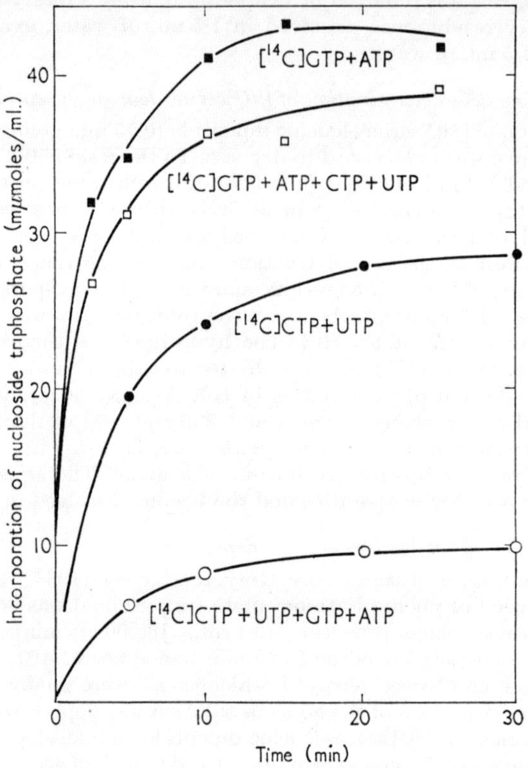

FIG. 1. Kinetics of incorporation of ribonucleoside triphosphates using poly d-TC:AG as template for RNA polymerase. The reaction mixture contained 0·4 mg of RNA polymerase, 40 mμmoles of poly d-TC:AG, and 160 mμmoles of each ribonucleoside triphosphate as indicated. For other details see Materials and Methods.

TABLE 1

Characteristics of incorporation of individual ribonucleoside 5′-triphosphates

Nucleoside triphosphate	Incorporation of labeled nucleoside triphosphate (mμmoles/ml.)		
	5 min	10 min	30 min
[¹⁴C]ATP + GTP	27·3	34·0	
[¹⁴C]GTP + ATP	31·3	36·3	
[¹⁴C]ATP + GTP + UTP + CTP	32·5	37·7	
[¹⁴C]GTP + ATP + UTP + CTP	34·9	41·7	
[¹⁴C]UTP + CTP		24·2	28·9
[¹⁴C]CTP + UTP		25·0	29·6
[¹⁴C]UTP + CTP + ATP + GTP		8·4	9·9
[¹⁴C]CTP + UTP + ATP + GTP		8·3	11·6

The reaction mixture contained 42 mμmoles of poly d-TC:AG, 0·4 mg of RNA polymerase and 160 mμmoles of nucleoside triphosphates as specified, in addition to the other components described under Materials and Methods, in 1·0 ml. of total volume. The incubation was carried out at 37°C. Specific activity of nucleoside triphosphate used: [¹⁴C]ATP and [¹⁴C]GTP, 4 μc/μmole; [¹⁴C]UTP and [¹⁴C]CTP, 2 μc/μmole.

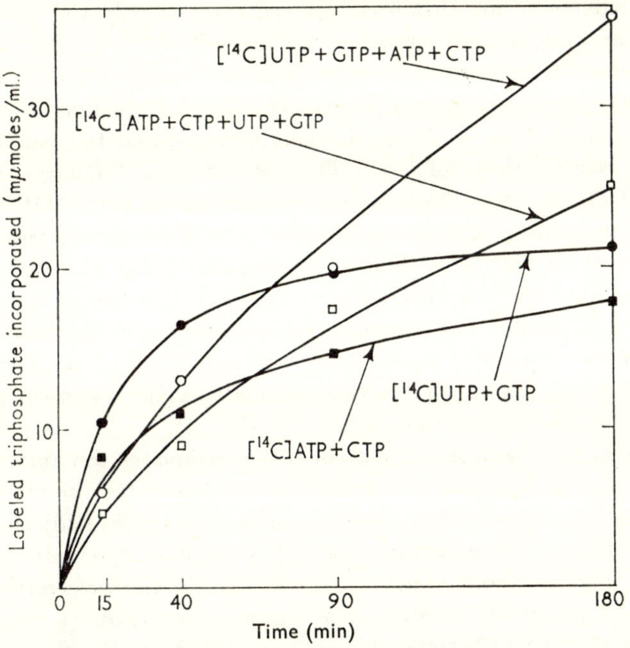

FIG. 2. Kinetics of incorporation of ribonucleoside triphosphates using poly d-TG:AC as template for RNA polymerase. The reaction mixture contained 0·19 mg of RNA polymerase, 4·0 mμmoles of poly d-TG:AC, and 160 mμmoles of each nucleoside triphosphate as indicated. For other details see Materials and Methods.

TABLE 2

Characteristics of incorporation of individual ribonucleoside 5′-triphosphates

Nucleoside triphosphate	Incorporation of labeled nucleoside 5′-triphosphates (mμmoles/ml.)			
	15 min	40 min	90 min	180 min
[14C]ATP + CTP	8·1	9·8	15·0	17·8
[14C]CTP + ATP	7·9	10·8	14·4	17·8
[14C]ATP + CTP + UTP + GTP	4·7	9·0	17·5	25·0
[14C]CTP + ATP + UTP + GTP	4·4	9·5	13·0	24·3
[14C]UTP + GTP	10·6	16·3	19·4	21·0
[14C]GTP + UTP	12·7	19·7	24·2	20·8
[14C]UTP + GTP + ATP + CTP	6·0	13·0	20·0	35·5
[14C]GTP + UTP + ATP + CTP	5·8	13·3	19·0	30·6

The reaction mixture contained 4·0 mμmoles of poly d-TG:AC, 0·19 mg of RNA polymerase and 160 mμmoles of nucleoside triphosphates as specified, in addition to the other components described under Materials and Methods. The incubation was carried out at 37°C. Specific activity of nucleoside triphosphate used: [14C]ATP and [14C]GTP, 4 μc/μmole; [14C]UTP and [14C]CTP, 2 μc/μmole.

strands, are sufficient; and that excellent syntheses of poly UC, poly UG, poly AG and poly AC are thus achieved. Further, as expected, the incorporation of each of the two triphosphates belonging to every set goes hand in hand (Tables 1 and 2).

Thus, in the case of experiments using poly d-TC:AG, the kinetics of incorporation of [^{14}C]ATP match those of [^{14}C]GTP incorporation, and the kinetics of [^{14}C]CTP incorporation match those of [^{14}C]UTP incorporation. Similarly, in experiments using poly d-TG:AC, the kinetics of incorporation of [^{14}C]ATP match those of [^{14}C]CTP incorporation, and the kinetics of [^{14}C]UTP match those of [^{14}C]GTP incorporation. With both DNA-like templates, this holds when two triphosphates are used or when all the four triphosphates are used. It is further seen (Figs 1 and 2) that there are differences in the amount of the copying of the two complementary strands, both when two triphosphates are used and when all the four triphosphates are used. Thus with poly d-TC:AG as template, under identical conditions, the synthesis of poly AG exceeds that of poly UC (Fig. 1).

Finally, while with both deoxypolynucleotide templates the amount of ribopolynucleotide synthesized exceeded the amount of the templates, the net synthesis in the experiment of Fig. 2 was much more extensive (about 30-fold) than that (about fourfold) obtained in the experiment of Fig. 1. In fact, it is noted that under the conditions of Fig. 2, ribopolynucleotide synthesis in the presence of all the four ribonucleoside triphosphates continued more or less linearly over a much longer period of time. These differences between the results in Fig. 1 and Fig. 2 are ascribed to the higher enzyme to deoxypolynucleotide template ratio used in the experiments of Fig. 2. (This conclusion is further supported below by the results recorded in Fig. 5.)

(ii) *Requirements of ribonucleoside triphosphate incorporation*

The results shown in Tables 3 and 4 further illustrate the interdependence of the incorporation of the different ribonucleoside triphosphates. Thus using poly d-TG:AC as template, incorporation of [^{14}C]ATP was specifically stimulated by CTP, and

TABLE 3

Interdependence of the incorporation of ribonucleoside 5′-triphosphates

	Incorporation of labeled nucleoside triphosphate (mμmoles/ml.)			
	[^{14}C]ATP 60 min	[^{14}C]GTP 60 min	[^{14}C]UTP 60 min	[^{14}C]CTP 60 min
− Other nucleoside triphosphate	0·8	2·1	2·2	0·6
+ ATP	—	1·5	2·9	53·5
+ GTP	0·8	—	69·0	0·7
+ UTP	1·6	79·0	—	0·5
+ CTP	66·5	2·3	1·9	—

The reaction mixture contained 20 mμmoles of poly d-TG:AC, 0·56 mg of RNA polymerase and 400 mμmoles of nucleoside triphosphate as specified, in addition to the other components described under Materials and Methods, in 1·0 ml. of total volume. The incubation was carried out at 37°C. Specific activity of nucleoside triphosphate used: [^{14}C]ATP and [^{14}C]GTP, 4μc/μmole; [^{14}C]UTP and [^{14}C]CTP, 2 μc/μmole.

TABLE 4

Interdependence of the incorporation of ribonucleoside 5'-triphosphates

	Incorporation of labeled nucleoside triphosphate (mμmoles/ml.)							
	[14C]ATP		[14C]GTP		[14C]UTP		[14C]CTP	
	10 min	40 min	10 min	40 min	10 min	40 min	10 min	40 min
− Other nucleoside triphosphate	0·49	0·92	1·0	16·1	0·64	1·1	0·54	1·0
+ ATP			28·7	39·7	0·66	1·6	0·52	0·96
+ GTP	23·7	29·7			0·69	0·91	0·54	1·6
+ UTP	0·51	1·1	0·84	9·1			9·7	19·4
+ CTP	0·52	0·85	0·84	7·2	14·8	19·8		

The reaction mixture contained 27 mμmoles of poly d-TC:AG, 0·4 mg of RNA polymerase and 160 mμmoles of nucleoside triphosphates as specified, in addition to the other components described under Materials and Methods, in 1·0 ml. of total volume. The incubation was carried out at 37°C. Specific activity of nucleoside triphosphate used: [14C]ATP and [14C]GTP, 4 μc/μmole; [14C]UTP and [14C]CTP, 2 μc/μmole.

[14C]CTP incorporation was specifically stimulated by ATP. Similarly, the incorporation of UTP and that of GTP were interdependent. Analogous results were obtained using poly d-TC:AG as the template. Thus, no significant incorporation of a ribonucleoside triphosphate occurred unless the second triphosphate, which was required for the transcription of a particular deoxypolynucleotide strand, was also present.

However, as seen in Fig. 3 and Table 4, when GTP alone was present and poly d-TC:AG was the template, its incorporation was noted after a rather long lag period.

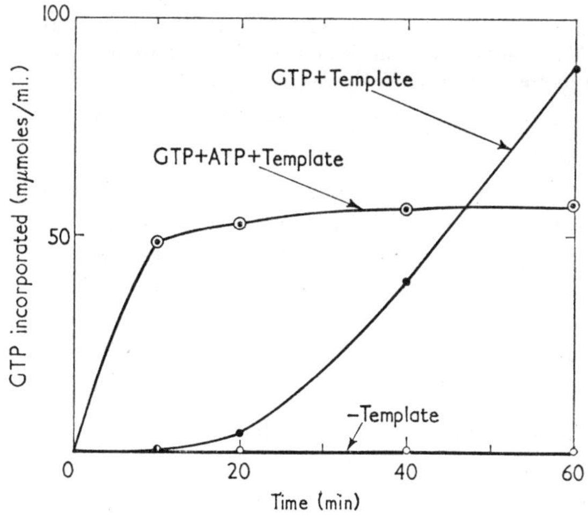

FIG. 3. GTP incorporation in the presence of poly d-TC:AG and RNA polymerase. The reaction mixture contained 0·4 mg of RNA polymerase, 42 mμmoles of poly d-TC:AG, 160 mμmoles of GTP, and 160 mμmoles of ATP where indicated in 1·0 ml. of total volume. The incubation was carried out at 37°C.

The nature of this incorporation has not been investigated further; but it should be noted that this reaction caused no interference in the synthesis of the required ribo-polynucleotide-containing alternating adenylate and guanylate units, because the synthesis of the latter reached completion well before the start of the anomolous incorporation of GTP.

(iii) *Further characteristics of ribopolynucleotide synthesis*

Figure 4 shows the influence of the concentration of RNA polymerase on the synthesis of poly UC and of poly AG with poly d-TC:AG as the template. Under the conditions used in the present work, a concentration of 0·4 mg/ml. of the enzyme gave maximum synthesis, and this is the amount used in most of the work with poly d-TC:AG template.

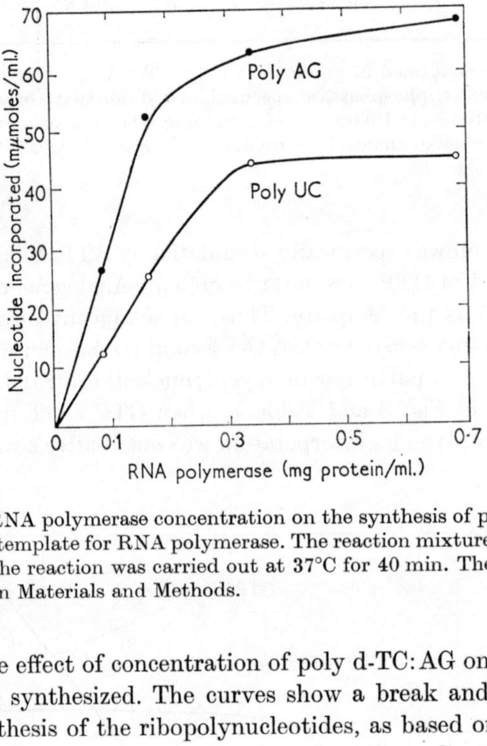

FIG. 4. Influence of RNA polymerase concentration on the synthesis of poly UC and of poly AG using poly d-TC:AG as template for RNA polymerase. The reaction mixture contained 27 mμmoles of poly d-TC:AG/ml. The reaction was carried out at 37°C for 40 min. The other conditions were the same as described in Materials and Methods.

Figure 5 shows the effect of concentration of poly d-TC:AG on the amount of poly UC and of poly AG synthesized. The curves show a break and it is clear that the extent of "net" synthesis of the ribopolynucleotides, as based on the amount of the template, is much higher at low concentrations (less than about 10 mμmoles/ml.) of the template.

(iv) *Nearest-neighbor frequency analysis of poly UC, poly AG, poly UG and poly AC*

The alternating sequence of the two component nucleotides in poly UC, poly AG, poly UG and poly AC was determined by preparation of these polymers using one of the two triphosphates ^{32}P-labeled in the α-position, subsequent alkaline hydrolysis and determination of the transfer of ^{32}P to ribonucleoside 2′(or 3′)-phosphates. Table 5 shows the conditions used for the preparation of the polymers (similar to those actually used in the first step of the *in vitro* peptide synthesis described below) and for the nearest-neighbor frequency analysis. The results obtained are given in Table 6. It is seen that in every case, the transfer of radioactivity to the expected

TABLE 5

Conditions for the preparation of ribopolynucleotides containing alternating nucleotide sequences and for nearest-neighbor frequency analysis

Template	Concentration of templates (mμmoles/ml.)	Substrates	Concentration of substrates (mμmoles/ml.)	Incubation conditions	Product	Hydrolysis conditions	Solvent for chromatography of mononucleotides
Poly d-TC:AG	40	[α-^{32}P]UTP + CTP	160 160	37°C, 30 min	Poly UC	25°C, 20 hr 0·3 M-NaOH	Solvent I
Poly d-TC:AG	40	[α-^{32}P]ATP + GTP	160 160	37°C, 15 min	Poly AG	25°C, 20 hr 0·3 M-NaOH	Solvent I
Poly d-TG:AC	4	[α-^{32}P]UTP + GTP	168 160	37°C, 40 min	Poly UG	25°C, 21 hr 0·3 M-NaOH	Solvent II
Poly d-TG:AC	4	[α-^{32}P]ATP + CTP	160 160	37°C, 90 min	Poly AC	25°C, 20 hr 0·3 M-NaOH	Solvent I

TABLE 6

Nearest-neighbor frequency analysis of poly UC, poly AG, poly UG and poly AC

Template	Substrates	Product	Radioactivity in isolated nucleotide				% Transfer	
			2′-(3′) UMP	2′-(3′) CMP	2′-(3′) AMP	2′-(3′) GMP	Found (%)	Theoretical (%)
Poly d-TC:AG	[α-^{32}P]UTP + CTP	Poly UC	70	6,229	—	—	99	100
Poly d-TC:AG	[α-^{32}P]ATP + GTP	Poly AG	—	—	357	47,880	99·3	100
Poly d-TG:AC	[α-^{32}P]UTP + GTP	Poly UG	13	—	—	3,921	99·7	100
Poly d-TG:AC	[α-^{32}P]ATP + CTP	Poly AC	—	20,601	214	—	99	100

The conditions for the nearest-neighbor frequency analysis are given in Table 5. For other details see Materials and Methods.

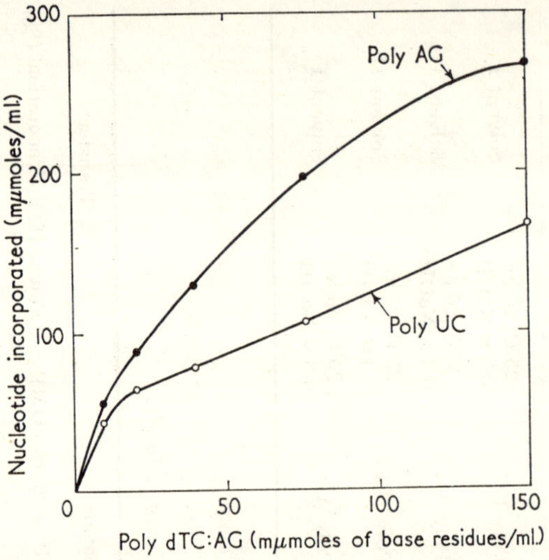

FIG. 5. Effect of concentration of poly d-TC:AG on the synthesis of poly UC and poly AG. In the case of poly UC synthesis, the reaction mixture contained 0·4 mg of RNA polymerase, 320 mμmoles each of UTP and CTP and other components described in Materials and Methods. The incubation was carried out at 37°C for 45 min. In the case of poly AG synthesis, 320 mμmoles each of ATP and GTP were used. Time of incubation was 15 min.

nucleoside 2′(3′)-phosphate was 99% or better. There is the possibility that the minor deviation from theoretical transfer was introduced by trace contamination of the ^{32}P-labeled nucleoside triphosphate by the second triphosphate of the pair. In any event, the results of nearest-neighbor frequency analysis show that the ribopolynucleotide products are made up to the extent of at least 99% of strictly alternating base sequences.

(v) *Chain-length determination of poly UC and of poly AG*

Poly UC and poly AG were prepared from the appropriate pairs of nucleoside triphosphates containing ^{14}C-label of equal specific activity (Nishimura *et al.*, 1964). The results of chain-length determination by alkaline hydrolysis are shown in Table 7.

TABLE 7

Chain-length determination of poly UC and of poly AG by alkaline hydrolysis

	Poly AG	Poly UC
Ap + Gp	4852	
A + G	43	
Up + Cp		2920
U + C		25
Chain-length	113	116

In the case of poly AG synthesis, 160 mμmoles of each of [^{14}C]ATP and [^{14}C]GTP of equal specific activity (4 μc/μmole) and 42 mμmoles of poly d-TC:AG were used. The incubation was carried out at 37°C for 30 min. In the case of poly UC synthesis, 160 mμmoles of each of [^{14}C]UTP and [^{14}C]CTP of equal specific activity (2 μc/μmole) were used. Time of the incubation was 15 min at 37°C.

The chain-length figures 113 and 116 are only approximate, because tailing of the major radioactive products (the mononucleotides) was observed on DEAE-cellulose paper chromatography and the radioactivity recorded for them is therefore on the low side; and, furthermore, the quantity of radioactive material recovered in the nucleoside region is too small to be reliable. It is clear, however, that the ribopolynucleotide products consist of long chains and are in a size range which is suitable for their use as messengers in the amino acid incorporation system.

(b) *Stimulation of the binding of [^{14}C]aminoacyl-sRNA to ribosomes by poly UC*

The results of binding experiments using poly UC are given in Fig. 6 and Table 8. Thus, poly UC stimulated the binding of [^{14}C]leucyl-sRNA (prepared from fractionated sRNA) and of [^{14}C]seryl-sRNA to ribosomes. While the results shown were

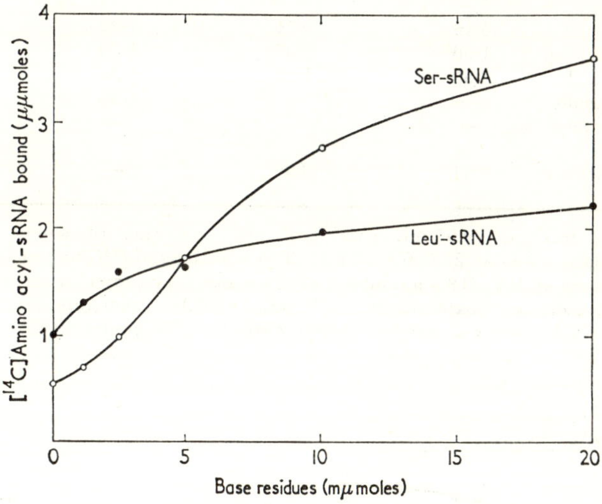

FIG. 6. Stimulation of the binding of [^{14}C]leucyl-sRNA and [^{14}C]seryl-sRNA to ribosomes by poly UC. Reaction mixtures contained each: ribosomes (2 O.D. units), [^{14}C]leucyl-sRNA (18·9 $\mu\mu$moles; 5900 cts/min), prepared as described in Materials and Methods, or [^{14}C]seryl-sRNA (28·0 $\mu\mu$moles; 5200 cts/min), prepared from unfractionated sRNA, in 0·05 ml. final volume. The incubations were at 25°C for 20 min. The concentration of magnesium ions was 0·02 M.

obtained at a concentration of 0·02 M-Mg^{2+} and incubation for 20 minutes at 25°C, the results using 0·01 M-Mg^{2+} at 37°C were similar. The binding of [^{14}C]phenylalanyl-sRNA and of [^{14}C]prolyl-sRNA to ribosomes was not stimulated by poly UC (Table 8). The binding of other aminoacyl-sRNA's has not been tested. It may further be noted from the last line of Table 6 that poly d-TC:AG was ineffective in stimulating the binding of any of the aminoacyl-sRNA's.

(c) *Serine–leucine co-polypeptide synthesis directed by poly UC*

(i) *Interdependence of serine and leucine incorporation*

Figure 7 shows the kinetics of incorporation of [^{14}C]serine and of [^{14}C]leucine in two separate experiments. It is clear that the incorporation of either of these amino acids requires the presence of the other amino acid. Although not shown in the

Figure, there was no incorporation of either [^{14}C]serine or [^{14}C]leucine in the absence of poly UC and of the second amino acid; the kinetics were superimposable on those shown in Fig. 7 for − poly UC + leucine (or + serine).

It should be noted that when serine–leucine co-polypeptide synthesis was not carried out under optimal conditions (as shown in Fig. 7) and the incorporation of each amino acid was about 0·5 to 0·8 mμmole/ml., the dependence of incorporation

TABLE 8

Stimulation of the binding of aminoacyl-sRNA's by poly UC

	[^{14}C]Aminoacyl-sRNA attached to ribosomes (μμmoles)			
	[^{14}C]Ser-sRNA	[^{14}C]Leu-sRNA	[^{14}C]Phe-sRNA	[^{14}C]Pro-sRNA
− Polymer	0·48	0·71	0·42	0·09
+ Poly UC, 5 mμmoles	1·66	1·57	0·39	0·10
10 mμmoles	2·83	1·65	0·40	0·08
+ Poly U, 10 mμmoles	—	—	6·05	—
+ Poly C, 10 mμmoles	—	—	—	0·71
+ Poly d-AG:TC, 5 mμmoles	0·51	0·90	—	—

Each incubation mixture (0·05 ml. total volume) contained ribosomes (1·0 O.D.) and 0·02 M-Mg^{2+}. Incubation was at 25°C for 20 min. The additions of [^{14}C]aminoacyl-sRNA's were as follows: [^{14}C]leucyl-sRNA (18·9 μμmoles; 5900 cts/min) prepared from fraction 114 of countercurrent distribution experiment; [^{14}C]seryl-sRNA (28·0 μμmoles; 5200 cts/min); [^{14}C]phenylalanyl-sRNA (10·8 μμmoles; 4900 cts/min); [^{14}C]prolyl-sRNA (20·3 μμmoles; 5100 cts/min).

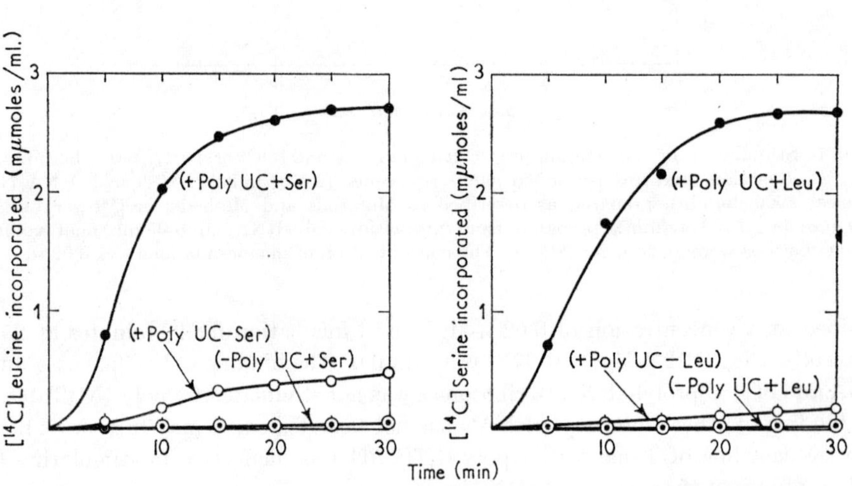

FIG. 7. Characteristics of the incorporation of [^{14}C]serine and [^{14}C]leucine into polypeptide in the presence of poly UC. The reaction mixture at the first stage contained 0·45 mg of RNA polymerase, 155 mμmoles of poly d-TC:AG, 320 mμmoles of UTP and 320 mμmoles of CTP per ml. After incubation at 37°C for 50 min, 168 mμmoles/ml. of poly UC had been synthesized. As a control experiment, the reaction mixture without adding poly d-TC:AG was incubated under the same conditions as above. These reaction mixtures were taken to the second stage as described in Materials and Methods. In the second-stage reaction, 3 μC of [^{14}C]leucine (15·1 mμmoles), 3 μC of [^{14}C]serine (24·5 mμmoles), 62 mμmoles of cold leucine and 62 mμmoles of cold serine were added where indicated.

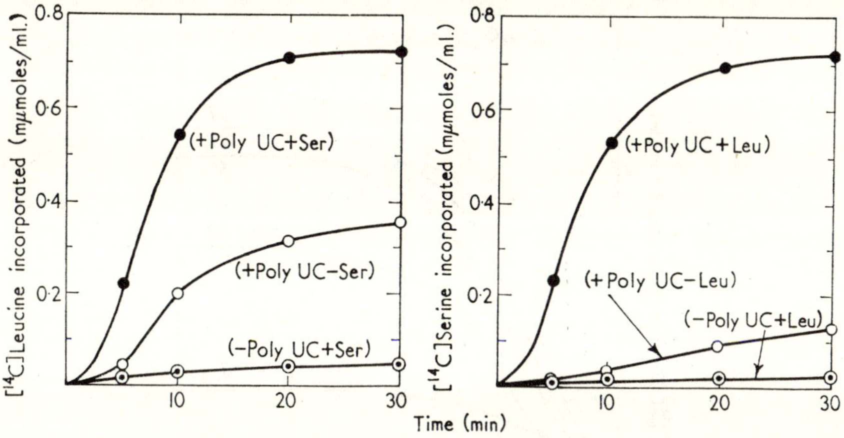

FIG. 8. Incorporation of [^{14}C]leucine and [^{14}C]serine using low concentration of poly UC template. The reaction mixture at the first stage contained 40 mμmoles of poly d-TC:AG/ml. After incubation at 37°C for 50 min, 43 mμmoles of poly UC were synthesized per ml. The reaction mixture was taken to the second stage as described in Materials and Methods. In the second-stage reaction, 1 μc of [^{14}C]leucine (5 mμmoles), 1 μc of [^{14}C]serine (8·1 mμmoles), 50 mμmoles of cold leucine and 50 mμmoles of cold serine were used where indicated.

of [^{14}C]leucine on the presence of serine was not so clearly demonstrable (Fig. 8). It is probable that this incorporation of [^{14}C]leucine was permitted by endogenous serine and that the product formed again was leucine–serine co-polypeptide. This conclusion was supported by the susceptibility of the product to acid hydrolysis. No evidence for the formation of [^{14}C]homopolyleucine was obtained.

Two other features of the experiment of Fig. 7 were: (1) the incorporation of [^{14}C]leucine was equivalent in molar proportion to that of [^{14}C]serine; and (2), when poly UC isolated from the RNA polymerase experiment (see Materials and Methods) was tested for the stimulation of incorporation of [^{14}C]leucine or [^{14}C]serine, the results were negative. The result could be ascribed to the increased exposure of poly UC to degradative enzymes during the isolation procedure. This aspect requires further study.

(ii) *Further characteristics of serine–leucine co-polypeptide synthesis*

Figure 9 shows the effect of concentration of poly UC on the synthesis of serine–leucine co-polypeptide. It is seen that the amino acid incorporation increased with increasing concentration of poly UC and that, even at the highest concentration (84 mμmoles/ml.) tried, the system was not completely saturated; although the effect of increase of the template in this range of concentration was not so marked.

Table 9 further describes the requirements of serine–leucine co-polypeptide synthesis. It is clear that: (1) the incorporation of serine and leucine is due to poly UC; (2) deoxypolynucleotide template does not directly serve to stimulate the incorporation of the amino acids; and (3) poly AG synthesized at the first stage does not stimulate the incorporation of serine and leucine. This last finding completely rules out the possibility that poly AG, which might have been formed during the conditions of the second-stage reaction (when ATP and GTP, RNA polymerase and poly d-TC:AG are present), brought about the incorporation of the above two amino acids.

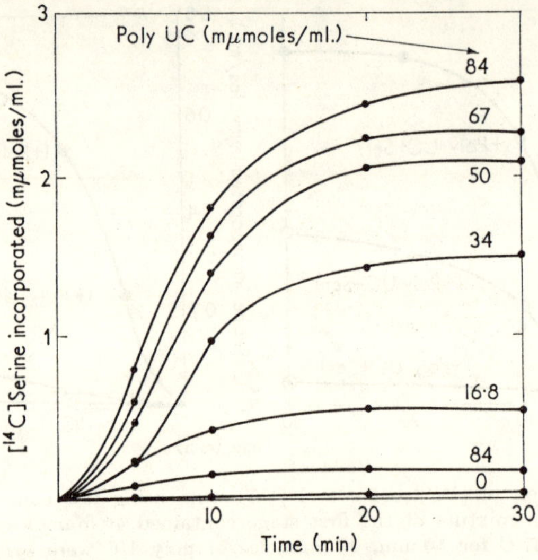

FIG. 9. Effect of concentration of poly UC on the synthesis of serine–leucine polypeptide. The conditions of the first-stage reaction were the same as described in Fig. 7. After the completion of poly UC synthesis, the reaction mixture was diluted, wherever necessary, by adding appropriate amounts of the buffer for RNA polymerase reaction (0·04 M-tris–HCl, pH 7·9, 0·004 M-$MgCl_2$, 0·001 M-$MnCl_2$ and 0·012 M-β-mercaptoethanol) to give the final concentrations of poly UC as indicated on the individual curves in the second-stage reaction. In the second-stage reaction, 36 mμmoles of [^{14}C]serine (2 μc) and 50 mμmoles of cold leucine were used per ml. of reaction mixture.

TABLE 9

Requirements for the synthesis of serine–leucine co-polypeptide

Components at first stage	[^{14}C]Serine incorporated at second stage (μμmoles/ml.)
RNA polymerase + UTP + CTP + poly d-TC:AG	495
RNA polymerase + UTP + CTP	17·0
Poly d-TC:AG + UTP + CTP	14·8
RNA polymerase + GTP + ATP + poly d-TC:AG	15·3
RNA polymerase + GTP + ATP	15·0

The reaction mixture at the first stage contained 40 mμmoles of poly d-TC:AG, 0·4 mg of RNA polymerase and 160 mμmoles of each of the nucleoside triphosphates as specified. Poly UC synthesized amounted to 78 mμmoles/ml. after 45 min incubation; poly AG synthesized was 124 mμmoles/ml. after 15 min incubation. The control reactions without the addition of RNA polymerase or poly d-TC:AG template were carried out under the same incubation conditions. In the second stage reaction, the reaction mixture contained 0·5 μc of [^{14}C]serine (4·1 mμmoles) and 125 mμmoles of cold leucine in addition to other components as described under Materials and Methods; time of incubation at second stage was 20 min. Assay was by measurement of radioactivity of material insoluble in hot trichloroacetic acid.

(iii) *Absence of incorporation of other amino acids in the presence of poly UC*

The possibility of incorporation of amino acids other than leucine and serine was tested by providing in each reaction mixture one labeled amino acid and the remaining 19 cold amino acids. The results are given in Table 10. It is seen that, within the sensitivity of the assay, no significant incorporation of any other amino acid was detected. It is therefore concluded that poly UC stimulates the incorporation of only serine and leucine.

TABLE 10

[^{14}C]L-amino acid	Incorporation of amino acid (cts/min/0·02 ml.) at second stage	
	− Poly d-TC:AG in stage I	+ Poly d-TC:AG in stage I
Serine	297	2946
Leucine	780	7067
Alanine	321	320
Arginine	463	437
Aspartic acid	110	110
Asparagine	54	52
DL-Cystine	73	55
Glutamic acid	74	124
Glutamine	16	22
Glycine	169	203
Histidine	137	138
Isoleucine	324	301
Lysine	475	422
Methionine	15	14
Phenylalanine	459	359
Proline	175	192
Threonine	402	200
Tryptophan	9	11
Tyrosine	331	303
Valine	505	535

The reaction mixture at the first stage contained 0·45 mg of RNA polymerase, 155 mμmoles of poly d-TC:AG, 320 mμmoles of UTP and 320 mμmoles of CTP in 1 ml. The control experiments omitted poly d-TC:AG in the first stage. Incubation of the first-stage reaction was at 37°C for 50 min. In the second-stage reaction, each reaction mixture contained one [^{14}C]labeled amino acid + 19 cold amino acids. The amount of the labeled amino acid was 1 μc/ml., except for methionine and tryptophan which were added in the amount of 0·05 μc; the amount of each cold amino acid added was 100 mμmoles/ml. Other components added were as described under Materials and Methods. The incubation at the second-stage reaction was at 37°C for 30 min. Assay was by measurement of radioactivity of material insoluble in hot trichloroacetic acid.

(d) *Characterization of serine–leucine co-polypeptide*

(i) *Kinetics of acidic hydrolysis of co-polypeptide to seryl-leucine dipeptide*

Serine–leucine co-polypeptide containing both amino acids labeled with ^{14}C was subjected to acidic hydrolysis (see Materials and Methods) and the hydrolysis was followed by paper chromatography in solvent III. Portions removed after 6, 11, 25 and 33 hours gave the patterns of radioactivity shown in Fig. 10. It is noted that even very early portions gave, presumably, oligopeptides which had high mobility in this solvent system. The starting polypeptide, in contrast, had zero mobility in this solvent. This is further shown in Fig. 11(a). After treatment with 10 N-HCl for five days, a single sharp band, which is shown below to correspond to seryl-leucine, was obtained (Fig. 11(b)).

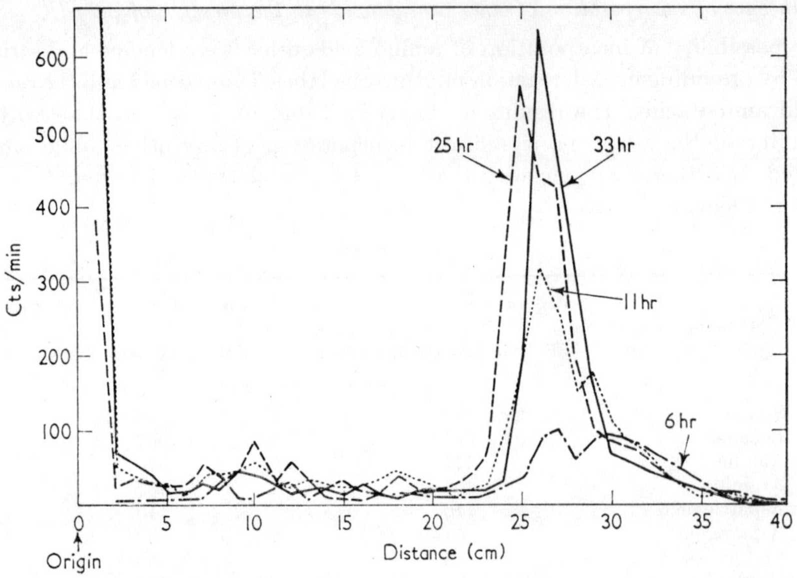

FIG. 10. Kinetics of acidic hydrolysis of [^{14}C]serine–leucine co-polypeptide. The polypeptide was kept in 10 N-HCl at 30°C and portions which were removed at different intervals were chromatographed in solvent III. The chromatogram was cut into 1-cm wide strips and counted for radioactivity in the scintillation counter.

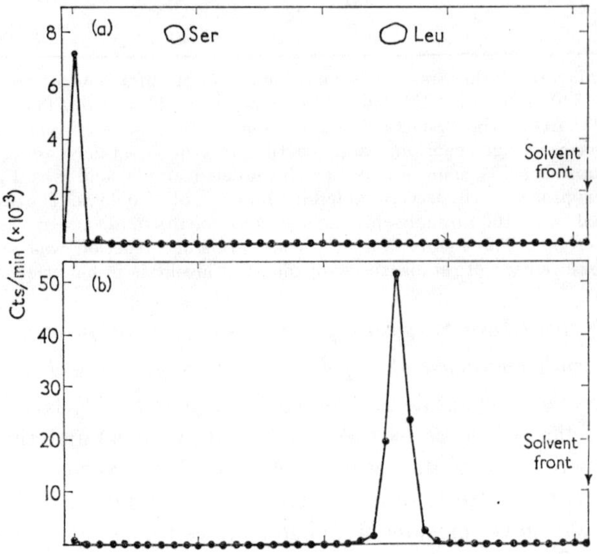

FIG. 11. Selective acidic hydrolysis of [^{14}C]serine–leucine co-polypeptide to serine–leucine dipeptide. (a) Co-polypeptide as isolated from amino acid incorporation experiment as described under Materials and Methods; (b) product (seryl-leucine) formed after 5 days at 30°C in 10 N-HCl. Chromatograms were developed in solvent III.

(ii) *Characterization of seryl-leucine dipeptide: homogeneity and amino end-group determination*

The [^{14}C]dipeptide obtained above was homogeneous as examined by paper chromatography in two additional solvents: solvent V, $R_F = 0{\cdot}11$, R_F of leucine $= 0{\cdot}23$; solvent VI, $R_F = 0{\cdot}80$, R_F values of leucine and serine $= 0{\cdot}90$ and $0{\cdot}42$, respectively.

The dipeptide was treated with 1-fluoro-2,4-dinitrobenzene and the product was chromatographed in solvent III. The pattern of radioactivity obtained (Fig. 12) showed completion of reaction with the reagent. In solvent VII, the dinitrophenyl-dipeptide had an R_F of 0·41; the R_F values for dinitrophenyl-leucine and dinitrophenyl-serine were 0·60 and 0·22, respectively.

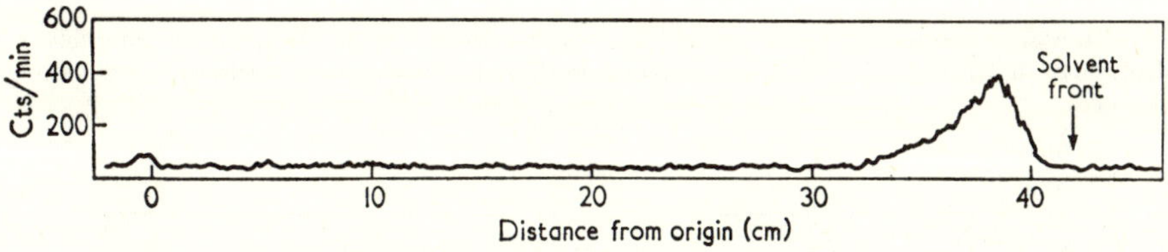

FIG. 12. Paper chromatography (solvent III) of the product of reaction of seryl-leucine dipeptide (Fig. 11(b)) with 1-fluoro-2,4-dinitrobenzene.

The dinitrophenyl-dipeptide thus obtained was subjected to complete hydrolysis and the dinitrophenyl-amino acid obtained was chromatographed along with markers of dinitrophenyl-serine and dinitrophenyl-leucine in solvent V. As seen in Fig. 13, the only ^{14}C-labeled product corresponded to dinitrophenyl-serine.

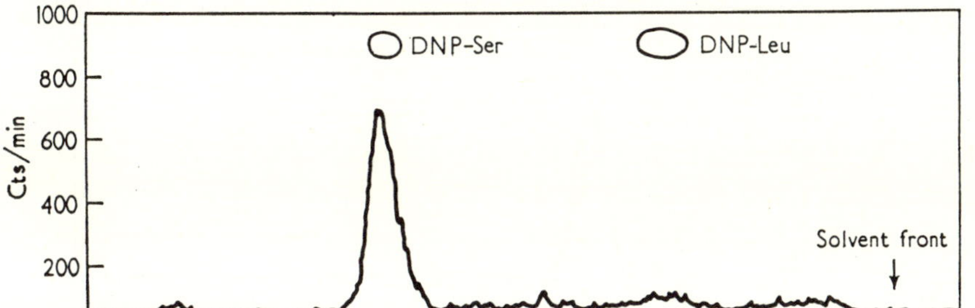

FIG. 13. Paper chromatography (solvent V) of dinitrophenyl-amino acid obtained on complete acidic hydrolysis of dinitrophenyl-seryl-leucine (Fig. 12). Dinitrophenyl-serine and dinitrophenyl-leucine were run alongside as markers.

Paper chromatography in solvent III of the free amino acid fraction along with marker serine and leucine gave the results shown in Fig. 14. Thus it is seen that the major radioactive product corresponded to leucine. The formation of a small amount of [^{14}C]serine was also noted; this probably arose from the partial breakdown of dinitrophenyl-serine during the conditions necessary for acidic hydrolysis of the peptidic linkage (Porter, 1957).

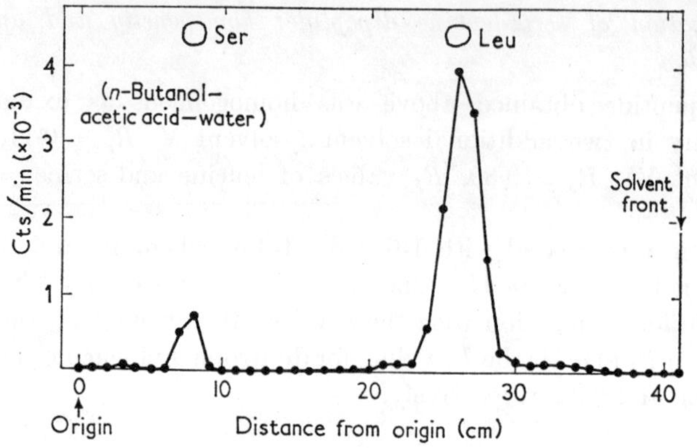

FIG. 14. Paper chromatography (solvent III) of free amino acids obtained on complete acidic hydrolysis of dinitrophenyl-seryl-leucine dipeptide (Fig. 12). Serine and leucine were run alongside as markers.

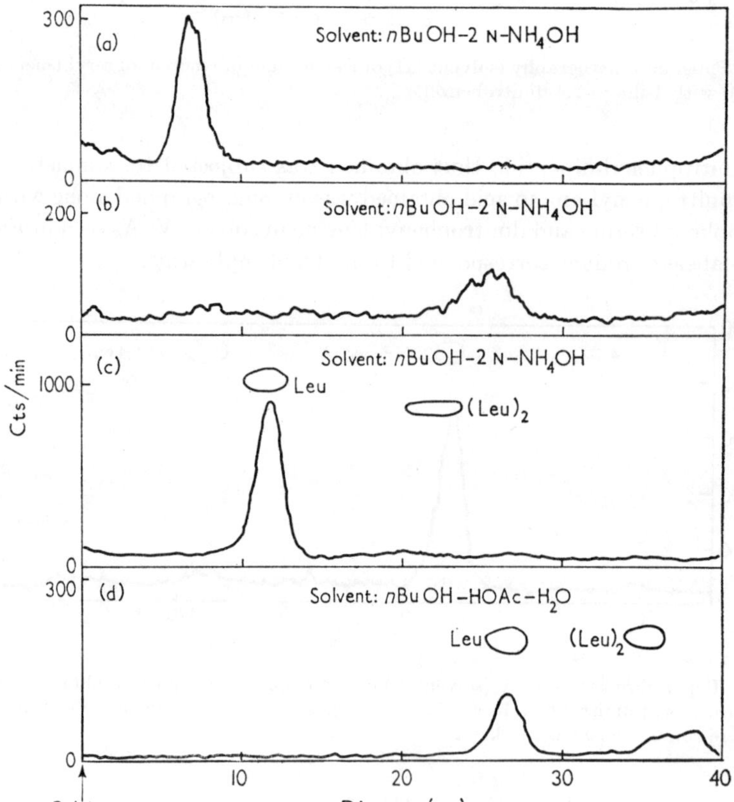

FIG. 15. Characterization of seryl-leucine dipeptide by Edman degradation. (a) Seryl-leucine dipeptide before treatment with phenyl isothiocyanate (paper chromatography in solvent V); (b) product formed on reaction of seryl-leucine dipeptide with phenyl isothiocyanate (paper chromatography in solvent V); (c) free amino acid released after thiohydantoin formation step in Edman degradation (paper chromatography in solvent V); (d) same as (c), except that chromatography was in solvent VIII.

(iii) *Characterization of seryl-leucine dipeptide: Edman degradation*

After reaction of the dipeptide with phenyl isothiocyanate, the products were chromatographed in solvent V. The pattern of radioactivity (Fig. 15(b)) showed completion of reaction with the reagents. (Figure 15(a) shows the position of the dipeptide before treatment with the reagent.)

The dipeptide derivative was treated with acid and the product was chromatographed in solvents V and VIII. As seen in Fig. 15(c), leucine was the only product formed after the degradative reaction. The radioactivity pattern in Fig. 15(d) again showed leucine as a product of reaction, but in addition showed material moving near to the solvent front, presumably the cyclization product of the N-terminal serine.

4. Discussion

The combined use of organochemical methods (Khorana *et al.*, 1965) and of enzymic methods has permitted the synthesis of long ribopolynucleotides of known repeating di- and trinucleotide sequences. The approach previously described (Nishimura *et al.*, 1964; Mehrotra & Khorana, 1965) involved the use of short synthetic deoxyribopolynucleotides containing repeating nucleotide sequences as templates for DNA-dependent RNA polymerase. The second approach has involved, as the first step, the use of short-chain deoxyribopolynucleotides as templates for DNA-polymerase, so as to obtain DNA-like polymers of defined sequences; and, in the second step, the use of the latter as templates for RNA polymerase. The present study of poly d-TC:AG and of poly d-TG:AC as templates for RNA polymerase has further demonstrated that the enzyme can "copy" only one strand of a double-stranded DNA molecule. This property has enabled the preparation of single-stranded ribopolynucleotides with repeating dinucleotide sequences; these are of interest as messengers in the *in vitro* synthesis of polypeptides of defined sequences.

The DNA-like polymers, poly d-TC:AG and poly d-TG:AC, offer opportunities for further studies of the mechanism of the action of RNA polymerase. With the demonstration that, in the presence of the two appropriate ribonucleoside 5′-triphosphates, either of the two strands may be copied, a very convenient system becomes available for the study of the mechanism of strand selection by the RNA polymerase. Furthermore, the DNA-like polymers may prove to be useful for further studies of the mode of inhibition of RNA synthesis by antibiotics such as the actinomycins.

The most important result of the study of poly UC as messenger in the bacterial cell-free system has been the demonstration of the synthesis of the co-polypeptide containing serine and leucine in alternating sequence. The first suggestions for this structure of the polypeptide came from the molar equivalence of the incorporation, and from the interdependence of the incorporation of the two amino acids. Conclusive evidence for the structural assignment was obtained in a variety of ways. Thus, the total polypeptidic material could be degraded quantitatively to the dipeptide, seryl-leucine, by specific cleavage at every serine–amino group of the polypeptide chain. It is emphasized that all of the original radioactivity in the polypeptide was accounted for by the dipeptide, no free serine or any other amino acid or peptide being produced. Seryl-leucine itself was characterized as pure by extensive chromatography, both as the free dipeptide and as its dinitrophenyl derivative. The identification of the dipeptide itself was accomplished by reaction with dinitrofluorobenzene followed by acidic hydrolysis. Dinitrophenyl-serine was the only dinitrophenyl-amino acid produced, and leucine was the major free amino acid released. The small amount of serine detected almost certainly came from the partial breakdown of dinitrophenyl-serine during the acidic hydrolysis of the dinitrophenyl-dipeptide. The amount formed is consistent with the rates of acidic breakdown for this dinitrophenyl-amino acid given by Porter (1957). The identification of seryl-leucine was further confirmed by applying the Edman degradation. In this degradation, after the removal of the terminal serine as the thiohydantoin, leucine was released as the free amino acid. The latter amino acid was identified and shown to be distinct from leucyl-leucine by paper chromatography. The total evidence therefore firmly established the structure of the dipeptide as seryl-leucine and hence that of the original polypeptide as a co-polypeptide containing serine and leucine in strictly alternating sequence.

The co-polypeptide appears to be very long, although no direct estimate of the size is available. The only evidence bearing on this point was that the total product formed in the experiment on amino acid incorporation had zero mobility on paper chromatograms (solvent III). In contrast, the oligopeptides formed even very early during partial acidic hydrolysis had very high mobility in the same solvent system and, in fact, the oligomers, larger than the dipeptide, travelled faster than the dipeptide. From this pattern of mobilities, it may be concluded that the chain length of the original material with zero mobility was high.

The incorporation of amino acids other than leucine and serine was looked for, but none was detected at the limits of the sensitivity of the experiment. The present results therefore indicate absence of ambiguity and, furthermore, the fact that a co-polypeptide of two amino acids was obtained strengthens confidence in the precision of the "read-out" mechanism in the *in vitro* system.

Of the two repeating codon sequences, UpCpU and CpUpC, the former has previously been assigned to serine on the basis of the stimulation of seryl-sRNA to ribosomes (Bernfield & Nirenberg, 1965) and this assignment is consistent with amino acid replacement data. The codon sequence, CpUpC, is now assigned to leucine. It should be noted that the trinucleotide, CpUpC, failed to stimulate the binding of leucyl-sRNA to ribosomes (Bernfield & Nirenberg, 1965).

Experiments to be reported subsequently (Jones, Nishimura & Khorana, manuscript in preparation) have shown that poly AG directs the synthesis of a co-polypeptide of arginine and glutamic acid; poly UG directs the synthesis of a co-polypeptide containing cysteine and valine; and poly AC directs the synthesis of a co-polypeptide containing threonine and histidine. The formation of co-peptides containing two amino acids in alternating sequence under the direction of ribopolynucleotides containing two nucleotides in alternating sequence is therefore a general feature of the amino acid incorporation system.

Finally, the present results provide direct proof that the amino acid code must consist of an odd number of letters. Taken together with the findings reported in the preceding paper, namely, the formation of homopeptides under the direction of a ribopolynucleotide containing a repeating trinucleotide sequence, the results almost certainly establish the three-letter and non-overlapping property of the amino acid code.

This work has been supported by grants from the National Science Foundation (no. GB–976), the National Cancer Institute of the National Institutes of Health, U.S. Public Health Service (no. CA–05178) and the Life Insurance Medical Research Fund (grant no. G–62–54).

REFERENCES

Bernfield, M. R. & Nirenberg, M. W. (1965). *Science*, **147**, 479.

Byrd, C., Ohtsuka, E., Moon, M. W. & Khorana, H. G. (1965). *Proc. Nat. Acad. Sci., Wash.* **53**, 79.

Chamberlin, M. & Berg, P. (1962). *Proc. Nat. Acad. Sci., Wash.* **48**, 81.

Chamberlin, M. & Berg, P. (1964). *J. Mol. Biol.* **8**, 708.

Desnuelle, P. & Casal, A. (1948). *Biochim. biophys. Acta*, **2**, 64.

Gray, W. R. & Hartley, B. S. (1963). *Biochem. J.* **89**, 379.

Holley, R. W., Apgar, J., Everett, G. A., Madison, J. T., Merrill, S. H. & Zamir, A. (1963). *Cold Spr. Harb. Symp. Quant. Biol.* **28**, 117.

Jacob, T. M., Ohtsuka, E., Moon, M. W., Narang, S. A. & Khorana, H. G. (1964). *Fed. Proc.* **23**, 531.

Khorana, H. G., Jacob, T. M., Moon, M. W., Narang, S. A. & Ohtsuka, E. (1965). *J. Amer. Chem. Soc.*, in the press.

Levy, A. L. (1954). *Nature*, **174**, 126.

Mans, R. J. & Novelli, G. D. (1961). *Arch. Biochem. Biophys.* **94**, 48.

Mehrotra, B. D. & Khorana, H. G. (1965). *J. Biol. Chem.* **240**, 1750.

Nirenberg, M. W. & Leder, P. (1964). *Science*, **145**, 1399.

Nishimura, S., Jacob, T. M. & Khorana, H. G. (1964). *Proc. Nat. Acad. Sci., Wash.* **52**, 1494.

Nishimura, S., Jones, D. S., Ohtsuka, E., Hayatsu, H., Jacob, T. M. & Khorana, H. G. (1965). *J. Mol. Biol.* **13**, 283.

Nishimura, S., Jones, D. S., Wells, R. D., Jacob, T. M. & Khorana, H. G. (1965). *Fed. Proc.* **24**, 409.

Ohtsuka, E., Moon, M. W. & Khorana, H. G. (1965). *J. Amer. Chem. Soc.*, in the press.

Porter, R. R. (1957). In *Methods in Enzymology*, ed. by S. P. Colowick & N. Kaplan, vol. 4, p. 221. New York: Academic Press.

Waley, S. G. & Watson, J. (1953). *Biochem. J.*, **55**, 328.

Weisblum, B., Benzer, S. & Holley, R. W. (1962). *Proc. Nat. Acad. Sci., Wash.* **48**, 1449.

Wells, R. D., Ohtsuka, E. & Khorana, H. G. (1965). *J. Mol. Biol.* in the press.

Wood, W. B. & Berg, P. (1962). *Proc. Nat. Acad. Sci., Wash.* **48**, 94.

Reprinted from SCIENCE, Vol. 149, 1965

Reports

Suppression in vitro: Identification of a Serine-sRNA as a "Nonsense" Suppressor

Abstract. *In a cell-free system, with RNA from a suppressible mutant of bacteriophage R17 as messenger, no functional coat protein of the bacteriophage is synthesized unless serine-accepting soluble RNA from the suppressor strain,* Escherichia coli *S26R1E, is present.*

The molecular mechanism by which the amber suppressor genes in *Escherichia coli* reverse the effects of mutations in other genes recently has been shown to be a misreading of the genetic code. The amber suppressor genes constitute one class of suppressors; there is evidence that other suppressors in *E. coli* (*1*, *2*) and in yeast (*3*) act in a similar manner.

Mutations suppressible by amber suppressor genes can occur in many different cistrons, regardless of the cistron's function. These amber mutations—mutations suppressible by amber suppressor genes—have been isolated in the rII region of bacteriophage T4 (*4*), in the entire T4 genome (*5*), in the *E. coli* structural gene specifying alkaline phosphatase (*6*), in the bacteriophage lambda (*7*), and in the RNA-containing (*8*) bacteriophage f2 (*9*).

Benzer and Champe (*10*) and Garen and Siddiqi (*6*) first suggested that the amber mutation produces a nonsense codon—a nucleotide triplet which does not code for any amino acid and thus interrupts polypeptide synthesis in a nonpermissive cell (one lacking an amber suppressor). They further postulated that this codon would be read as an amino acid code word in a permissive (suppressor-containing) host, thereby permitting protein synthesis to continue. This hypothesis received strong support from the work of Brenner and his colleagues, who showed that amber mutants in the head protein of bacteriophage T4D produce a normal yield of released chain fragments in a nonpermissive *E. coli* host, but make a mixture of about 60 percent complete protein and 40 percent fragments in the permissive host (*11*).

Four genetically distinguishable amber suppressor genes in *E. coli* have been identified (*12*). The action of one of these, the *su-A* suppressor gene contained in strain S26R1E of Garen and Siddiqi, has been investigated extensively. In the T4D head protein (*13*), in alkaline phosphatase (*12*), and in f2 coat protein (*14*), a glutamine or tryptophan residue in the wild-type protein is always replaced by serine in the mutant protein produced in the *su-A*–containing host. Thus, the amber nonsense codon has arisen through mutations from triplets coding for glutamine and tryptophan. An alteration of the protein-synthesizing machinery of the permissive host has changed its ability to recognize the nonsense codon, so that part of the time it is recognized as a code word for serine.

Since amber suppression is based on the misreading of the code in the permissive cell, an alteration of any component that guarantees the accuracy of reading might result in suppression. Four possibilities are: (i) an altered aminoacyl synthetase which transfers an amino acid to a class of sRNA molecules which normally recognizes the nonsense triplet, but does not bind any amino acid; (ii) a new species of sRNA specific for a particular amino acid, yet capable of recognizing the nonsense codon; (iii) mutant ribosomes which distort the natural mRNA-sRNA complex, thereby permitting the recognition of a nonsense codon by a normally occurring sRNA species (*2*); and (iv) the presence of base analogs or an inaccurate RNA polymerase, which would produce an altered messenger (for example, fluoracil inserted in place of uracil) (*4*).

We show here that the component active in suppression in strain S26R1E is a serine-accepting sRNA species not present in the nonpermissive isogenic strain S26. This result was made possible by the development of an assay, in vitro, for suppression by S26R1E of an amber mutation in the coat-protein cistron of the RNA-containing bacteriophage R17. The RNA isolated from this mutant directs the synthesis of functional coat protein in a cell-free protein-synthesizing system derived entirely from the permissive host S26R1E. No functional protein is produced in a system derived from S26. However, a mixed system, consisting of serine-accepting sRNA from S26R1E added to an otherwise nonpermissive system, produces functional coat protein.

Preparation of the assay system. Escherichia coli bacterial strains S26 and S26R1E, an isogenic pair obtained from A. Garen, are nonpermissive (Su−) and permissive (Su+), re-

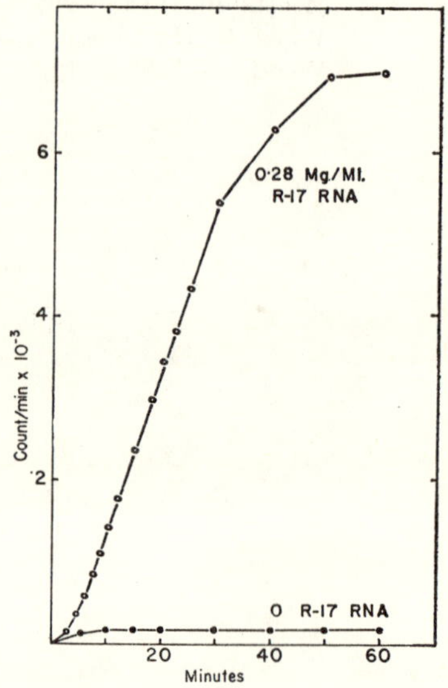

Fig. 1. Kinetics of lysine incorporation in vitro in the presence and absence of R17 RNA. The 1-ml reaction mixture contained 0.25 ml of "preincubated" *E. coli* B S-30, 0.8 mg of stripped sRNA from *E. coli* B, and 0.5 μc of C^{14}-lysine (1.93 $\times 10^7$ count min^{-1} $\mu mole^{-1}$). Each point represents the radioactive material in 50 μl of reaction mixture, which was precipitable by hot trichloroacetic acid (TCA).

Table 1. Capacity of sRNA's to accept amino acids; results in mμmole of amino acids per milligram of sRNA. The conditions used for these assays were those of Berg *et al.* (*26*).

Amino acid	sRNA		
	Su+	Serine-accepting	Leucine-accepting
Alanine	2.21	0.024	0.025
Arginine	2.14	.026	.022
Isoleucine	0.895	.015	.083
Leucine	3.51	.028	3.26
Serine	1.81	1.64	0.027

spectively. *Escherichia coli* strain C600G, a permissive host used as an indicator strain by Campbell (*7*), was obtained from M. Meselson and converted to F+ so that it would support the growth of R17. *Escherichia coli* B and *E. coli* Hfr_1 are nonpermissive strains. Strain Hfr was obtained with a stock of R17 from A. Graham.

For the preparation of *E. coli* extracts (S-30) (*8*), cells were grown in a glycerol-casamino acid medium (*15*). Cells were harvested during exponential growth, washed twice in tris-magnesium buffer (0.01*M* magnesium acetate and 0.01*M* tris, *p*H 7.5), and frozen until used. The S-30 extracts were prepared according to the method of Nirenberg and Matthaei (*16*) except for two modifications: (i) NH_4Cl was used in place of KCl and (ii) 1.5 ml of tris-magnesium buffer per gram of cells was used instead of 3.0 ml per gram.

The S-30 extracts were "preincubated" (*8*) with an ATP-generating system, chilled, and immediately used in incorporation experiments. For cell-free protein synthesis, the reaction mixtures contained per milliliter: one-fourth volume of preincubated S-30, 0.003 mmole ATP, 0.002 mmole GTP, 0.005 mmole phosphoenolpyruvate, 20 μg pyruvate kinase, 4 × 10^{-5} mmole of each amino acid, 0.010 mmole magnesium actetate, 0.075 mmole of NH_4Cl, 0.05 mmole tris (*p*H 7.8), 0.01 mmole glutathione, stripped sRNA (*8*), and R17 RNA. The temperature of incubation was 36°C.

Total sRNA was prepared by the procedure of Zubay (*17*). The biological activity of all sRNA species except those protected by the attachment of a chosen amino acid to the sRNA terminal adenylic acid residue was destroyed by oxidation with periodate. Stripped Su+ sRNA was charged with (esterified to) either serine or leucine, saturating amounts of Su− activating enzymes being used. The reaction was stopped after 20 minues at 35°C by precipitation with ethanol. The precipitate was suspended in distilled water and dialyzed for 2 hours at 4°C. The RNA solution was then centrifuged at 15,000 rev/min for 20 minutes in a Serval SS-2 centrifuge to remove denatured protein. The supernatant was made 0.1*M* with sodium acetate, *p*H 4.7. A 12-fold excess of sodium periodate was added to the sRNA solution, and the oxidation was allowed to proceed for 30 minutes at 30°C in the dark. The reaction was stopped by the addition of excess glucose, and the mixture was precipitated with ethanol. The precipitate was dissolved in distilled water, and the solution was dialyzed for 24 hours. To test the capacity of the oxidized sRNA preparations to accept amino acid, the sRNA's were first stripped by incubating with 0.1*M* tris, *p*H 8.8, at 37°C for 3 hours. The oxidation procedure was adopted to assure protection of sRNA charged with serine. P. C. Zamecnik pointed out to us that the normal procedure for periodate oxidation is not successful because of the oxidation of the amino and hydroxyl groups of serine and the subsequent lability of the oxidation product.

Bacteriophage R17 was grown and purified according to the procedure of Gesteland and Boedtker (*15*). Titers of 1 to 3 × 10^{12} pfu/ml (*8*) were routinely obtained. The R17 RNA was extracted with phenol and then precipitated from the extract with ethanol. The ratio of the absorbancy at 260 mμ to that at 280 mμ of the R17 RNA so prepared is between 1.9 and 2.0.

For the isolation of R17 amber mutants, a concentrated phage suspension was incubated in 0.2*M* $NaNO_2$, *p*H 4.6, at 37°C for 8 to 9 hours. After dilution to stop the reaction, the suspension of phage was plated on S26R1E; individual plaques were picked, and each was tested for growth on S26. After nitrous acid treatment of a phage suspension which originally contained 10^{13} viable particles per milliliter, 10^5 particles per milliliter survived. Among the survivors, about 1 to 3 in 500 were mutants.

For the growth of the R17 mutant am11B, a mutant stock containing 2 × 10^{11} pfu/ml, 0.8 percent revertants, was used to infect 16-liter cultures at a phage–bacteria ratio of about 3 to 1, and an optical density of 0.6 to 0.8. Stocks thus obtained all contain 5 to 7 × 10^{11} pfu/ml, with 1 to 2 percent revertants. The phage particles were purified in the same manner as wild-

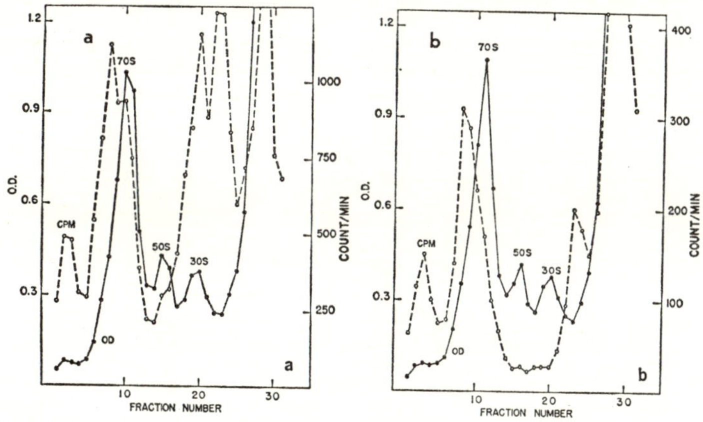

Fig. 2. Sucrose-gradient analysis of the total amino acid–incorporating system after 20-minute incubation with R17 RNA and either C^{14}-lysine and C^{14}-arginine or H^3-histidine. The 200-μl reaction mixture contained 50 μl of preincubated *E. coli* B S-30, 104 μg of R17 RNA, 160 μg of stripped *E. coli* B sRNA. (*a*) The reaction mixture contained 0.2 μc of C^{14}-lysine (7.25 × 10^7 count min^{-1} $\mu mole^{-1}$) and 0.2 μc of C^{14}-arginine (7.75 × 10^7 count min^{-1} $\mu mole^{-1}$); (*b*) the reaction mixture contained 5 μc of H^3-histidine (5.25 × 10^8 count min^{-1} $\mu mole^{-1}$). A 100-μl portion of the reaction mixture was layered on a 5-ml linear sucrose gradient (5 to 20 percent) and centrifuged at 38,000 rev/min for 2.3 hours at −12°C. Each fraction was first assayed for optical density at 260 mμ and then treated with 6 percent TCA at 90°C for 20 minutes to prepare the samples for radioactivity analysis. Of the acid-insoluble (hot TCA) radioactive material originally put on the gradient, 85 percent was recovered. The optical density of the 30*S* peaks in this and succeeding experiments is higher than might be expected because of the absorption of phage RNA.

type phage and the percentage of revertants remained constant throughout this procedure.

Amino acids labeled with C^{14} or H^3 were obtained from New England Nuclear Corporation, Boston, Mass. The specific activities of the C^{14} amino acids in $\mu c/\mu$mole were: alanine, 100; arginine, 239; isoleucine, 221; lysine, 210; serine, 120; H^3-histidine had a specific activity of 1.1 mc/μmole. Pyruvate kinase, ribonuclease, and deoxyribonuclease were obtained from Worthington Biochemical Corporation.

Development of the assay for amber suppression. The kinetics of C^{14}-lysine incorporation in vitro, with a preincubated S-30 extract in the presence and absence of added R17 RNA, are shown in Fig. 1. Under optimum conditions, there is a 30- to 50-fold stimulation of amino acid incorporation with the addition of R17 RNA.

If we assume that the product synthesized in vitro has an amino acid composition similar to that of R17 coat protein (*18*), then we can calculate that approximately 260 amino acids per ribosome are being incorporated into protein. This corresponds to two molecules of phage coat protein per ribosome in the reaction mixture. From Fig. 1, one can see that preincubation virtually eliminates protein synthesis mediated by endogenous *E. coli* messenger. All experiments reported here were performed with a preincubated S-30 homogenate.

After 20 minutes of incubation with R17 RNA as messenger, newly synthesized polypeptide sedimented in six discrete regions, with sedimentation constants of approximately 112*S*, 85*S*, 70*S*, 30*S*, 20*S*, and less than 4*S* (Fig. 2*a*).

The radioactivity which is associated with fractions sedimenting at 70*S*, 85*S*, and 112*S* represents newly made chains attached to ribosomes. The 85*S* and 112*S* peaks arise from the combination of an R17 RNA chain with one and two ribosomes, respectively; the 70*S* peak represents ribosomes from which the RNA messenger has been detached or degraded, but which have not released their polypeptide product. All the more slowly sedimenting material consists of protein products released from ribosomes. Particularly relevant is the newly synthesized material which sediments at 30*S*.

We believe that this material consists of complete coat protein for the following reasons: (i) autoradiograms of peptides resulting from digestion of the 30*S* material by trypsin reveal a peptide pattern (fingerprint) identical to that obtained from coat protein (*19*); (ii) sucrose-gradient analysis of cell-free reaction mixture in which the radioactive label was exclusively in histidine reveals (Fig. 2*b*) that no radioactivity is present in the 30*S* peak. This is in agreement with the fact that the coat protein of R17, like that of f2 and MS2, does not contain histidine (*18*).

The rapid rate of sedimentation of this coat protein synthesized in vitro suggests that it consists either of a coat protein aggregate or of protein subunits bound to the R17 RNA which, in the reaction mixture, normally sediments at about 30*S*. Mild

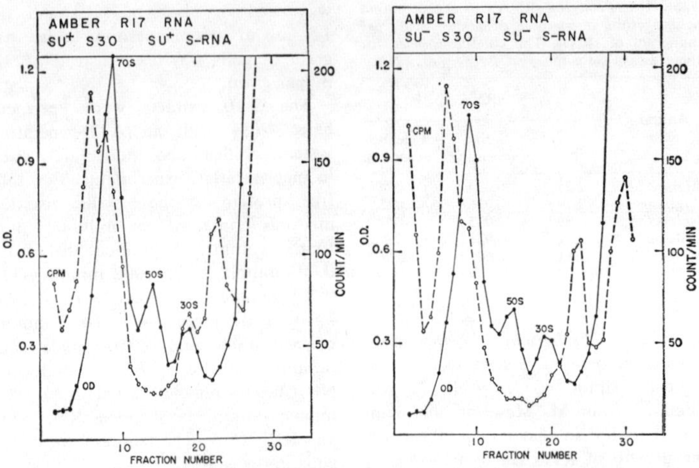

Fig. 3. Sucrose-gradient analyses of amino acid incorporation directed by am11B RNA. Each reaction mixture contained per 200 μl: 100 μg of am11B RNA containing less than 2 percent contamination from wild type, 0.2 μc of C^{14}-isoleucine (8.4×10^7 count min^{-1} $\mu mole^{-1}$), and either 50 μl of preincubated Su^+ S-30 and 160 μg of stripped Su^+ sRNA (left) or 50 μl of preincubated Su^- S-30 and 160 μg of Su^- stripped sRNA (right). A 100 μl sample of the reaction mixture was layered on a 5-ml sucrose gradient after a 20-minute period of incubation at 36°C. The gradients were centrifuged at 38,000 rev/min for 2.4 hours at −12°C. Approximately 85 percent of the acid-insoluble (hot TCA) radioactive material that had been put on the gradients was recovered.

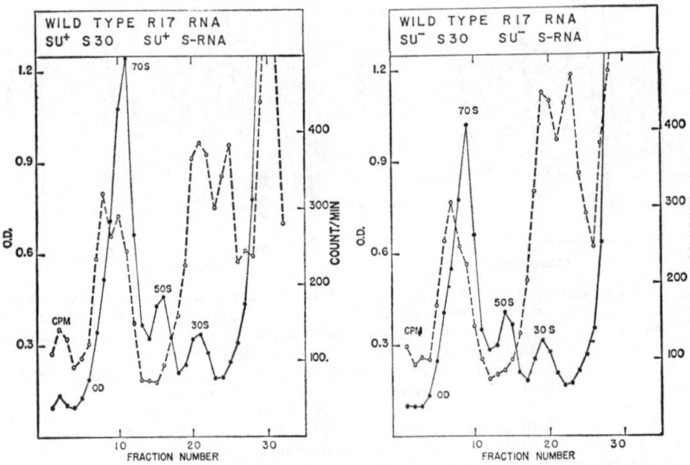

Fig. 4. Sucrose-gradient analyses of amino acid incorporation directed by wild-type RNA in systems derived from Su^+ and Su^- bacteria. These experiments were run in parallel with, and under the same conditions as, those described in Fig. 3. In each 200-μl reaction mixture there were 104 μg of wild-type R17 RNA. On the left is the result of incorporation in the Su^+ system, on the right, that in the Su^- system.

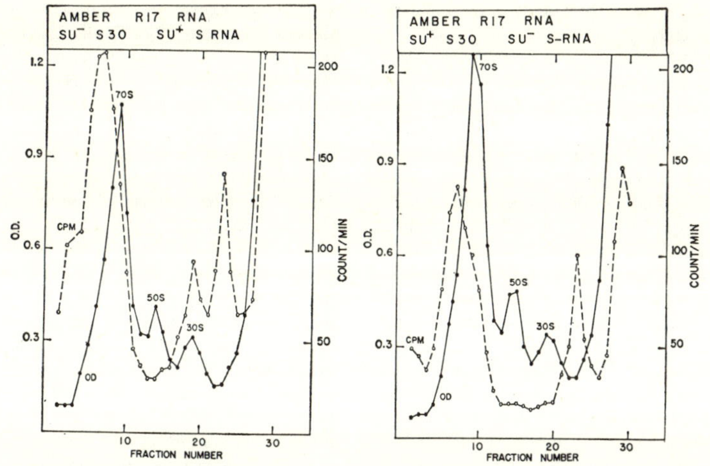

Fig. 5. sRNA mixing experiment. The sucrose-gradient analysis of protein synthesis in vitro with am11B RNA as messenger was run in parallel with, and under the same conditions as, the experiment shown in Fig. 3. The only difference was that to the reaction mixture containing 50 μl of preincubated Su$^+$ S-30, was added 160 μg of stripped Su$^-$ sRNA (right) and to the reaction mixture containing Su$^-$ S-30 was added stripped Su$^+$ sRNA (left).

treatment with ribonuclease (0.2 μg/ml for 3 minutes at 0°C) of an incubation mixture prior to its being layered on a sucrose gradient shifts the newly synthesized 30*S* material to the top of the gradient, thereby demonstrating that it was previously bound to free R17 RNA molecules.

The nature of the 20*S* peak is not yet clear. It definitely is not coat protein, since (i) it contains histidine and (ii) its autoradiogram is different from, and far more complex than, that of coat protein (*19*). Thus it must arise either from a mixture of different proteins or from a protein of high molecular weight relative to that of coat protein. It is most tempting to believe that the 20*S* material might be R17-specific RNA synthetase.

Notani, Engelhardt, Konigsberg, and Zinder (*14*) have unambiguously demonstrated that a number of their amber mutations in f2 are in the coat-protein cistron. All of the coat-protein mutants (su-3 class) grow on strain S26R1E, but not on strain S26 or on strain C600. In contrast, a number of other mutants (su-1 class), identified by complementation tests as belonging to another cistron, grow on strain C600 as well as on S26R1E. A search was undertaken for R17 amber mutants (*20*) which had the same plating and complementation properties as the su-3 group of f2. Among those found, one, am11B, was chosen for studies in vitro. Parallel with the experiments in vitro, we have undertaken chemical experiments to demonstrate directly a difference in the amino acid composition of its coat protein. Fingerprinting of tryptic digests, a separation of the tryptic peptides by electrophoresis in one dimension followed by chromatography in the second dimension, reveals the disappearance of one normal peptide and the appearance of a new one compared to wild-type R17 coat protein. This result gives us reason to believe that am11B is a coat-protein mutant. The results of the studies in vitro that are described next confirm this identification.

When am11B RNA is used as messenger in the cell-free system, there are no gross differences either in the kinetics or in the final amount of protein synthesis between the Su+ (S26R1E) and Su− (S26) systems. Under conditions in which synthesis mediated by wild-type RNA is identical in the two systems, there is a 15 to 20 percent greater incorporation of amino acids in the Su+ than in the Su− system when RNA from mutant phage is used.

The reaction mixtures are supplemented with stripped sRNA (about 1 mg/ml) isolated from Su+ and Su− bacteria. This addition stimulates protein synthesis threefold, regardless of whether the mRNA used is from wild-type or am11B bacteriophage. The influence of added sRNA on the degree of amino acid incorporation is an important factor in the success of experiments described below.

Sucrose gradient analyses of Su+ and Su− reaction mixtures, incubated with am11B RNA, are shown in Fig. 3. Each reaction mixture contained

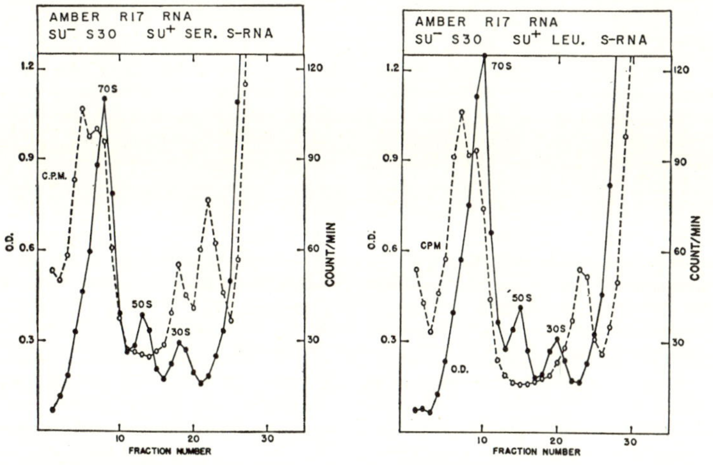

Fig. 6. Addition of Su$^+$ serine- or Su$^+$ leucine-accepting sRNA's to an otherwise nonpermissive protein-synthesizing system directed by am11B RNA. In each 200-μl reaction mixture there were 95 μg of am11B RNA, 50 μl of preincubated Su− S-30, 120 μg of stripped Su$^-$ sRNA, 0.2 μc of C^{14}-isoleucine (8 × 10^7 count min^{-1} μmole^{-1}), and approximately 5 μg of Su$^+$ sRNA having only serine-accepting activity (left) or leucine-accepting activity (right). A 100-μl sample of the reaction mixture was put on a 5-ml sucrose gradient after a 20-minute period of incubation at 36°C. Approximately 85 percent of the acid-insoluble (hot TCA) radioactive material, which had been put on the gradients, was recovered.

added sRNA which was homologous to the S-30 used. A difference between the two gradients is apparent. The 30*S* peak, previously identified as R17 coat protein synthesized in vitro and bound to R17 RNA, is absent in the gradient of the Su$^-$ reaction mixture (right side of Fig. 3). This observation is consistent with the hypothesis that am11B is a mutation in the R17 cistron specifying coat protein. As a control, the experiment was performed in parallel tubes with wild-type R17 RNA as messenger (Fig. 4). Since both bacterial strains permit the growth of wild-type phage, both the 30*S* and 20*S* proteins should be synthesized in vitro in either system when wild-type RNA is used as messenger. This is, in fact, observed.

We have repeated all of these experiments, with identical results, using different preparations of S-30's. sRNA's, R17 wild-type and am11B RNA's, and also using amino acids other than isoleucine as label. These results confirm prior indications that suppression occurs during messenger RNA translation; and most important, the presence or absence of the 30*S* peak constitutes as assay for suppression in vitro.

The amount of 30*S* protein synthesized in the Su$^+$ system is reduced relative to that of the 20*S* protein when am11B RNA is used as messenger. This reduction can be used as a measure of the efficiency with which the Su$^+$ system can suppress the amber mutation in vitro. The level observed is consistent with the observation that suppression of amber mutations in the T4 head protein by the *su-A* suppressor gene is approximately 60 percent efficient in vivo (*11*).

With this assay for suppression in vitro, we are in the position, by means of mixing experiments, to discover which component or components of the protein-synthesizing system are responsible for the suppression of the amber mutation. Experiments in which Su$^+$ sRNA was added to the incubation mixture containing Su$^-$ S-30 (Fig. 5, left), and vice versa (Fig. 5, right), were run in parallel with those shown in Figs. 3 and 4. Since addition of Su$^+$ sRNA to the S-30 prepared from Su$^-$ cells permits synthesis of functional coat protein (protein capable of binding to the added R17 RNA), and since, conversely, the Su$^+$ S-30 cannot produce 30*S* protein in the presence of Su$^-$ sRNA, suppression must be caused by a difference in the respective sRNA pools in Su$^+$ and Su$^-$ bacteria.

Weigert and Garen (*12*) and Notani *et al.* (*14*) have shown that suppression by S26R1E results in the insertion of serine into mutant protein, usually in place of a glutamine residue in the wild type. This fact, coupled with our finding that suppression in this strain is due to a difference between its sRNA pools and those of the isogenic strain S26, leads us to predict that the specific sRNA acting as the suppressor is a new or altered serine-accepting sRNA capable of inserting serine at the nonsense site. To test this hypothesis, we prepared Su$^+$ sRNA which contained only biologically active serine-accepting sRNA's. This we achieved by charging stripped Su$^+$ sRNA with serine, using an unfractionated, nucleic acid–free mixture of amino acid activating enzymes (*21*) prepared from S26 (Su$^-$) bacteria. As a control, a second portion of Su$^+$ sRNA was charged with leucine.

The two sRNA preparations were then treated with sodium periodate to destroy any uncharged sRNA by oxidation of the 2′- and 3′-hydroxyl groups on the ribose of the terminal adenylic acid residue of sRNA (*22*). To test the effectiveness of this method in protecting only a specific sRNA, the oxidized sRNA was stripped, and the amino acid–accepting activity for each of five different amino acids was examined and compared to that of the original Su$^+$ sRNA (Table 1). Protection of the serine- and leucine-accepting sRNA's is better than 90 percent, whereas other sRNA species were destroyed to the extent that their residual accepting activity was only 1 percent. The anomaly with isoleucine could be due either to contamination by C^{14}-leucine of the C^{14}-isoleucine used to test the isoleucine-accepting activity, or to contamination by C^{12}-isoleucine of the C^{12}-leucine employed for protection. The addition of sRNA's which have the 2′- and 3′-hydroxy groups of ribose oxidized to aldehydes does not inhibit protein synthesis in vitro.

The effect of adding either Su$^+$ serine sRNA or Su$^+$ leucine sRNA to the protein-synthesizing system mediated by am11B RNA is shown in Fig. 6. The reaction mixtures contained Su$^-$ S-30, 0.9 mg per milliliter of Su$^-$ sRNA, and 0.5 mg per milliliter of either Su$^+$ serine-accepting or Su$^+$ leucine-accepting sRNA. The results clearly implicate Su$^+$ serine-sRNA as the requirement for suppression of amber mutations by S26R1E. The experiment was repeated with wild-type RNA instead of am11B RNA as messenger, and no appreciable difference between the serine and leucine gradients was observed.

Molecular models for suppression. The C^{14}-labeled polypeptide sedimenting in the 30*S* region of the sucrose gradient has been identified by fingerprinting as phage coat protein. Its high sedimentation value is due to the binding of subunits of phage coat protein, synthesized in vitro, to the phage RNA, which was added as messenger. This formation of a 30*S* complex between input phage RNA and protein synthesized in vitro allows us to examine selectively the synthesis of intact coat-protein molecules and thus to develop an assay for suppression in a cell-free system.

The assay consists of testing the capacity of the amino acid incorporating system to read the nonsense codon in the coat-protein cistron of the R17 amber mutant am11B as a serine codon, and thereby to permit the synthesis of intact phage coat protein. We have shown that the only requirement for this translation of the amber codon is the addition of an Su$^+$ serine-accepting sRNA to the otherwise totally Su$^-$ system.

Based on studies of the pattern of mutation to, and reversion from, nonsense, Brenner *et al.* and Garen (*23*) have proposed that the amber triplet is UAG (*8*). This triplet could, in fact, be unreadable, or it could be a specific signal to terminate a completed polypeptide chain. In the first case, there would be no sRNA species able to recognize the amber codon, and chain termination might be merely a means of freeing ribosomes for the synthesis of new protein. In the second case, one might imagine an sRNA species capable of recognizing the UAG triplet, but unable to bind any amino acid.

With these possibilities in mind, two types of models can be proposed for the reading of the amber codon by a specific serine-accepting sRNA. These can be designated as (i) precise-reading models and (ii) ambiguous-reading models.

One precise-reading model could be based on a mutation in a serine-accepting sRNA anticodon, enabling it to recognize the amber codon. Since there is a UCG serine codon (*24*), the change of a single base in the corresponding serine-accepting sRNA would permit it to pair exactly with the amber codon UAG. However, that species of sRNA capable of recognizing the UCG serine codon would now be eliminated unless there existed more than one cistron for the UCG-specific sRNA. Furthermore, this model does not explain the release of a large proportion of polypeptide fragments in the Su+ cell, unless we include the additional hypothesis that an amber-specific chain-terminating sRNA exists.

A second precise-reading model is based on the assumption that an sRNA species exists that normally recognizes the amber codon as chain-terminating. A modification of this sRNA could enable it to accept serine. The production of a mixture of complete and incomplete protein would be a consequence of this model if the charging of the altered amber sRNA with serine were inefficient—that is, if not all such sRNA molecules became charged.

In the case of the ambiguous-reading model, no assumption about the nature of chain termination is necessary. Suppression could be caused by a modification of a serine-accepting sRNA, which destroys the specificity of its attachment to a codon without altering its anticodon. For example, modification of the overall structure of the serine-accepting sRNA species which recognizes the serine codon UCG might permit this sRNA species to recognize either the nonsense (UAG) or the serine (UCG) codon. This model requires no additional hypotheses to explain the production of a mixture of fragments and completed chains in the Su+ cell.

The amber suppressor mutation, in the second and third models, could be a mutation either in the sRNA cistron or in a gene coding for an enzyme which could modify the serine-accepting or amber-specific sRNA species assumed to act as the suppressor. Likely candidates for such a task are enzymes which produce the unusual bases in sRNA (for example, methylated bases, pseudouracil, or dihydrouracil). Our results do not allow us to distinguish between enzymatic and genetic modification of the specific sRNA involved. It is possible, therefore, that the new serine-accepting sRNA is only the indirect product of the *su-A* suppressor gene.

Finally, the conditions selected for the mixing experiment (Fig. 6) enable us to determine whether the suppressor mutation is dominant in vitro. Since, in these experiments, the incubation mixtures contained a twofold excess of Su− serine-accepting sRNA over Su+ serine-accepting sRNA, we conclude that the *su-A* suppressor is dominant. This is consistent with the findings of Signer, Beckwith, and Brenner (*25*) that a cell heterozygous for amber suppressor gene is permissive.

MARIO R. CAPECCHI
GARY N. GUSSIN
Biological Laboratories, Harvard University, Cambridge, Massachusetts

References and Notes

1. C. Yanofsky, D. R. Helinski, B. C. Maling, in *Cold Spring Harbor Symp. Quant. Biol.* **28**, 11 (1961); L. Gorini and E. Kataja, *Proc. Nat. Acad. Sci. U.S.* **51**, 487 (1964); J. R. Beckwith, *Biochim. Biophys. Acta* **76**, 162 (1963).
2. J. Davies, W. Gilbert, L. Gorini, *Proc. Nat. Acad. Sci. U.S.* **51**, 883 (1964).
3. D. C. Hawthorne and R. K. Mortimer, *Genetics* **47**, 1085 (1960); T. R. Manney, *ibid.* **50**, 109 (1964).
4. S. Benzer and S. P. Champe, *Proc. Nat. Acad. Sci. U.S.* **47**, 1025 (1961).
5. R. H. Epstein, A. Bolle, C. M. Steinberg, E. Kellenberger, E. Boy de la Tour, R. Chevalley, R. S. Edgar, M. Susman, G. H. Denhardt, A. Lielausis, in *Cold Spring Harbor Symp. Quant. Biol.* **28**, 375 (1963).
6. A. Garen and O. Siddiqi, *Proc. Nat. Acad. Sci. U.S.* **48**, 1121 (1962).
7. A. Campbell, *Virology* **14**, 22 (1961).
8. Abbreviations and definitions: RNA, ribonucleic acid; mRNA, messenger RNA; sRNA, soluble RNA; ATP, adenosine 5′-triphosphate; GTP, guanosine 5′-triphosphate; pfu, plaque forming units; S-30, a bacterial homogenate from which whole cells, cell-wall fragments, and cell-membrane fragments have been removed by centrifugation at 30,000*g* for 30 minutes; "preincubation," refers to incubation of the S-30 in the absence of labeled amino acids to eliminate amino acid incorporation directed by endogenous *E. coli* messenger RNA; "charged sRNA," is sRNA to which an amino acid is attached by a phosphate-ester linkage between the terminal adenylic acid of sRNA and the carboxyl group of the amino acid; "stripped sRNA," is sRNA from which the amino acid has been removed by hydrolysis of the phosphate-ester linkage; UAG and UCG are symbols representing the sequence of nucleotides in specific codons, the letters represent the nucleotides uridine, adenosine, guanosine, and cytidine 5′-monophosphate; TCA, trichloroacetic acid.
9. N. D. Zinder and S. Cooper, *Virology* **23**, 152 (1964).
10. S. Benzer and S. P. Champe, *Proc. Nat. Acad. Sci. U.S.* **48**, 1114 (1962).
11. S. Brenner and A. O. W. Stretton, *J. Cellular Comp. Physiol.* **64**, Suppl. 1, 43 (1964).
12. M. G. Weigert and A. Garen, *J. Mol. Biol.*, in press.
13. A. O. W. Stretton and S. Brenner, *ibid.*, in press.
14. G. W. Notani, D. L. Engelhardt, W. Konigsberg, N. D. Zinder, *ibid.*, in press.
15. R. F. Gesteland and H. Boedtker, *J. Mol. Biol.* **8**, 496 (1964).
16. M. W. Nirenberg and J. H. Matthaei, *Proc. Nat. Acad. Sci. U.S.* **47**, 1588 (1961).
17. G. Zubay, *J. Mol. Biol.* **4**, 347 (1962).
18. M. D. Enger and P. Kaesberg, *Proc. Nat. Acad. Sci. U.S.*, in press.
19. M. R. Capecchi, manuscript in preparation.
20. G. N. Gussin, in preparation.
21. E. M. Martin, C. Yegian, G. Stent, *Z. Vererbungslehre* **94**, 303 (1963).
22. P. C. Zamecnik, M. L. Stephenson, J. P. Scott, *Proc. Nat. Acad. Sci. U.S.* **46**, 811 (1960).
23. S. Brenner, A. O. W. Stretton, S. Kaplan, *Nature*, in press; A. Garen, personal communication.
24. M. W. Nirenberg, P. Leder, M. Bernfield, R. Brimacombe, J. Trupin, F. Rottman, C. O'Neal, *Proc. Nat. Acad. Sci. U.S.*, in press.
25. E. R. Signer, J. R. Beckwith, S. Brenner, in preparation.
26. P. Berg, F. H. Bergmann, E. J. Ofengand, M. Dieckmann, *J. Biol. Chem.* **236**, 1726 (1961).
27. We thank Professors J. D. Watson and W. Gilbert for their encouargement and helpful discussion. Supported in part by grants GM-09541-03 and -04 from the U.S. Public Health Service.

1 June 1965

Reprinted from NATURE, Vol. 217, 1968

Amber Suppression: a Nucleotide Change in the Anticodon of a Tyrosine Transfer RNA

by
HOWARD M. GOODMAN
JOHN ABELSON
ARTHUR LANDY
S. BRENNER
J. D. SMITH
MRC Laboratory of Molecular Biology, Cambridge

In certain mutants a single base change alters the meaning of a messenger codon in such a way that, instead of spelling out an amino-acid, it spells out chain termination. Mutants in a quite different gene, called a suppression gene, allow the chain-terminating triplet to be read as an amino-acid. Experiments have shown that this is caused by a mutated tRNA which carries a single base change in its anticodon. This allows it to read the chain-terminating codon as if it spelt tyrosine.

IN *Escherichia coli* the amber codon UAG is one of the three "nonsense" triplets (reviewed in ref. 1) that code as signals for termination of messenger RNA translation. Strains of *E. coli*, which allow translation of UAG as an amino-acid and subsequent propagation of the poly-peptide chain, result from mutation in one of several suppressor genes. There are two classes of UAG suppressors: first, ochre suppressors which permit both UAG and the related chain terminating triplet UAA to be read with low efficiency, and, second, the relatively efficient amber suppressors which are specific for UAG. It has been shown for three amber suppressors that suppression is caused by the presence of an altered *t*RNA in the su^+ cell[2-5]. Although there was no doubt that the suppressor itself was a *t*RNA molecule, these experiments did not show how the suppressor gene acts. It could, for example, control the synthesis of a *t*RNA not present in su^- cells, or of an enzyme which modifies *t*RNA. More simply it could specify the sequence of a *t*RNA. We will show that the amber suppressor su_{III} gene is the structural gene for one of the tyrosine *t*RNAs of *E. coli* and that the mutation su^-_{III} to su^+_{III} results in a base change in the anticodon of this *t*RNA.

The nature of the mutational change is shown by comparison of the nucleotide sequences of the su^+_{III} tyrosine *t*RNA and the corresponding *t*RNA specified by the su^-_{III} allele. The su_{III} suppressor *t*RNA is a small part of the total tyrosine *t*RNA in the cell. The rest of the tyrosine *t*RNA consists of two species (I and II). To overcome this difficulty we used a technique designed greatly to increase the amount of su_{III} gene product in the cell by increasing the number of copies of the gene. The su_{III} gene, which maps very near to the attachment site of phage φ80, was placed on a defective transducing φ80. Infection of *E. coli* with the transducing phage resulted in a phage DNA pool containing many copies of the su_{III} gene. In this way, the amount/cell of the su_{III} gene product, su_{III} tyrosine *t*RNA, was greatly increased. This technique also allows selective isotopic labelling of the *t*RNA for sequence determination.

Transducing Phage φ80 carrying the su_{III} Gene

In the experiments described here, a particular defective transducing phage φ80 dsu^+_{III} (V) was used[4]. The su^+_{III} gene was isolated as a spontaneous mutant from an su^-_{III} strain. The homologous phage, transducing the wild type allele, φ80 dsu^-_{III}, was derived as a recombinant from heterozygous strains lysogenic for φ80 dsu^+_{III}. Thus φ80 dsu^-_{III} and φ80 dsu^+_{III} are identical except for the single spontaneous mutation which produced the su^+_{III} allele.

Synthesis of su_{III} Tyrosine tRNA after Infection by φ80 dsu_{III}

Infection of *E. coli CA*274 (su^- lac^- amber) with φ80 dsu^+_{III} resulted in the appearance of suppressor activity as measured by the suppression of an amber mutation in the β galactosidase gene. This was closely paralleled by an increase in the amount of tyrosine *t*RNA in the cell (Fig. 1). No increase occurs in the relative proportion of any other *t*RNA.

Ribosome binding experiments with the *t*RNA synthesized after infection with φ80 dsu^+_{III} showed that the su^+_{III} strains contain a tyrosine *t*RNA which recognizes the amber codon. The corresponding *t*RNA in cells infected with φ80 dsu^-_{III} recognizes the normal tyrosine codons UAU and UAC[4]. Although these experiments do not prove that the suppressor gene specifies a tyrosine *t*RNA, this, however, is likely because φ80 dsu^+_{III} does contain a structural gene for tyrosine *t*RNA as shown by specific hybridization of tyrosine *t*RNA with the DNA from this phage[6].

Nucleotide Sequences of su^+_{III} and su^-_{III} Tyrosine *t*RNAs

If the su_{III} gene is the structural gene for a tyrosine *t*RNA, we expect that the mutation from the su^-_{III} to su^+_{III} allele will have produced a nucleotide change in the su_{III} *t*RNA. To test this idea we have determined the nucleotide sequence of the su^+_{III} tyrosine *t*RNA and compared it with that of the su^-_{III} *t*RNA.

The su^+_{III} *t*RNA labelled with phosphorus-32 was prepared as previously described[6] or by a slightly modified procedure (see legend of Fig. 1). The selective labelling of su_{III} tyrosine *t*RNA after infection was evident from the high proportion of *t*RNA labelled with phosphorus-32 eluting with tyrosine *t*RNA during purification on the DEAE 'Sephadex' column. Pooled fractions from this region often contained about 70 per cent radiochemically pure tyrosine *t*RNA and were sufficiently homogeneous

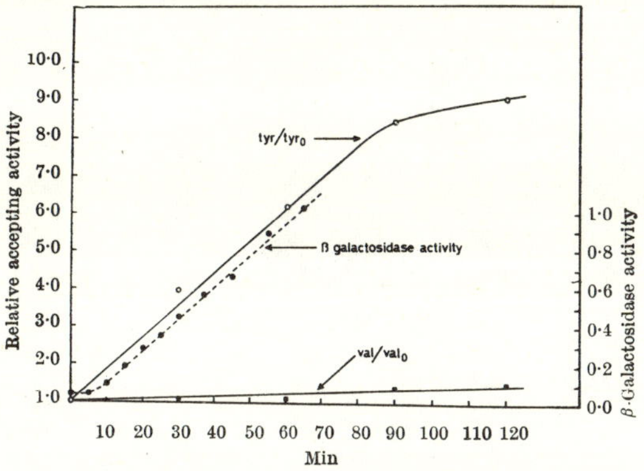

Fig. 1. Increase of tyrosine *t*RNA and suppressor activity in *E. coli CA*274 *su⁻ lac⁻* amber following infection by φ80 dsu^{+}_{III}. *E. coli CA*274 was grown to 2×10^8 cells/ml. in LP medium[6]. The cells were collected by centrifugation and concentrated ten-fold in preconditioned LP medium. $MgSO_4$ was added to 0·02 M and the cell suspension was irradiated with ultraviolet light for 90 s. This dose kills 70 per cent of the cells. The culture was then infected with 30 φ80 dsu^{+}_{III} particles/bacterium. After adsorption for 7 min the cell suspension was diluted ten-fold in preconditioned media. Two separate experiments are shown. In one experiment, the suppression of an amber mutation in the β-galactosidase gene was measured by assaying enzyme activity at various times after dilution into the growth medium. One mM of isopropylthio-galactoside was added to induce the β-galactosidase. In the second experiment, aliquots of the infected culture were removed, *t*RNA was extracted, and tested for its ability to be acylated with several different amino-acids. Fifty μg/ml. of chloramphenicol added at 25 min prevented lysis which normally occurs at 60 min after dilution. In experiments where the *t*RNA is labelled with ^{32}P, 0·05 m Ci/ml. is added 5 min after dilution of the infected cells and the culture is collected after 2 h.

for some of the nucleotide sequence work. In other cases, these fractions were further purified by either of two chemical methods which depend on the ability of the *t*RNA to be acylated with tyrosine[7,8].

The sequence of the su^{+}_{III} tyrosine *t*RNA labelled with phosphorus-32 was determined using the methods of Sanger *et al.*[9,10]. The end-products of complete digestion with T_1 ribonuclease or pancreatic ribonuclease were separated on a two-dimensional electrophoretic system and their sequences determined by further enzymatic digestion. These are listed in Table 1. The combination of products from the two enzyme digests gave only two significant overlaps—GAAGGTψCG and CCAAAGGGAGC.

To overlap the remaining end-products, larger fragments were obtained by partial enzyme digestion and separated as described before[11]. The sequences of these partial fragments were determined from the complete products of their digestion with T_1 ribonuclease and pancreatic ribonuclease. About one hundred large fragments were obtained by partial digestion with various enzymes and their sequences were determined. Fig. 2 shows a selected set of fragments from which the complete sequence was deducted. Fig. 3*a* shows the su^{+}_{III} tyrosine *t*RNA sequence arranged in the "clover leaf"

Table 1. END-PRODUCTS OF RIBONUCLEASE DIGESTION OF su^{+}_{III} TYROSINE-*t*RNA

T_1 ribonuclease	Pancreatic ribonuclease
7 G	20 C
2 AG	8 U
UG	ψ
pGp	AoH
CCG	2 AC
CAG	GC
AAG	AU
$CG_{2'OMe}G$	GAC
CCAAAG	AGAC
TψCG	$G_{2'OMe}GC$
U*UCCG	AA*Aψ
UCAUCG	GAGC
ACUUCG	GU
AAUCCUUCCCCCACCACCA–OH	GAAU
ACUCUAA*AψCUG	AAAGGGAGC
	GAAGGT
	GGGGU*
	pGGU

pattern characteristic of the yeast *t*RNA structures. For comparison, the sequence of yeast tyrosine *t*RNA as determined by Madison, Everett and Kung[12] is shown in Fig. 3*b*. Note that in su^{+}_{III} tyrosine *t*RNA, CUA is in the position expected for the anticodon.

There is only a single nucleotide difference between the su^{+}_{III} and su^{-}_{III} tyrosine *t*RNAs and this is best illustrated by comparison of their T_1 ribonuclease digestion products. Fig. 4 compares the fingerprints of the T_1 ribonuclease digestion products of the two *t*RNAs.

Comparison of the T_1 products shows that oligonucleotide 17, ACUCUAA*AψCUG, in the su^{+}_{III} *t*RNA is replaced in the su^{-}_{III} *t*RNA by the two oligonucleotides ACUG and UAA*AψCUG. For the moment we will ignore the presence in the su^{+}_{III} fingerprint of oligonucleotide 17a (which is related to 17); its significance will be discussed later. There are no differences in any other T_1 ribonuclease digestion products. In the su^{-}_{III} *t*RNA the sequence corresponding to oligonucleotide 17 (su^{+}_{III}) must therefore be ACUGUAA*AψCUG which is split by T_1 ribonuclease to give ACUG and UAA*AψCUG. The sequence GUA in su^{-}_{III} *t*RNA is thus changed to CUA in the su^{+}_{III} *t*RNA. GUA is the expected anticodon for the dual recognition of the tyrosine codons UAU and UAC (ref. 13) and in fact GψA is found in the expected anticodon position of yeast tyrosine *t*RNA (Fig. 3*b*) which recognizes both these codons[14]. CUA is the anticodon expected for the unique recognition of the amber codon UAG. In the complete sequence (Fig. 3*a*) of su^{+}_{III} *t*RNA, CUA occupies the anticodon position. These results show that the mutational event giving rise to the su^{+}_{III} allele can be accounted for by a G to C change in the structural gene for a tyrosine *t*RNA.

G in the Anticodon is Normally Modified

In these experiments, the su^{-}_{III} *t*RNA had an unmodified G in the first position of the anticodon. Normally, however, the G in the anticodon of the su^{-}_{III} tyrosine *t*RNA is

```
pG-G-U-G-G-G-G-U̇-U-C-C-C-G
                                       OMe
                                        |
  U-G-G-G-G-U̇-U-C-C-C-G-A-G-C-G-G-C-C-A-A-A-G

                          C-C-A-A-A-G-G-G-A-G-C-A-G

                A-G-C-A-G-A-C-U-C-U-A-Ȧ-A-Ψ-C-U-G-C-C-G-U-C-A-U-C-G

      U-C-A-U-C-G-A-C-U-U-C-G-A-A-G-G-T-Ψ-C-G-A-A-U-C-C-U-U-C-C-C-C-C-A-C-C-A-C-C-A—OH

                                       OMe
                                        |
pG-G-U-G-G-G-G-U̇-U-C-C-C-G-A-G-C-G-G-C-C-A-A-A-G-G-G-A-G-C-A-G-A-C-U-C-U-A-Ȧ-A-Ψ-C-U-G-C-C-G-U-C-A-U-C-G-A-C-U-U-C-G-A-A-G-G-T-Ψ-C-G-A-A-U-C-C-U-U-C-C-C-C-C-A-C-C-A-C-C-A—OH
```

Sequence of E. coli Su$^{+}_{III}$ Tyrosine - tRNA

Fig. 2. Examples of overlapping oligonucleotides used to determine the sequence of su^{+}_{III} tyrosine *t*RNA. The five oligonucleotides were separated after partial T_1 ribonuclease digestion and their sequences determined as described in the text.

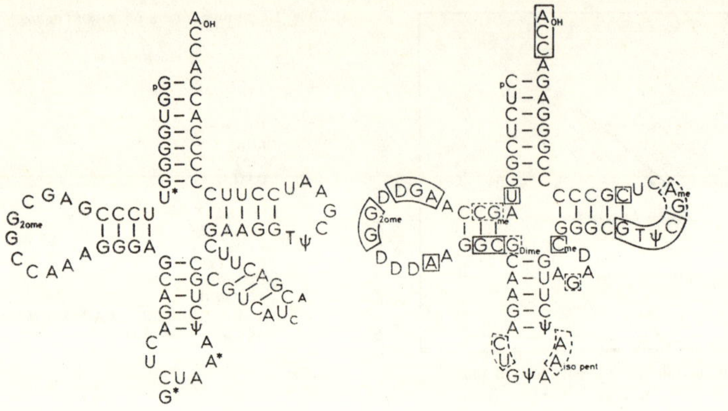

Su⁺ Tyrosine tRNA E.coli (a)

Tyrosine tRNA Yeast (b)

Fig. 3. Clover leaf models of *E. coli* and yeast tyrosine *t*RNAs. The sequence of the su^+_{III} tyrosine *t*RNA is shown with the differences that occur in su^-_{III}, and species I and II (*a*). For identification of the minor bases see Fig. 4 and the text. The su^-_{III} and species I tyrosine *t*RNAs have identical sequences and differ from su^+_{III} in having a G* in the anticodon. Species II *tyr t*RNA has the same anticodon sequence containing G* but is different in two positions where the T_1 ribonuclease product UCAUCG is replaced by UCACAG. The latter oligonucleotide has been independently identified by B. P. Doctor and S. Nishimura. The sequence of the yeast tyrosine *t*RNA is taken from Madison *et al.*[12] (*b*). The sequences enclosed in solid boxes are common to all the five yeast *t*RNA structures so far determined (neglecting differences in modification), while those enclosed in dotted boxes are common to most of the sequences.

a modified guanine derivative (G*). In su_{III} tyrosine *t*RNA synthesized in conditions where the cells were starved before infection[6] we have observed that A* and $G_{2'OMe}$ remain largely unmodified. The same is true of the G residue in the anticodon. In su^-_{III} *t*RNA prepared from cells not starved before infection (legend to Fig. 1), the G residue in the anticodon is partly or completely modified to G*. This is not split by T_1 ribonuclease so that the T_1 ribonuclease product containing the anticodon is ACUG*UAA*AψCUG. The same anticodon G*UA also occurs in the two principal tyrosine *t*RNA components of *E. coli* (Fig. 3*a*, I and II), and therefore it was possible to isolate small quantities of non-radioactive G^*_p to study its properties. G^*_p has an ultraviolet absorption spectrum very similar to that of G_p. It differs from G_p in having an additional basic group with a pK_a between 5 and 6. Neither T_1 nor pancreatic ribonuclease splits the anticodon at the position of G^*_p and hence in a pancreatic ribonuclease digest it is obtained as the dinucleotide $G^*_pU_p$. From ribosome binding studies on *E. coli* tyrosine *t*RNAs it is clear that G* has the same codon recognition properties as G. The nature of the modification (which we suspect is on the imidazole part of the purine ring) and its biological significance remain unknown.

Heterogeneity of su^+_{III} Tyrosine *t*RNA

In the su^+_{III} *t*RNA T_1 ribonuclease fingerprint we noted the presence of an additional oligonucleotide 17a (Fig. 4). This has the sequence ACUG*UAA*AψCUG and is present in amounts varying between 20 per cent and 40 per cent of the su^+_{III} anticodon-containing T_1 product ACUCUAA*AψCUG. The sum of the two is in molar yield in the *t*RNA. Thus the tyrosine *t*RNA synthesized after infection by φ80 dsu^+_{III} contains in addition to su^+_{III} *t*RNA a smaller proportion of a tyrosine *t*RNA with the su^- anticodon sequence. As will be shown later, the tyrosine *t*RNA in *E. coli* can be separated into two kinds of molecules. One of these, tyrosine *t*RNA I, has the same sequence as the su^- *t*RNA. The other, tyrosine *t*RNA II, differs by two nucleotides. Fingerprints of the su^+_{III} tyrosine *t*RNA exclude the presence of tyrosine *t*RNA II as the source of the contaminating su^- anticodon sequence. We have considered four possible explanations for this. (i) The φ80 dsu^+_{III} carries an additional tyrosine *t*RNA gene determining the structure of tyrosine *t*RNA species I. This is not supported by hybridization experiments between φ80 dsu^+_{III} DNA and tyrosine *t*RNA, which show that the transducing phage carries a single *t*RNA gene[6]. (ii) The φ80 dsu^+_{III} phage preparations always contain a proportion of φ80 dsu_{III} transducing phage. Such phages could be produced by abnormal excisions occurring in the induction of the heterogenote. We have no good method of assessing the frequency of such events although the general phenomenon is known to occur[15]. (iii) The transcription of genes near the φ80 attachment site is specifically promoted on infection by the φ80 transducing phage. (iv) The presence of many copies of the structural gene for the su_{III} tyrosine *t*RNA specifically induces the expression of one class of chromosomal tyrosine *t*RNA genes.

At present we cannot say which of these mechanisms produces the effect we have observed. Experiments are in progress which should distinguish between them.

Tyrosine *t*RNAs in Uninfected *E. coli*

There are two principal tyrosine *t*RNAs in *E. coli* which can be separated by column chromatography on DEAE 'Sephadex'[16]. These are referred to as I and II in the order they elute from the column. They are present in proportions of approximately 40 per cent (I) and 60 per cent (II). We have determined the sequences of I and II from *CA*265 su^+_{III} uniformly labelled with phosphorus-32. The labelled tyrosine *t*RNAs were isolated by a combination of chromatography on DEAE 'Sephadex', and either reversed phase partition chromatography[17] or chromatography on benzoylated DEAE cellulose[18]. The sequences were derived in the same way as for su^+_{III} *t*RNA. Certain overlaps of the complete T_1 and pancreatic digestion

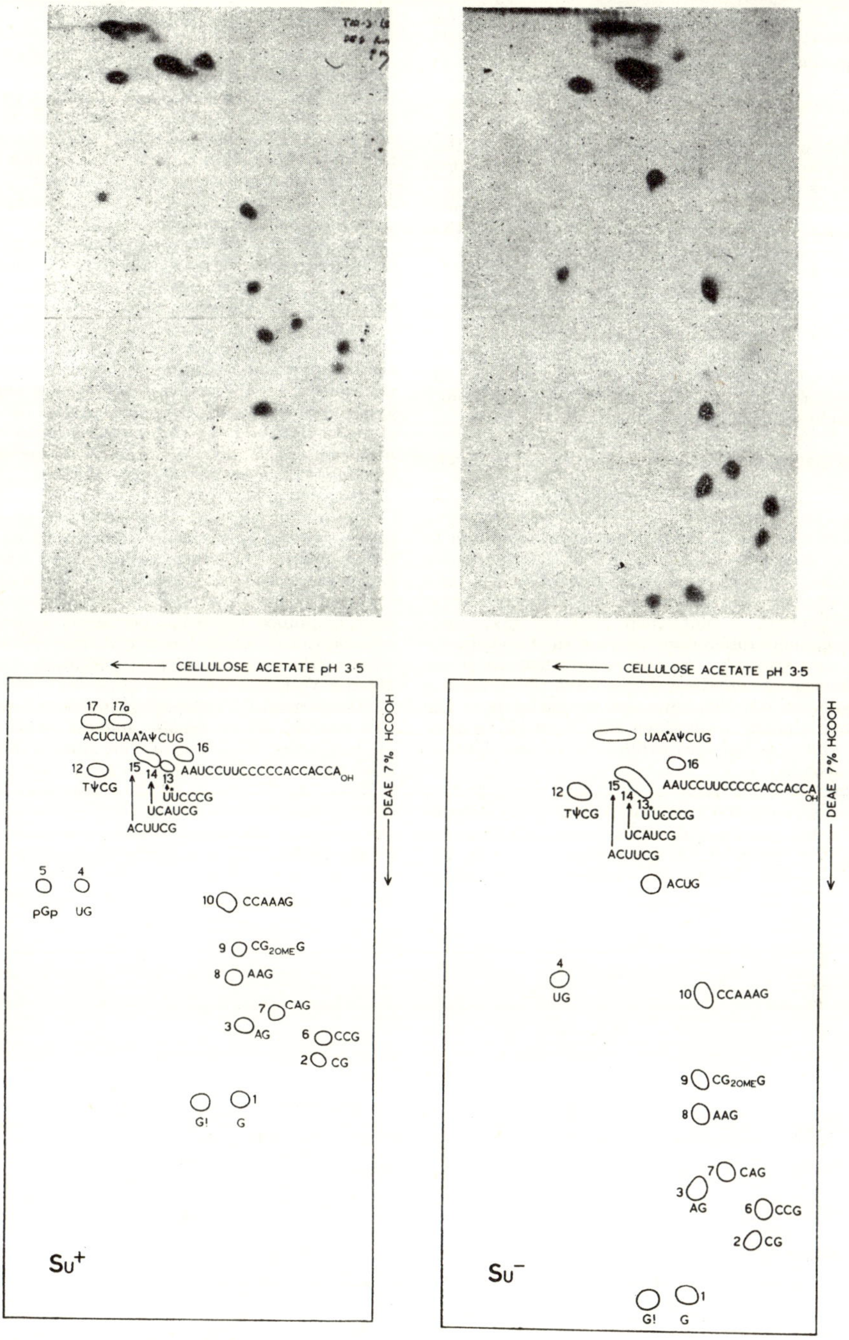

(*a*) (*b*)

Fig. 4. Two dimensional electrophoretic separation of a ribonuclease T_1 digest of su^{+}_{III} tyrosine *t*RNA labelled with ^{32}P. *a*, The end-products of ribonuclease T_1 digestion were separated by two dimensional electrophoresis first on cellulose acetate in pyridine–acetate buffer, *p*H 3·5, and then on DEAE–cellulose paper in 7 per cent formic acid (v/v). The lower diagram shows the identification of the nucleotides. The three hexanucleotides 13, 14 and 15 can be resolved after removal of their terminal 3′-phosphates with phosphomonoesterase. The appearance of CG is a consequence of incomplete 2′-*O*-methylation of $CG_{2'OMe}G$ giving rise to CG plus G after digestion with ribonuclease T_1. The hexanucleotide, U*UCCCG, always occurs in low yield, a fact that has not yet been adequately explained. pGp streaks badly and has partially migrated off the paper, making it difficult to detect. The two spots to the right and slightly above UG are minor contaminants which disappear on further purification of the *t*RNA. U* is a sulphur-containing nucleotide, tentatively identified as 4-thiouridylic acid. A* is derived from adenylic acid and, because it can be labelled with (^{35}S) sulphate and (^{14}C)-methyl methionine, it presumably contains both sulphur and a methyl group. *b*, Two dimensional electrophoretic separation of a T_1 ribonuclease digest of su^{-}_{III} tyrosine *t*RNA labelled with ^{32}P produced after infection with φ80 dsu^{-}_{III}. In this separation spot 16 has trailed but is normally present in molar yield. The time of electrophoresis on DEAE paper was longer than in *a*.

products in the first sixteen residues have not yet been obtained, and the sequence in this region is assigned by analogy with that of the su^+_{III} *t*RNA.

Both tyrosine *t*RNAs I and II contain the anticodon G*UA and differ only in a sequence of two nucleotides in the loop between the anticodon and TψCG arms (Fig. 3*a*). Tyrosine I has a sequence apparently identical with that of su^-_{III} tyrosine *t*RNA so that in *E. coli* there are only two basic types of tyrosine *t*RNA sequences.

Thus it is clear that the two different sequences (I and II) must be specified by separate genes. If tyrosine *t*RNA I and su^-_{III} were determined by the same gene, then we would expect that in an su^+_{III} strain all of tyrosine *t*RNA I would be converted to su^+_{III} *t*RNA. When tyrosine *t*RNA from *E. coli CA*275 su^+_{III} is fractionated on a DEAE 'Sephadex' column, assays for su^+_{III} *t*RNA show that it is located between the peaks of tyrosine *t*RNAs I and II (Fig. 5). This is confirmed by fingerprinting. We conclude that there is a separate gene for su^-_{III} *t*RNA even though it has the same sequence as tyrosine I.

Comparison between Yeast and *E. coli* Tyrosine *t*RNAs

It is interesting to compare the sequences of the yeast and *E. coli* tyrosine *t*RNAs. The shape and hydrogen bonding pattern of the clover leaf model are common to the two organisms. As indicated on the diagram, most of the specific sequence similarities between the two tyrosine *t*RNAs are also shared by all the other yeast *t*RNA structures[19-22], so that when these are taken into account the sequence differences between the two tyrosine *t*RNAs are very striking. An interesting feature is the complete absence of dihydro-uridylic acid (D) from the *E. coli* tyrosine *t*RNA. The greatest similarity is in the anticodon loop and the lower three base pairs of the anticodon arm, which, except for modified bases, are identical in the two *t*RNAs.

At this stage it is not useful to speculate which part of the structures serve as the amino-acyl synthetase recognition site. Neither yeast nor *E. coli* tyrosine *t*RNA can be acylated by the heterologous synthetase[23], so the two transfer RNAs may have quite different recognition sites.

Discussion

The mutation giving rise to one su^+_{III} gene changes the anticodon of a minor tyrosine *t*RNA species from GUA to CUA. This change enables the *t*RNA to recognize the amber codon UAG instead of the tyrosine codons UAU and UAC. It seems likely that the mutations leading to the su^+_{I} and su^+_{II} amber suppressor genes have similarly affected structural genes for serine and glutamine *t*RNA. These *t*RNAs could have anticodons related to the amber anticodon by a single base change.

The ochre suppressor gene su_c maps in the same position as the su_{III} gene[24] and also causes the insertion of tyrosine[25]. The su_c gene may therefore be another allele of the su_{III} gene. Amber suppressor genes may only arise where there are multiple genes for a particular *t*RNA species. In the case of tyrosine *t*RNA we have provided evidence that there are at least three structural genes. If there are no more than three genes our results suggest that the genes are expressed at different rates. We estimate that tyrosine *t*RNAs I, II and *su* are present in the approximate ratios 25 : 60 : 15. The need for this differential control is obscure because the three *t*RNAs have nearly identical sequences.

One might predict that multiple amber suppressor genes could arise corresponding to the multiple tyrosine *t*RNA genes. Perhaps not all of these could be obtained as the large amount of tyrosine II *t*RNA in the cell might make an amber suppressor mutant of this gene lethal. We have examined three independently isolated amber suppressor strains which insert tyrosine. All these map very close to the original su^+_{III}. In addition, Garen's *su*3 (ref. 26) and Yanofsky's *su Ymel* (personal communication from G. Dennert and U. Henning) insert tyrosine and map in this region. These results suggest either that only one of the tyrosine *t*RNA genes can mutate to su^+_{III} or that at least two of the genes are located close to each other.

The G to C change in the nucleotide sequence of the *su* tyrosine *t*RNA specifically changes its codon recognition but does not affect any of the other functional properties of the *t*RNA. This formally designates this position as part of the anticodon and also shows that at least the first position of the anticodon is not necessary for recogni-

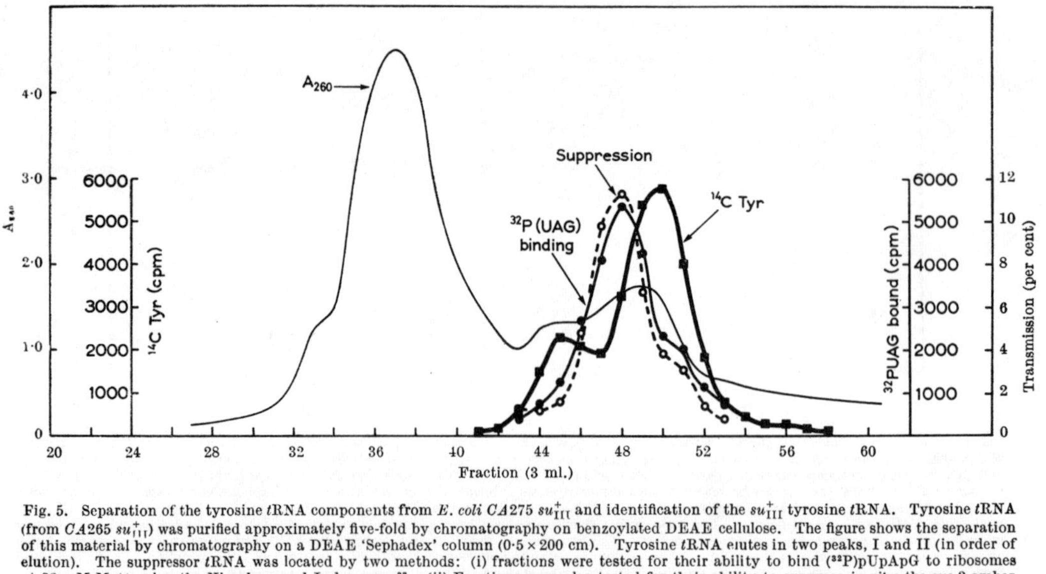

Fig. 5. Separation of the tyrosine *t*RNA components from *E. coli CA*275 su^+_{III} and identification of the su^+_{III} tyrosine *t*RNA. Tyrosine *t*RNA (from *CA*265 su^+_{III}) was purified approximately five-fold by chromatography on benzoylated DEAE cellulose. The figure shows the separation of this material by chromatography on a DEAE 'Sephadex' column (0·5 × 200 cm). Tyrosine *t*RNA elutes in two peaks, I and II (in order of elution). The suppressor *t*RNA was located by two methods: (i) fractions were tested for their ability to bind (^{32}P)pUpApG to ribosomes at 30 mM Mg^{++} using the Nirenberg and Leder assay[27]. (ii) Fractions were also tested for their ability to suppress *in vitro* the *sus* 3 amber mutation of the coat protein of the phage *f*2. This assay was performed by Dr Harvey Lodish using a modification[28] of the method of Engelhardt *et al.*[3]. The measure of suppression is the percentage of transmission, defined as the ratio of completed coat protein molecules : the total number of polypeptide chains initiated, at a concentration of 8 μg/ml. of suppressor *t*RNA added.

tion of amino-acyl synthetase. To identify the function of other parts of the molecule, different mutants of this gene must be studied. These can be obtained by isolating su^- revertants of the su^+ gene. Any mutation which gives rise to a non-functional *t*RNA will have the su^- phenotype. Garen, Garen and Wilhelm[26] have isolated su^- revertants of the su^+_{I} gene. These map in five closely linked sites indicating that different parts of the su^+_{I} gene may be mutated to give the su^- phenotype.

With L. Barnett we have isolated a number of su^- revertants of the su^+_{III} gene. We are in the process of characterizing the structural and functional changes in the su^- tyrosine *t*RNA caused by these mutations. This should provide insight into the structural basis for function in *t*RNA molecules.

We thank Dr F. Sanger, Dr G. G. Brownlee and Mr B. G. Barrell for their help. Dr S. Zadrazil helped to characterize the minor nucleotides. We thank the Radiochemical Centre, Amersham, for preparing labelled *t*RNA and Mrs R. Fishpool, Miss E. Holgate and Mr G. Chambers for technical assistance. One of us (A. L.) was supported by US Public Health Service and American Cancer Society postdoctoral fellowships; another (J. A.) by a US Public Health Service postdoctoral fellowship; and another (H. M. G.) by the Helen Hay Whitney Foundation.

Received February 15, 1968.

[1] Weigert, M. G., Galluci, E., Lanka, E., and Garen, A., *Cold Spring Harbor Symp. Quant. Biol.*, **31**, 145 (1966); Stretton, A. O. W., Kaplan, S., and Brenner, S., ibid., **31**, 173 (1966).
[2] Capecchi, M., and Gussin, G., *Science*, **149**, 417 (1965).
[3] Engelhardt, D. L., Webster, R. E., Wilhelm, R. C., and Zinder, N. D., *Proc. US Nat. Acad. Sci.*, **54**, 1791 (1965).
[4] Smith, J. D., Abelson, J. N., Clark, B. F. C., Goodman, H. M., and Brenner, S., *Cold Spring Harbor Symp. Quant. Biol.*, **31**, 479 (1966).
[5] Gesteland, R. F., Salser, W., and Bolle, A., *Proc. US Nat. Acad. Sci.*, **58**, 2036 (1967).
[6] Landy, A., Abelson, J., Goodman, H. M., and Smith, J. D., *J. Mol. Biol.*, **29**, 457 (1967).
[7] von Portatius, H., Doty, P., and Stephenson, M. L., *J. Amer. Chem. Soc.*, **83**, 3351 (1961).
[8] Katchalsky, E., Yankofsky, S., Novogradsky, A., Galenter, Y., and Littauer, U. Z., *Biochim. Biophys. Acta*, **123**, 641 (1966).
[9] Sanger, F., Brownlee, G. G., and Barrell, B. G., *J. Mol. Biol.*, **13**, 373 (1965).
[10] Brownlee, G. G., and Sanger, F., *J. Mol. Biol.*, **23**, 337 (1967).
[11] Brownlee, G. G., Sanger, F., and Barrell, B. G., *J. Mol. Biol.* (in the press).
[12] Madison, J. T., Everett, G. A., and Kung, H. K., *Science*, **153**, 531 (1966).
[13] Crick, F. H. C., *J. Mol. Biol.*, **19**, 548 (1966).
[14] Doctor, B. P., Loebel, J. E., and Kellogg, D. A., *Cold Spring Harbor Symp. Quant. Biol.*, **31**, 543 (1966).
[15] Appleyard, R. K., *Genetics*, **39**, 440 (1954).
[16] Nishimura, S., Harada, F., Narushima, U., and Seno, T., *Biochim. Biophys. Acta*, **142**, 133 (1967).
[17] Kelmers, A. D., Novelli, G. D., and Stulberg, M. P., *J. Biol. Chem.*, **240**, 3979 (1965).
[18] Gillam, I., Millward, S., Blew, D., von Tigerstrom, M., Wimmer, E., and Tener, G. M., *Biochemistry*, **6**, 3043 (1967).
[19] Holley, R. W., Apgar, J., Everett, G. A., Madison, J. T., Marquisee, M., Merrill, S. H., Penswick, J. R., and Zamir, A., *Science*, **147**, 1462 (1965).
[20] Zachau, H. G., Dutting, D., and Feldman, H., *Ang. Chemie*, **5**, 122 (1966).
[21] Bayer, A. A., Venkstern, T. V., Mirsabekov, A. D., Krutilina, A. I., Li, L., and Axelrod, V. D., *Molecular Biology (USSR)*, **1**, 754 (1967).
[22] Raj Bhandary, U. L., Chang, S. H., Stuart, A., Faulkner, R. D., Hoskinson, R. M., and Khorana, H. G., *Proc. US Nat. Acad. Sci.*, **57**, 751 (1967).
[23] Doctor, B. P., and Mudd, J., *J. Biol. Chem.*, **238**, 3677 (1963).
[24] Signer, E. R., Beckwith, J. R., and Brenner, S., *J. Mol. Biol.*, **14**, 153 (1965).
[25] Weigert, M. G., Lanka, E., and Garen, A., *J. Mol. Biol.*, **23**, 401 (1967).
[26] Garen, A., Garen, S., and Wilhelm, R. C., *J. Mol. Biol.*, **14**, 167 (1965).
[27] Nirenberg, M., and Leder, P., *Science*, **145**, 1399 (1964).
[28] Lodish, H. F., Abelson, J., and Clark, B. F. C., *Abst. of Fed. Europ. Biochem. Soc., Fourth Meeting, Oslo* (1967).

Reprinted from JOURNAL OF MOLECULAR BIOLOGY, Vol. 19, 1966

Codon—Anticodon Pairing:

The Wobble Hypothesis

F. H. C. CRICK

Medical Research Council, Laboratory of Molecular Biology

Hills Road, Cambridge, England

(*Received 14 February 1966*)

It is suggested that while the standard base pairs may be used rather strictly in the first two positions of the triplet, there may be some wobble in the pairing of the third base. This hypothesis is explored systematically, and it is shown that such a wobble could explain the general nature of the degeneracy of the genetic code.

Now that most of the genetic code is known and the base-sequences of sRNA molecules are coming out, it seems a proper time to consider the possible base-pairing between codons on mRNA and the presumed anticodons on the sRNA.

The obvious assumption to adopt is that sRNA molecules will have certain common features, and that the ribosome will ensure that all sRNA molecules are presented to the mRNA in the same way. In short, that the pairing between one codon–anticodon matching pair will to a first approximation be "equivalent" to that between any other matching pair.

As far as I know, if this condition has to be obeyed, and if all four bases must be distinguished in any one position in the codon, then the pairing in this position is *highly likely* to be the standard one; that is:†

G ==== C

and A ==== U

or some equivalent ones such as, for example,

I ==== C

and A ==== T

since this is the only type of pairing which allows all four bases to be distinguished in a strictly equivalent way.

We now know enough of the genetic code to say that in the *first two* positions of the codon the four bases are clearly distinguished; certainly in many cases, and probably in all of them. I thus deduce that the pairings in the first two positions are likely to be the standard ones.

† Throughout this paper the sign ==== is used to mean "pairs with". If two bases are equivalent in their coding properties, this is written $\begin{matrix}U\\C\end{matrix}$ or $\left.\begin{matrix}U\\C\end{matrix}\right\}$

However, what we know about the code has already suggested two generalizations about the third place of the codon. These are:

(1) U} † this already appears true in about a dozen cases out of the possible 16,
C} and there are no data to suggest any exceptions.

(2) A} probably true in about half of the possible 16 cases, but the evidence
G} suggests it may perhaps be incorrect in several other cases.

The detailed experimental evidence is rather complicated and will not be discussed here. (For details of the code see, for example, Nirenberg *et al.*, 1965; and Söll *et al.*, 1965.) It suffices that these rules *may* be true, as suggested by Eck (1963) a little time ago. Alternatively, only the first one may be true.

This naturally raises the question: Does *one* sRNA molecule recognize more than one codon, e.g. both UUU *and* UUC. Some evidence for this was first presented by Bernfield & Nirenberg (1965). They showed that *all* the sRNA for phenylalanine can be bound by poly U, although this sRNA also recognizes the triplet UUC, at least in part. More recent evidence along these lines is presented in Söll *et al.* (1966) and Kellogg *et al.* (1966). Again I do not wish to discuss here the evidence in detail, but simply to ask: If one sRNA codes both XYU and XYC, how is this done?

Now if we do not know anything about the geometry of the situation, it might be thought that almost any base pairs might be used, since it is well known that the bases can be paired (i.e. form at least two hydrogen bonds) in many different ways. However, it occurred to me that if the first two bases in the codon paired in the standard way, the pairing in the third position might be *close* to the standard ones.

We therefore ask: How many base pairs are there in which the glycosidic bonds occur in a position close to the standard one? Possible pairs are:

$$G ==== A \qquad (1)$$

In my opinion this will not occur, because the NH_2 group of guanine cannot make one of its hydrogen bonds, even to water (see Fig. 1).

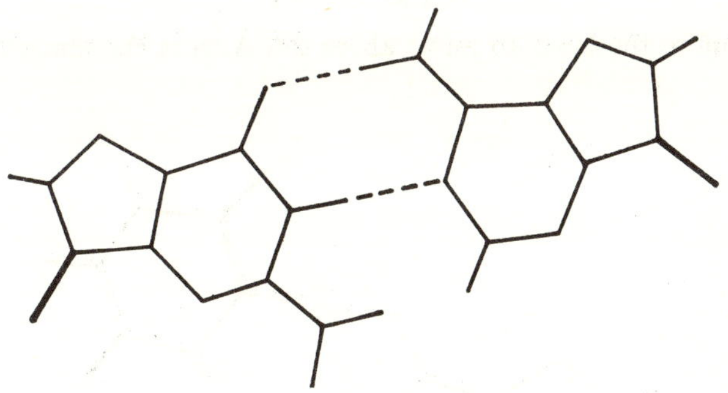

FIG. 1. The unlikely pair guanine–adenine.

$$U ==== C \qquad (2)$$

This brings the two keto groups rather close together and also the two glycosidic bonds, but it may be possible (see Fig. 2).

† This symbol implies that both U and C code the same amino acid.

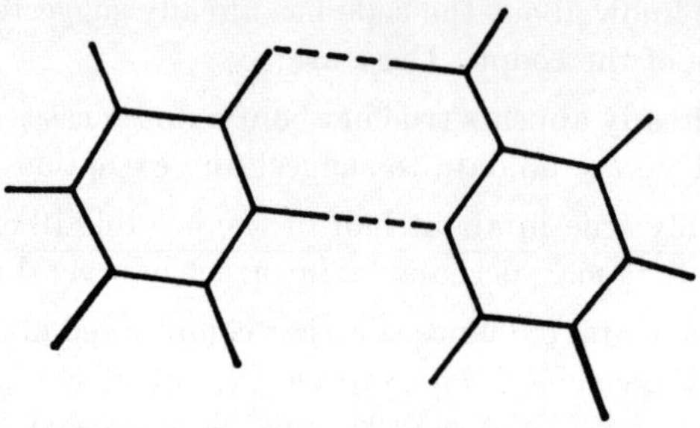

FIG. 2. The close pair uracil–cytosine.

U ==== U (3)

Again rather close together (see Fig. 3).

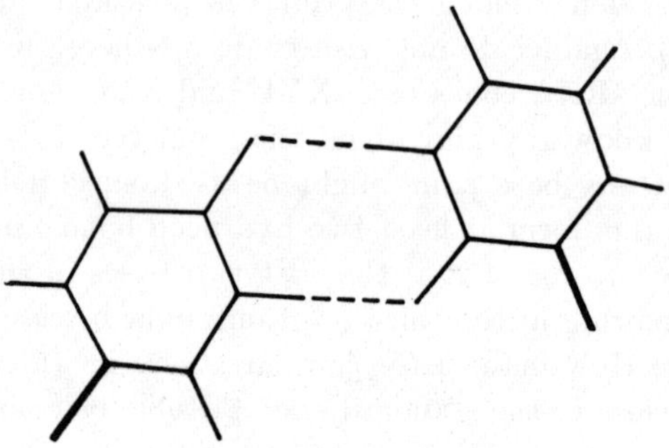

FIG. 3. The close pair uracil–uracil.

G ==== U (4)

or I ==== U

These only require the bond to move about 2·5 Å from the standard position (see Fig. 4).

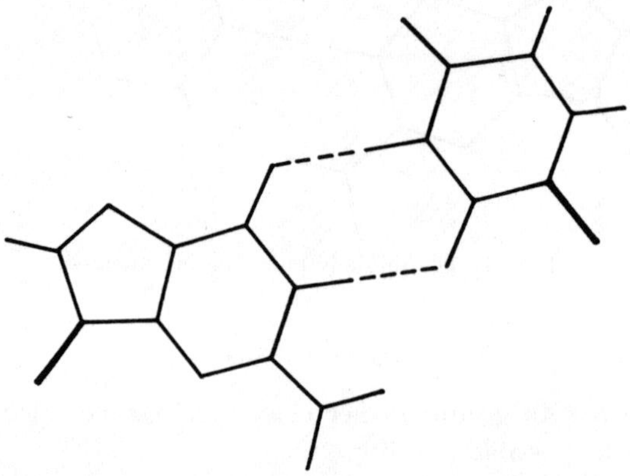

FIG. 4. The pair guanine–uracil (the pair inosine–uracil is similar).

$$I = = = = A \qquad (5)$$

This is perfectly possible. Poly I and poly A will form a double helix. The distance between the glycosidic bonds is increased (see Fig. 5).

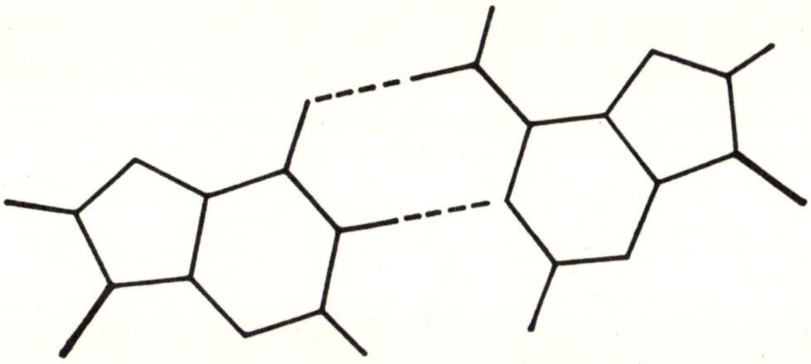

FIG. 5. The pair inosine–adenine.

As far as I know, these are all the possible solutions if it is assumed that the bases are in their usual tautomeric forms.

I now postulate that in the base-pairing of the third base of the codon there is a certain amount of play, or wobble, such that more than one position of pairing is possible.

As can be seen from Fig. 6, there are seven possible positions which might be reached by wobbling. However, it by no means follows that all seven are accessible, since the molecular structure is very likely to impose limits to the wobble. We should therefore strictly consider all possible *combinations of allowed positions*. There are 127 of these, but most of them are trivial. If we adopt the rule that *all four bases* on the codon (in the third position) must be recognized (that is, paired with) we are left with 51 different combinations. This is too many for easy consideration, but fortunately we can eliminate most of them by only accepting combinations which do not violate the broad features of the code. If we assume:

(a) that all four bases must be recognizable;

(b) that the code must *in some cases* distinguish between

$\left.\begin{matrix}U\\C\end{matrix}\right\}$ and $\left.\begin{matrix}A\\G\end{matrix}\right\}$ as it appears to do for the pairs

Phe	Tyr	His	Asn	Asp
Leu	C.T.†	Gln	Lys	Glu

(not all of which are likely to be wrong)

then by strictly logical argument it can be shown both that the standard position must be used, and that the three positions on the left of Fig. 6 cannot be used.

This leaves us with only four possible sites to consider one of which—the standard one—must be included. There are therefore only seven possible combinations. I have examined all these, but I shall restrict myself here to the case in which all four positions are used, as this is structurally the most likely and also seems to give the code (called code 4 in the note privately circulated) which best fits the experimental data.

† C.T., Chain termination.

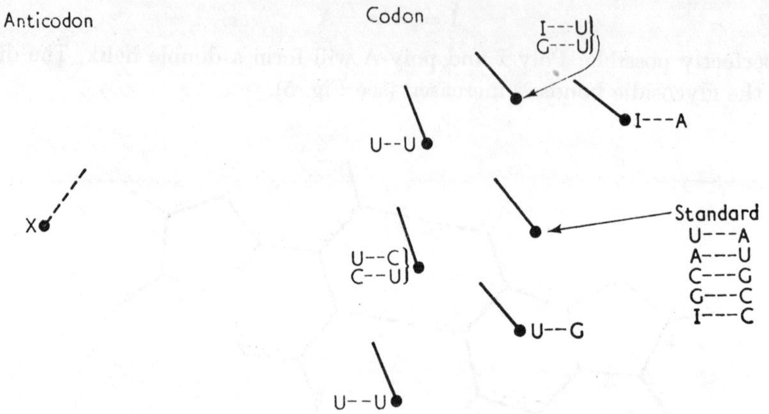

FIG. 6. The point X represents the position of the C_1' atom of the glycosidic bond (shown dotted) in the anticodon. The other points show where the C_1' atom and the glycosidic bond fall for the various base pairs. (Pairs with inosine in the codon have been omitted for simplicity.) The wobble code suggested uses the four positions to the right of the diagram, but not the three close positions.

The rules for pairing between the third base on the codon and the corresponding base on the anticodon are set out in Table 1. It can be seen that these rules make several strong predictions:

(1) it is not possible to code for either C alone, or for A alone.

For example, at the moment the codon UGA has not been decisively allocated. Wobble theory states that UGA might either:

(a) code for cysteine, which has UGU and UGC; or
(b) code for trypotophan, which has UGG; or
(c) not be recognized.

TABLE 1

Pairing at the third position of the codon

Base on the anticodon	Bases recognized on the codon
U	A, G
C	G
A†	U
G	U, C
I	U, C, A

† It seems likely that inosine will be formed enzymically from an adenine in the nascent sRNA. This may mean that A in this position will be rare or absent, depending upon the exact specificity of the enzyme(s) involved.

However it does *not* permit UGA to code for any amino acid other than cysteine or tryptophan. This rule could also explain why no suppressor has yet been found which suppresses only *ochre* mutants (UAA), although suppressors exist which suppress both *ochre* and *amber* mutants (UA^{A}_{G}).

(2) If an sRNA has inosine in the place at the relevant position on the anticodon (i.e. enabling it to pair with the third base of the codon), then it must recognize U, C and A in the third place of the codon. Conversely, those amino acids coded only by XY^{U}_{C} (such as Phe, Tyr, His, etc.) cannot have inosine in that place on their sRNA.

(3) Wobble theory does not state exactly how many different types of sRNA will actually be found for any amino acid. However if an amino acid is coded for by all four bases in the third position (as are Pro, Thr, Val, etc.), then wobble theory predicts that there will be at least two sRNA's. These can have the recognition pattern:

$$\left.\begin{matrix}U\\C\end{matrix}\right\} \text{ plus } \left.\begin{matrix}A\\G\end{matrix}\right\}$$

or

$$\left.\begin{matrix}U\\C\\A\end{matrix}\right\} \text{ plus } G$$

Note that the sets actually used for any amino acid may well vary from species to species.

The Anticodons

At this point it is useful to examine the experimental evidence for the anticodon. In the sRNA for alanine from yeast, Holley *et al.* (1965) have the following sequences:

——— pUpUpIp Gp CpMeIpΨp ———

position ——— 36 37 38 ———

Zachau and his colleagues (Dütting, Karan, Melchers & Zachau, 1965) have for one of the serine sRNA's from yeast:

——— pΨpUpIpGpApA⁺pΨp ———

(A^+ stands for a modified A)

For the valine sRNA from yeast, Ingram & Sjöquist (1963) have shown that the only inosine occurs in the sequence:

——— pIpApCp ———

Holley *et al.* (1965) have already pointed out that IGC is a possible anticodon for alanine, and the additional evidence makes it almost certain to my mind that this is correct, and that the anticodons are as given in the Table below†:

† *Note added 26 April 1966.* Drs J. T. Madison, G. A. Everett and H. Kung (personal communication) have completed the sequence of the tyrosine sRNA from yeast. The sequence strongly suggests that the anticodon in this case is GΨA, corresponding to the known codons UA^{U}_{C}. Since Ψ can form the same base pairs as U, this is in excellent agreement with the previous data.

Yeast sRNA		
	Anticodon	Codon
Ala	I G C	G C ?
Ser	I G A	U C ?
Val	I A C	G U ?

remembering that the pairing proposed between codon and anticodon is *anti*-parallel. Thus I confidently predict: the anticodon is a triplet at (or very near) positions 36–37–38 on every sRNA, and that the *first two bases* in the codon pair with this (in an anti-parallel manner) *using the standard base pairs.*

However, inosine does not occur in every sRNA. In particular Holley *et al.* (1963) (and personal communication) have reported that the tyrosine sRNA has two peaks, neither of which contains inosine. Moreover, Sanger (personal communication) tells me that there is rather little inosine in the total sRNA from *E. coli.*

Testing the Theory

Two obvious tests present themselves:

(1) To find which triplets are bound by any one type of sRNA. This is being done by Khorana and his colleagues (Söll *et al.*, 1966), and also by Nirenberg's group (Kellogg, Doctor, Loebel & Nirenberg, 1966). The difficulty here is to be sure that the sRNA used is pure, and not a mixture.

(2) To discover unambiguously the position of the anticodon on sRNA, and to find further anticodons. This will certainly happen as our knowledge of the base sequence of sRNA molecules develops. The absence of inosine from any anticodon is obviously of special interest.

In conclusion it seems to me that the preliminary evidence seems rather favourable to the theory. I shall not be surprised if it proves correct.

I thank my colleagues for many useful discussions and the following for sending me material in advance of publication: Dr M. W. Nirenberg, Dr H. G. Khorana, Dr G. Streisinger, Dr W. Holley, Dr J. Fresco, Dr H. G. Zachau, Dr C. Yanofsky, Dr H. G. Wittmann, Dr H. Lehmann and Dr J. D. Watson.

REFERENCES

Bernfield, M. R. & Nirenberg, M. W. (1965). *Science*, **147**, 479.

Dütting, D., Karan, W., Melchers, F. & Zachau, H. G. (1965). *Biochim. biophys. Acta*, **108**, 194.

Eck, R. V. (1963). *Science*, **140**, 477.

Holley, R. W., Apgar, J., Everett, G. A., Madison, J. T., Marquisee, M., Merrill, S. H., Penswick, J. R. & Zamir, A. (1965). *Science*, **147**, 1462.

Holley, R. W., Apgar, J. Everett, G. A., Madison, J. T., Merrill, S. H. & Zamir, A. (1963). *Cold Spr. Harb. Symp. Quant. Biol.* **28**, 117.

Ingram, V. M. & Sjöquist, J. A. (1963). *Cold. Spr. Harb. Symp. Quant. Biol.* **28**, 133.

Kellogg, D. A., Doctor, B. P., Loebel, J. E. & Nirenberg, M. W. (1966). *Proc. Nat. Acad. Sci., Wash.* **55**, 912.

Nirenberg, M., Leder, P., Bernfield, M., Brimacombe, R., Trupin, J., Rottman, F. & O'Neal, C. (1965). *Proc. Nat. Acad. Sci., Wash.* **53**, 1161.

Söll, D., Jones, D. S., Ohtsuka, E., Faulkner, R. D., Lohrmann, R., Hayatsu, H., Khorana, H. G., Cherayil, J. D., Hampel, A. & Bock, R. M. (1966). *J. Mol. Biol.* **19**, 556.

Söll, D., Ohtsuka, E., Jones, D. S., Lohrmann, R., Hayatsu, H., Nishimura, S. & Khorana, H. G. (1965). *Proc. Nat. Acad. Sci., Wash.* **54**, 1378.

Reprinted from NATURE, Vol. 206, 1965

BASE COMPOSITION OF NONSENSE CODONS IN *E. coli*

EVIDENCE FROM AMINO-ACID SUBSTITUTIONS AT A TRYPTOPHAN SITE IN ALKALINE PHOSPHATASE

By MARTIN G. WEIGERT and PROF. ALAN GAREN
Department of Molecular Biology and Biophysics, Yale University, New Haven, Connecticut

CERTAIN mutations produce a codon which appears incapable, in some strains of *E. coli*, of coding for any amino-acid. Such a mutation, first identified in *T*4 phage and *E. coli*[1,2], is called a 'nonsense mutation', and the resulting codon a 'nonsense codon'. A nonsense mutation drastically alters the normal course of protein synthesis, by blocking the complete synthesis of the polypeptide chain specified by the gene in which the mutation occurs[3]. A codon which is nonsense in one strain of *E. coli* may code for an amino-acid in another strain if an appropriate suppressor gene is present[1-4].

In this article we present evidence on the base composition of a nonsense codon. This evidence consists of different amino-acid substitutions which occur at one position in the alkaline phosphatase molecule as a result of reversions of a nonsense mutation in the phosphatase structural gene. We find that the codon assignments reported for these amino-acids[5] can be related by single base changes only to a nonsense codon with the RNA base composition UAG.

The procedure for isolating phosphatase-positive revertants from phosphatase-negative nonsense mutants has been reported elsewhere[4]. Mutagenesis with ethyl methane sulphonate (EMS) was performed by spreading about 10^9 mutant cells on agar plates containing a medium selective for phosphatase-positive revertants and placing a small drop of the mutagen on the plate[4] (this procedure was developed by Drs. C. Levinthal and E. Lin). After incubation for 3 days at 37° C, revertant colonies appeared in a cluster surrounding the mutagen spot. For mutagenesis with *N*-methyl-*N*′-nitro-*N*-nitrosoguanidine, the same procedure was used, except that a drop of a saturated solution of this mutagen (instead of EMS) was applied to the agar plate. For nitrous acid mutagenesis, the cells were treated for 35 min at 37° C in a solution of acetate buffer at *p*H 5·0 containing 0·1 M sodium nitrite, after which the cells were grown overnight in nutrient broth and plated for revertants.

The procedures for purifying alkaline phosphatase, for finger-printing its tryptic peptides, and for determining

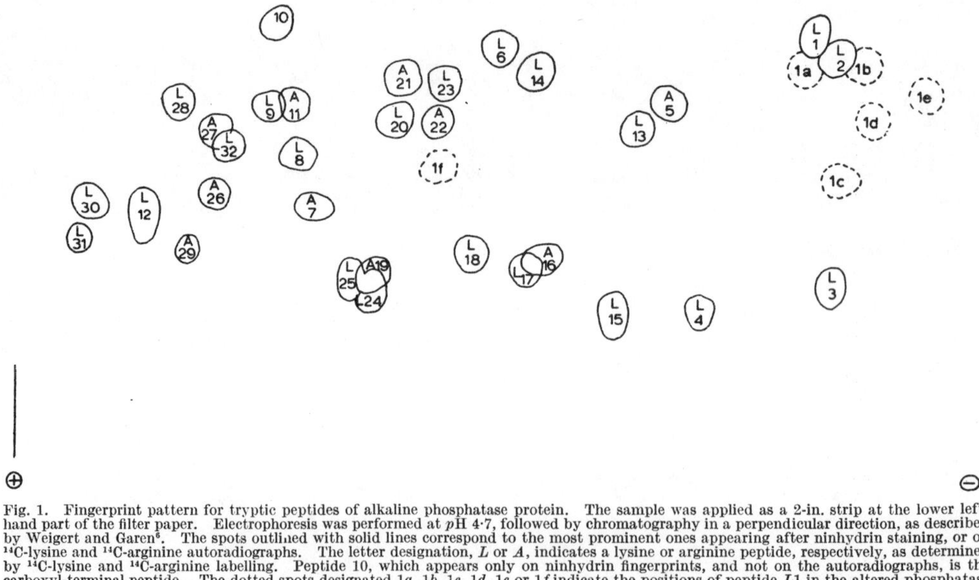

Fig. 1. Fingerprint pattern for tryptic peptides of alkaline phosphatase protein. The sample was applied as a 2-in. strip at the lower left-hand part of the filter paper. Electrophoresis was performed at *p*H 4·7, followed by chromatography in a perpendicular direction, as described by Weigert and Garen[6]. The spots outlined with solid lines correspond to the most prominent ones appearing after ninhydrin staining, or on ^{14}C-lysine and ^{14}C-arginine autoradiographs. The letter designation, *L* or *A*, indicates a lysine or arginine peptide, respectively, as determined by ^{14}C-lysine and ^{14}C-arginine labelling. Peptide 10, which appears only on ninhydrin fingerprints, and not on the autoradiographs, is the carboxyl-terminal peptide. The dotted spots designated 1*a*, 1*b*, 1*c*, 1*d*, 1*e* or 1*f* indicate the positions of peptide *L*1 in the altered phosphatase molecules isolated from revertants. The amino-acid changes in the altered peptides involve the following substitutions for the original tryptophan residue (Table 1): 1*a*, tyrosine; 1*b*, leucine; 1*c*, glutamine; 1*d*, serine; 1*e*, lysine or arginine (the shifted peptide does not contain the new basic resi due since it was released as a free amino-acid by trypsin, as described in Table 1); 1*f*, glutamic acid

Table 1. AMINO-ACID COMPOSITION OF PEPTIDE 1 FROM REVERTANTS OF THE NONSENSE MUTANT *H*12

Strain	Mutagenic agent [a]	Electrophoretic character of alkaline phosphatase [b]	Peptide analysed [c]	Amino-acid residues [d]								Amino-acid substitution [i]
				Lys	Try [e]	Gly	Leu	Ala	Glu [f]	Tyr [g]	Ser [h]	
Standard P^+		Normal	1	1·00	+	0·98	0·90					
			1+2	2·00	+	1·96	2·96	2·04				
Revertants												
*R*1*a*	Spontaneous	Normal	1+2	1·00	–	1·99	4·10	1·91				Try→Leu
*R*3*a*	Spontaneous	Normal	1	1·00	–	1·03	1·02				0·90	Try→Ser
*R*4*a*	Spontaneous	Normal	1	1·00	–	1·12	1·23		0·61			Try→Glu.NH_2
*R*5*a*	Spontaneous	Negative	1	1·00	–	0·83	1·04		0·82			Try→Glu
*R*11*b*	Spontaneous	Normal	1+2	2·00	+	1·93	3·26	2·06				Try→Try
*R*13*b*	Spontaneous	Normal	1+2	2·00	+	1·64	3·02	2·20				Try→Try
*R*14*b*	Spontaneous	Normal	1	1·00	–	1·07	0·97		0·95			Try→Glu.NH_2
*R*16	Nitrous acid	Normal	1+2	2·00	–	1·93	2·99	1·74		0·57		Try→Tyr
*R*17*a*	Nitrous acid	Normal	1	1·00	–	1·05	1·10		0·85			Try→Glu.NH_2
*R*18*b*	Nitrous acid	Normal	1+2	2·00	–	1·80	2·80	1·76		0·52		Try→Tyr
*R*20	Nitrous acid	Normal	1+2	2·00	–	1·85	3·02	1·70		0·55		Try→Tyr
*R*22	Nitrous acid	Normal	1	1·00	–	0·89	0·98				0·84	Try→Ser
*R*25*a*	Nitrous acid	Normal	1+2	2·00	–	1·84	3·03	2·00		0·68		Try→Tyr
*R*30	EMS	Normal	1	1·00	–	0·80	1·02		0·91			Try→Glu.NH_2
*R*34	EMS	Normal	1	1·00	–	0·94	0·97		0·70			Try→Glu.NH_2
*R*39	EMS	Normal	1	1·00	–	0·99	1·05		0·92			Try→Glu.NH_2
*R*61	EMS	Normal	1+2	2·00	+	2·11	3·14	2·20				Try→Try
*R*67	EMS	Normal	1+2	2·00	+	1·78	3·26	1·80				Try→Try
*R*47	NG	Normal	1	1·00	–	0·87	0·86		0·84			Try→Glu.NH_2
*R*48	NG	Positive	1	1·00	–	0·72	0·91					Try→Lys or Arg
*R*52	NG	Positive	1	1·00	–	0·83	1·07					Try→Lys or Arg

Table 1. *a*, EMS is ethyl methane sulphonate, and NG is *N*-methyl-*N'*-nitro-*N*-nitrosoguanidine. The mutagenic procedures are described in materials and methods. *b*, The electrophoretic tests were done by starch-gel electrophoresis[7]. The designation as negative indicates that the enzyme migrated more rapidly than the standard enzyme towards the positive electrode; a designation as positive indicates the converse. *c*, For some of the enzymes, peptide 1 separated on the fingerprint from peptide 2 and could be analysed alone. For the others, the two peptides were too close for separate analyses and therefore were combined for a single analysis. The separation of peptide 1 of the standard P^+ enzyme was achieved by electrophoresis at *p*H 1·9 (ref. 6). *d*, Details on the sensitivity and accuracy of the amino-acid analysis are reported in another publication[6]. *e*, Tryptophan was determined by a specific staining test on the fingerprint paper[6]. *f*, The distinction between glutamic acid and glutamine was based on the electrophoretic behaviour of the enzyme. Since the original amino-acid was tryptophan, it was concluded that, if the enzyme was not changed electrophoretically, the substitution was glutamine, and, if there was a change towards electronegativity, the substitution was glutamic acid. This criterion was consistent with the position of the altered peptide in the fingerprint pattern: a glutamine substitution did not change the electrophoretic mobility of peptide 1, while a glutamic acid substitution made the peptide more electronegative. *g*, There was destruction of tyrosine during hydrolysis, amounting to about 23 per cent. The values in this Table have not been corrected for this loss. *h*, There was destruction of serine during hydrolysis, amounting to 9 per cent. The values in this Table have not been corrected for this loss. *i*, The substitution of lysine or arginine for tryptophan was inferred from the finding that the revertant enzyme was more electropositive than the standard enzyme and showed no new amino-acid replacing the lost tryptophan. A change to a basic amino-acid should subject the peptide to cleavage by trypsin. If the new basic residue in peptide 1 was located terminally or in the position adjacent to the lysine normally present, it should be released as a free amino-acid. There is preliminary evidence from the Edman degradation procedure[10] that the position of the tryptophan residue in the standard peptide 1 is at the amino-end of the peptide (Weigert and Lanka, unpublished results).

the amino-acid composition of the peptides, have been described elsewhere[6].

Most of the reversions of the nonsense mutation *H*12 cause an alteration in the alkaline phosphatase molecule produced by the revertant strain. The site of the alteration is in the tryptic peptide 1. The position of this peptide is shifted in the phosphatase fingerprint patterns for several revertant strains, as shown in Fig. 1. The amino-acid composition data for peptide 1 as determined for the phosphatases from twenty-one revertant strains are presented in Table 1. Some of the reversions were of spontaneous origin, and the others were induced either by nitrous acid, ethyl methanesulphonate or *N*-methyl-*N'*-nitro-*N*-nitrosoguanidine. With seventeen of the revertant strains, peptide 1 showed an amino-acid substitution which, in every case, involved the loss of the tryptophan residue in this peptide. A total of six different amino-acids were found to substitute for tryptophan: glutamine, glutamic acid, tyrosine, serine, leucine and either lysine or arginine. Each of these substitutions is compatible with the enzymatic function of alkaline phosphatase.

Of special interest is the finding that substitution of serine for tryptophan can result from reversion. The same substitution was also observed in a suppressed strain of this nonsense mutant[6]. Although serine is substituted at the same position in both the revertant and the suppressed mutant strains, different codons for serine must be involved.

Phosphatase-negative nonsense mutants are totally incapable of synthesizing alkaline phosphatase[2]. Restoration of enzyme synthesis can be achieved in two ways: either by introducing into the mutant strain a suppressor gene that acts on nonsense mutations[4], or by inducing another mutation (reversion) in the phosphatase structural gene of the mutant. In another communication[6] we reported amino-acid substitutions in alkaline phosphatase resulting from suppression of nonsense mutations. In the present experiments amino-acid substitutions produced by reversions of a nonsense mutation have been identified.

It is known that reversions of *mis*sense mutations can occur outside the codon in which the original mutation is located[8], leading to the formation of a protein with amino-acid substitutions at two positions in the molecule. For nonsense mutations, we expect reversions to occur exclusively within the nonsense codon, since this codon must be transformed into one that codes for an amino-acid if protein synthesis is to proceed to completion. The reverted nonsense codon need not, however, be identical to the original codon present in the parental strain. In terms of protein structure, reversions of one nonsense mutation should affect only one position in the protein molecule, possibly with different amino-acid substitutions occurring at that position. The results reported in Table 1 for reversions of one alkaline phosphatase nonsense mutation, *H*12, show that six different amino-acids can be substituted for tryptophan at one position in the phosphatase molecule. No other effect attributable to reversions of this mutation has been detected in the fingerprint patterns of the phosphatase molecules. It should be noted that amino-acid substitutions at the same tryptophan position can also result from suppression of this nonsense mutation[6], providing further evidence that this is the position affected by the original mutation.

The amino-acid substitution data in Table 1 place certain restrictions on the possible base composition of the nonsense codon in the mutant studied. Consider the RNA triplet codon assignments that have been reported for the amino-acids involved[5], as listed in Table 2. We shall assume that these assignments are complete for each amino-acid. The problem is to designate a triplet which can be derived by a single base change from the RNA codon originally present at the nonsense position (the UGG codon for tryptophan), and which can be

Table 2

Amino-acid	RNA codons
Try	UGG
Glu.NH_2	UAC UAG AAG AAC
Glu	UAG AAG ACG
Tyr	UAC UUA
Ser	UUC UCC UCG ACG
Leu	UUG UUC UCC UUA
Lys	AAA UAA
Arg	UCG CCG AAG ACG

The coding assignments are taken from Speyer *et al.* (ref. 5) and Nirenberg *et al.* (ref. 5), and include all the RNA triplets listed for each amino-acid. Only the composition, and not the sequence, is indicated for each triplet.

transformed by a single base change into any of the codons for the known amino-acid substitutions at that position. If the substitution involving a basic amino-acid, which could be either lysine or arginine, is assumed to be arginine, there are three possibilities for the composition of the nonsense codon, UAG, UCG or AGG. If lysine is the basic amino-acid substituted, UAG emerges as the unique nonsense codon.

Another way of deducing the composition of the nonsense codon is from the mutagenic specificity of hydroxylamine, the mutagen used to produce the *H*12 nonsense mutation[4]. Hydroxylamine appears to mutate cytosine to thymine in DNA (ref. 10). According to the information in Tables 1 and 2, the original DNA codon affected by the nonsense mutation is ACC, corresponding to the RNA codon UGG for the tryptophan residue at that position. This codon should mutate to ATC with hydroxylamine, yielding as the RNA nonsense codon UAG.

The present experiments involve one of many nonsense mutations in the structural gene for alkaline phosphatase. Do nonsense mutations at other sites generate the same nonsense codon, or is there more than one such codon that responds to the same set of suppressor genes? Evidence on this point will be presented in a separate report.

This work was supported by research grants from the U.S. National Science Foundation and U.S. National Institutes of Health.

[1] Benzer, S., and Champe, S. P., *Proc. U.S. Nat. Acad. Sci.*, **48**, 1114 (1962).

[2] Garen, A., and Siddiqi, O., *Proc. U.S. Nat. Acad. Sci.*, **48**, 1121 (1962).

[3] Sarabhai, A. S., Stretton, A. O. W., Brenner, S., and Bolle, A., *Nature*, **201**, 13 (1964).

[4] Garen, S., Garen, A., and Wilhelm, R. C., *J. Mol. Biol.* (in the press).

[5] Speyer, J. F., Lengyel, P., Basilio, C., Wahba, A. J., Gardner, R. S., and Ochoa, S., *Cold Spring Harbor Symp. Quant. Biol.*, **28**, 559 (1963). Nirenberg, M. W., Jones, O. W., Leder, P., Clark, B. F. C., Sly, W. S., and Pestka, S., *ibid.*, **28**, 549 (1963).

[6] Weigert, M., and Garen, A., *J. Mol. Biol.* (in the press).

[7] Levinthal, C., Signer, E. R., and Fetherholf, K., *Proc. U.S. Nat. Acad. Sci.*, **48**, 1230 (1962).

[8] Garen, A., Levinthal, C., and Rothman, F., *J. Chim. Phys.*, **58**, 1068 (1961). Helinski, D. R., and Yanofsky, C., *J. Biol. Chem.*, **238**, 1043 (1963).

[9] Freese, E., Bautz-Freese, E., and Bautz, E., *J. Mol. Biol.*, **3**, 133 (1961); Freese, E., Bautz, E., and Freese, E. B., *Proc. U.S. Nat. Acad. Sci.*, **47**, 845 (1961); Schuster, H., *J. Mol. Biol.*, **3**, 447 (1961); Tessman, I., Poddar, R. K., and Kumar S., *J. Mol. Biol.*, **9**, 352 (1964).

[10] Konigsberg, W., and Hill, R. J., *J. Biol. Chem.*, **237**, 2547 (1962).

Reprinted from NATURE, Vol. 206, 1965

GENETIC CODE: THE 'NONSENSE' TRIPLETS FOR CHAIN TERMINATION AND THEIR SUPPRESSION

By Dr. S. BRENNER, F.R.S., Dr. A. O. W. STRETTON and Dr. S. KAPLAN
Medical Research Council, Laboratory of Molecular Biology, Cambridge, England

THE nucleotide sequence of messenger RNA is a code determining the amino-acid sequences of proteins. Although the biochemical apparatus which translates the code is elaborate, it is likely that the code itself is simple and consists of non-overlapping nucleotide triplets. In general, each amino-acid has more than one triplet corresponding to it, but it is not known how many of the sixty-four triplets are used to code for the twenty amino-acids. Triplets which do not correspond to amino-acids have been loosely referred to as 'nonsense' triplets, but it is not known whether these triplets have an information content which is strictly null or whether they serve some special function in information transfer.

The evidence that there are nonsense triplets is mainly genetic and will be reviewed later in this article. A remarkable property of nonsense mutants in bacteria and bacteriophages is that they are suppressible; wild-type function can be restored to such mutants by certain strains of bacteria carrying suppressor genes. It was realized early that this implies an ambiguity in the genetic code in the sense that a codon which is nonsense in one strain can be recognized as sense in another. The problem of nonsense triplets has become inextricably connected with the problem of suppression and, in particular, it has proved difficult to construct a theory of suppression without knowing the function of the nonsense triplet.

In this article we report experiments which allow us to deduce the structure of two nonsense codons as UAG and UAA. We suggest that these codons are the normal recognition signals in messenger RNA for chain termination and, on this basis, propose a theory of their suppression.

Nonsense mutants and their suppressors. One class of suppressible nonsense mutants which has been widely examined includes the subset I ambivalent *r*II mutants[1], the suppressible mutants of alkaline phosphatase[2], the *hd* or *sus* mutants of phage λ[3], and many of the *amber* mutants of bacteriophage *T*4 (ref. 4). These mutants have been isolated in various ways and the permissive (su^+) and non-permissive (su^-) strains used have been different. When isogenic bacterial strains, differing only in the *su* locus, are constructed, it can be shown that all these mutants respond to the same set of suppressors. They are therefore of the same class and we propose that all these mutants should be called *amber* mutants.

We may now consider the evidence that these mutants contain nonsense codons. Garen and Siddiqi[2] originally noted that *amber* mutants of the alkaline phosphatase of *E. coli* contained no protein related immunologically to the enzyme. Benzer and Champe[1] also showed that the mutants exert drastic effects and suggested that *amber* mutants of the *r*II gene interrupt the reading of the genetic message. In the *r*II genes, a deletion, *r*1589, joins part of the A cistron to part of the B cistron; complementation tests show that this mutant still possesses B activity although it lacks the A function. It may therefore be used to test the effects of mutants in the A cistron on the activity of the B. Double mutants, composed of an A *amber* mutant together with *r*1589 did not have B activity on the su^- strain; this effect is suppressed by the su^+ strain which restores B activity. This result is explained by our finding that each *amber* mutant of the head protein produces a characteristic fragment of the polypeptide chain in su^- bacteria[5]. More recently we have shown that the polarity of the fragment is such that it can only be produced by the termination of the growing polypeptide chain at the site of the mutation[6]. In su^+ strains, both the fragment and a completed chain are produced, and the efficiency of propagation depends on which of the

Wild Type	Ala.Gly.(Val,Phe)Asp.Phe.Gln.Asp.Pro.Ile.Asp.Ile.Arg . . .
H 36 on su⁻	Ala.Gly.Val.Phe.Asp.Phe.
H 36 on su^+_I	Ala.Gly.Val.Phe.Asp.Phe.
	and
	Ala.Gly.Val.Phe.Asp.Phe.Ser.Asp.Pro.Ile.Asp.Ile.Arg . . .

Fig. 1. Amino-acid sequences of the relevant region of the head protein in wild-type *T4D* and in the *amber* mutant *H*36 on su^- and su^+_I strains

suppressors is carried by the strain. In su^+_I, the efficiency of propagation is about 65 per cent[7], and the completed chain contains a serine at a position occupied by a glutamine in the wild type[6] as shown in Fig. 1.

In addition to the *amber* mutants, there are other mutants, called *ochre* mutants, which are suppressed by a different set of suppressors[8]. *Ochre* mutants of the A cistron of *r*II abolish the B activity of *r*1589. This effect is not suppressed by *amber* suppressors, which shows that the *ochre* mutants are intrinsically different from the *amber* mutants.

Thus, nonsense mutants may be divided into two types, *amber* and *ochre* mutants, depending on their pattern of suppression. In Table 1, which is abstracted from a larger set of results[8], it can be seen that strains, carrying the *amber* suppressors su^+_I, su^+_{II}, su^+_{III} and su^+_{IV}, suppress different but overlapping sets of *amber* mutants, but do not suppress any *ochre* mutants. This is the feature which distinguished the two classes of mutants from each other. Table 1 also shows that *ochre* mutants are suppressed by one or more of the strains carrying the *ochre* suppressors su^+_B, su^+_C, su^+_D and su^+_E. These suppressors are also active on various *amber* mutants, and we have not yet been able to isolate suppressors specific for *ochre* mutants. Table 1 also shows that the suppressor strains can be differentiated by the set of mutants they suppress.

Table 1. SUPPRESSION OF *r*II MUTANTS BY su^+ STRAINS OF *E. coli Hfr H*(λ)

		Amber suppressors				*Ochre* suppressors			
		su^+_I	su^+_{II}	su^+_{III}	su^+_{IV}	su^+_B	su^+_C	su^+_D	su^+_E
Amber mutants									
*r*IIA	*HD*120	+	+	+	+	+	0	+	0
	*N*97	+	poor	+	0	poor	poor	+	0
	*S*116	0	0	+	0	0	+	+	0
	*N*19	+	poor	+	0	poor	0	+	0
	*N*34	+	+	+	+	+	+	+	0
*r*IIB	*HE*122	+	+	+	0	poor	0	+	0
	*HB*74	+	+	+	+	+	+	+	+
	*X*237	+	poor	+	0	0	+	+	0
	*HB*232	+	+	+	+	+	0	+	+
	*X*417	+	+	+	+	0	poor	+	0
Ochre mutants									
*r*IIA	*HD*147	0	0	0	0	+	+	+	+
	*N*55	0	0	0	0	+	+	+	0
	*X*20	0	0	0	0	+	0	+	0
	*N*21	0	0	0	0	+	0	0	+
*r*IIB	*UV*375	0	0	0	0	+	+	+	+
	360	0	0	0	0	+	0	+	+
	375	0	0	0	0	+	+	+	0
	*HF*208	0	0	0	0	+	0	0	+
	*N*29	0	0	0	0	0	0	+	0

Unlike some of the *amber* suppressors, all the *ochre* suppressors are weak[7,8]. This has made the isolation of *ochre* mutants of the head protein impossible. We therefore do not know the molecular consequences of *ochre* mutants, but we shall assume that, like the *amber* mutants, they too result in chain termination.

If we accept that both types of nonsense mutants result in termination of the polypeptide chain, we have to ask: at which level of information transfer is this effect exerted ? We have recently shown that it is likely that chain termination occurs as part of protein synthesis, since both types of mutants vanish when the phase of reading of the genetic message is altered[9]. This leads us to conclude that *amber* and *ochre* mutants produce different triplets which have to be read in the correct phase and which are recognized as signals for chain termination.

Decoding of amber and ochre triplets. We have done two types of experiments which allow us to deduce the structures of the *amber* and *ochre* triplets. First, we studied the production and reversion of *r*II *amber* and *ochre* mutants using chemical mutagens. We show that the two triplets are connected to each other and that we can define their possible nucleotide compositions. Next, we investigated head protein *amber* mutants, to define the amino-acids connected to the *amber* triplet. Comparison of these results with known amino-acid codons allows us to deduce the structures of the *amber* and *ochre* triplets.

The experiments with *r*II mutants depend on the specificity of the mutagenic agent, hydroxylamine. This reacts only with cytosine in DNA, and although the exact structure of the product has not yet been defined, the altered base (called U′) appears to act like T with high efficiency, producing base-pair transitions of the G—C → A—T type[10]. Hence, response of any particular site of the DNA to hydroxylamine is evidence for the existence of a G—C pair at that site. Usually, phage particles are treated with hydroxylamine and, since these contain double-stranded DNA, the alteration of C occurs on only one of the strands. In any given gene, only one of the strands is transcribed into messenger RNA (ref. 11). This is the *sense* strand; it carries the genetic information proper and contains a nucleotide sequence which is the inverse complement of the sequence of the messenger RNA. The other strand, the *antisense* strand, has the same sequence as the messenger. In a phage, treated with hydroxylamine, the altered base, U′, could be on either the sense or the antisense strand of the DNA. Since the *r*II genes express their functions before the onset of DNA replication[12], only sense strand changes will register a phenotypic effect in the first cycle of growth; changes on the antisense strand, while still yielding altered DNA progeny, will go unexpressed (Fig. 2).

We can now examine the reversion properties of *amber* and *ochre* mutants. Champe and Benzer[12] studied reversion of a large number of *r*II mutants using different mutagens. They noted that no *amber* mutant was induced to revert by hydroxylamine. Some of the mutants they studied can now be identified as *ochre* mutants, and their results show that *ochre* mutants are equally insensitive to hydroxylamine. However, in their experiments, the treated mutant phages were plated directly on the bacterial strain which restricts the growth of *r*II mutants. They could therefore detect *sense* strand changes only, and any mutation on the antisense strand would not have been expressed. Strictly speaking, then, their results tell us that neither the *amber* nor the *ochre* triplet contains a C on the sense strand of the DNA, or, if any one does, it is connected by a C → U change to another nonsense codon. To extend this result and to recover all possible mutational changes, we grew the mutagenized phages in *E. coli B*, in which the *r*II functions are unnecessary, and then plated the progeny on strain *K* to measure reversion frequency. Table 2 shows that *amber* and

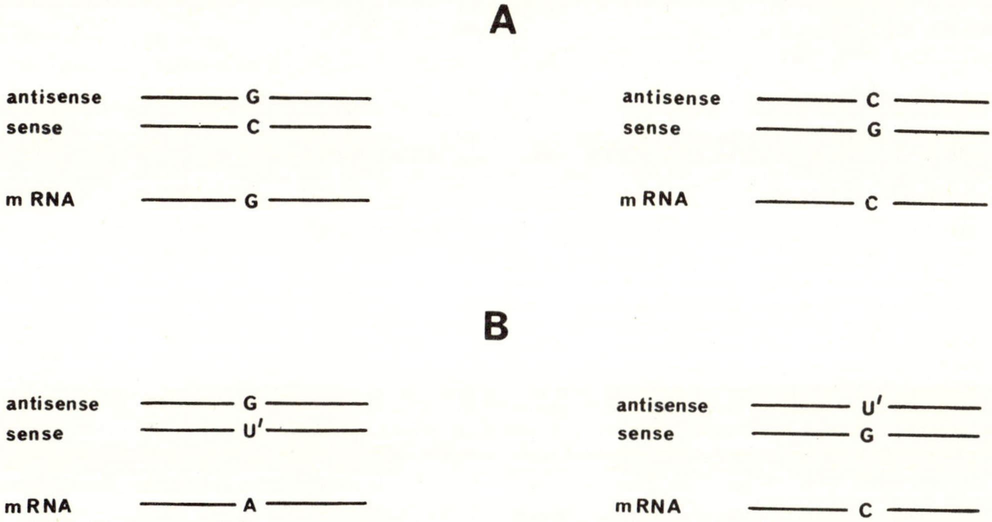

Fig. 2. Diagram illustrating the expression of the two types of G–C pairs (*A*) before and (*B*) after treatment with hydroxylamine

Table 2. Reversion of Amber and Ochre Mutants after Allowing DNA Replication

		Reversion index × 10^{-7} Control	NH_2OH
amber mutants	*S*116	0·04	0·03
	*HD*26	0·1	0·2
	*S*24	0·4	0·9
	*S*99	0·1	0·1
	*N*19	0·15	0·13
	*HD*59	0·0	0·05
	*HB*232	0·1	0·4
ochre mutants	*UV*375	0·8	2·0
	360	0·6	0·8
	*X*27	0·3	0·5
	375	0·8	0·9
	*X*511	0·2	0·3
	*UV*256	1·0	230

Phages were incubated in a solution of M NH_2OH in 2 M NaCl and 0·05 M sodium phosphate (*p*H 7·5) for 2 h at 37°. The reaction was terminated by dilution into acetone broth. About 10^8 phage particles were used to infect a culture of *E. coli* B which was grown to lysis, and the progeny assayed on *E. coli B* and *E. coli K*12(λ). The reversion index is the *K*/*B* ratio. The control was treated in the same way except that hydroxylamine was omitted. The mutant *UV*256, which is not an *ochre* or an *amber* mutant, was used to check the efficacy of the mutagenic treatment.

ochre mutants are not induced to revert by hydroxylamine, and we conclude that in neither mutant, does the triplet in the DNA contain G—C pairs, or, if a G—C pair is present, that triplet is connected by a G—C → A—T transition to another nonsense codon. In other words, subject to the last important reservation, we can conclude that the codons on the messenger RNA contain neither G nor C.

However, we next discovered that *ochre* mutants can be converted into *amber* mutants by mutation. Since *ochre* mutants are not suppressed by *amber* suppressors, plating on strains carrying such suppressors selects for *amber* revertants. Wild-type revertants also grow, but the two can be distinguished by testing revertant plaques on the su^- strain. Twenty-six *r*II *ochre* mutants have been studied and, of these, 25 have been converted into *amber* mutants. A sample of the results is given in Table 3, which shows that the mutation is strongly induced by 2-aminopurine, as strongly as the reversion of the *ochre* mutant to wild type. Other experiments, not reported here, show that the mutations of *ochre* mutants both to the *amber* and to the wild type are also induced by 5-bromouracil, but the induction is weaker than with 2-aminopurine. These results prove that the *amber* and *ochre* triplets differ from each other by only one nucleotide base, and must have the other two bases in common. 2-Aminopurine is a base analogue mutagen inducing the transition A—T ⇌ G—C in both directions[13]. This tells us that one of the triplets has a G—C pair in the DNA. The experiment reported in Table 4 shows that *ochre* mutants cannot be induced to mutate to *amber* mutants with hydroxylamine, even after the treated phages have been grown in *E. coli B*. This shows that it is the *amber* triplet which has the G—C pair and the *ochre* which contains the A—T pair.

Although the insensitivity of the mutants to reversion induction by hydroxylamine might suggest that they contain A—T base pairs only, the conversion of *ochre* mutants to *amber* mutants shows that, in one position, the *amber* mutant contains a G—C pair. The other two bases must be common to both triplets, but we cannot conclude that both are A—T pairs. In fact, both could be G—C pairs and the triplets may be connected to other nonsense triplets by G—C → A—T changes. However, we know that *amber* and *ochre* mutants can be induced by hydroxyl-

Table 3. Mutation of *ochre* Mutants to *amber* Mutants

*r*IIA cistron	Reversion index × 10^{-7} Spontaneous (wild type + *ambers*)	2-Aminopurine wild type	*amber*
*N*55	0·5	830	280
*X*20	0·05	100	2,100
*X*372	0·1	300	2,100
*X*352	0·06	340	1,500
*HD*147	0·3	370	50
*r*IIB cistron			
*X*511	0·2	710	65
*N*17	0·2	610	80
*SD*160	0·6	380	390
*N*29	1·0	3,900	330
*AP*53	0·7	350	15

Cultures of *E. coli B* in minimal medium with and without 2-aminopurine (600 μg/ml.) were inoculated with about 100 phages and grown to lysis. These were plated on *E. coli B* and on *E. coli K*12(λ) su^+_I. About 50 induced revertants were tested on *E. coli K*12(λ) su^- to measure the relative frequencies of *amber* revertants.

Table 4. Induction of the *ochre*→*amber* Mutation

		Reversion index × 10^{-7} r^+	*amber*
360	Control	0·6	0·2
	Hydroxylamine	0·8	0·3
	2AP	200	1,200
*UV*375	Control	0·8	0·4
	Hydroxylamine	2·0	1·0
	2AP	660	140
*X*27	Control	0·3	< 0·1
	Hydroxylamine	1·0	0·5
	2AP	7·0	73
375	Control	0·8	0·4
	Hydroxylamine	2·0	1·0
	2AP	1,400	1,700

Hydroxylamine treatment and growth of the mutagenized phages, and 2-aminopurine induction, were carried out as described in Tables 2 and 3.

amine from wild type. This proves that both triplets have at least one common A—T pair.

We now present an experiment which shows that the *amber* triplet has two A—T base pairs, and which also establishes the orientation of the pairs with respect to the two strands of DNA. Let us suppose that the *amber* triplet in the messenger RNA contains a U. This corresponds to an A in the sense strand of DNA of the *amber* mutant, implying that the wild-type DNA contains a G in this strand and a C in the antisense strand. When the wild-type DNA is treated with hydroxylamine to alter this C the change is not effective and normal messenger is still made (Fig. 2, right). On the other hand, if the *amber* triplet in the messenger contains an A, the mutant will be induced by the action of hydroxylamine on a C in the sense strand of the wild-type phage DNA, and provided that the U′ produced acts identically to U in messenger synthesis, mutant messenger will be made. This argument has been tested by the following experiment. Wild-type $T4r^+$ phages were treated with hydroxylamine to induce *r* mutants to a frequency of 1 per cent. In set B, the phages were then grown on *E. coli B*, in which the *r*II functions are not required, to recover all mutants. In set K, the phages were grown through *E. coli K*12(λ) su^-, to eliminate from the population all phages with an immediate mutant expression. *Amber* and *ochre* mutants were then selected and mapped. Table 5 summarizes the results. About the same number of *r*I mutants were recorded in each set, and since these mutants show no difference in growth on the two bacterial strains, this shows that the results may be compared directly. It will be seen that *amber* mutants at the sites, *N*97, *S*116, *S*24, *N*34, *X*237 and *HB*232, recur many times in set B, but are absent or rarely found in set K. At other *amber* sites, such as *HB*118, *HB*129, *EM*84 and *AP*164, mutants occur with approximately equal frequency in both sets. The first class fulfils the expectation for a C → U change on the sense strand, while the second class must arise by C → U changes on the antisense strand. This shows that the *amber* triplet in the messenger contains both an A and a U. The same should be true of the *ochre* mutants. However, as shown in Table 5, *ochre* mutants are not as strongly induced by hydroxylamine as are *amber* mutants, and we cannot separate the two classes with the same degree of confidence. Nevertheless, since we have already shown that the mutants are connected, it follows that the *ochre* triplet must also contain an A and a U. We conclude that the *amber* and *ochre* triplets are, respectively, either (UAG) and (UAA), or (UAC) and (UAU). If we had a strain which suppressed *ochre* mutants only, we could specify the third base by studying the induction of the *amber* → *ochre* change with hydroxylamine.

Table 5. HYDROXYLAMINE INDUCTION OF *amber* AND *ochre* MUTANTS

A. No. of mutants isolated

	Set B	Set K
*r*I	2,010	1,823
Leaky or high reverting *r*II	720	508
non-suppressible *r*II	1,144	433
amber	319	121
ochre	83	82
Total	4,276	2,967

B. Recurrences found at different sites

Amber mutants	No. found at each site		*Ochre* mutants	No. found at each site	
Site	Set B	Set K	Site	Set B	Set K
A cistron			A cistron		
*HB*118	27	15	*HD*147	2	0
*C*204	1	0	*HF*220	1	0
*N*97	44	1	*HF*240	1	0
*S*116	31	2	*N*55	19	19
*N*11	3	3	*X*20	9	8
*S*172	9	5	*HF*219	1	0
*S*24	44	3	*HF*245	1	0
*HB*129	14	25	*N*31	3	5
*S*99	12	16	*HM*127	0	1
*N*19	15	9	*N*21	2	2
*N*34	8	0			
B cistron			B cistron		
*HE*122	1	0	360	11	10
*EM*84	29	21	*UV*375	2	0
*HB*74	16	5	*N*24	6	4
*X*237	14	2	375	2	5
*AP*164	28	12	*N*17	5	3
*HB*232	21	1	*HF*208	1	2
*X*417	1	0	*N*7	4	12
*HD*231	1	1	*N*12	5	7
			*X*234	0	2
			*X*191	1	0
			*HE*267	5	0
			*AP*53	2	2

*T*4Br^+ was treated with M hydroxylamine (see Table 2) for 2 h at 37° C. Survival was 50 per cent, and the frequency of *r* and mottled plaques, 1 per cent. $1 \cdot 2 \times 10^7$ phage particles were adsorbed to 10^9 cells of *E. coli B* (set B) and to *E. coli K*12(λ) su^- (set K). After 8 min, the infected bacteria were diluted a thousand-fold into 2 litres of broth, incubated for 35′ and lysed with $CHCl_3$. The burst sizes in both sets were 60. *r* mutants were isolated from each set using less than 2 ml. to ensure that the mutants selected had mostly arisen from independent events. These were picked and stabbed into B and K, and *r*I mutants and leaky mutants discarded. The *r*II mutants were then screened on su^+III and su^+B to select for *amber* and *ochre* mutants which were then located by genetic mapping.

Fortunately, we can resolve the ambiguity by determining the amino-acids to which the *amber* triplet is connected by mutation. In particular, we note that it should be connected to two and only two amino-acid codons by transitions, corresponding, in fact, to the two types of origin of the mutants described here. The third codon to which it is connected by a transition is the *ochre* triplet. As mentioned earlier, the head *amber* mutant *H*36 has arisen from glutamine (Fig. 1). This mutant was induced with hydroxylamine. We have evidence that two other mutants, *E*161 and *B*278 induced by 2-aminopurine and 5-bromouracil respectively, have arisen from tryptophan. In a recent study of two *amber* mutants of the alkaline phosphatase, Garen and Weigert[14] found one mutant to arise from glutamine and the other from tryptophan; and Notani *et al.*[15] have found an *amber* mutant to arise from glutamine in the RNA phage *f*2. In addition, we have examined 2-aminopurine induced revertants of 10 different head *amber* mutants. Ten to 12 independently induced revertants of each of the mutants have been screened for tryptophan containing peptides by examining the ^{14}C-tryptophan labelled protein. Among a total of 115 revertants, 62 are to tryptophan. Determination of glutamine involves sequence analysis and takes more time. So far, among the remaining 53 revertants, glutamine has been identified in one revertant of *H*36. These results suggest that the two amino-acids connected to the *amber* triplet are glutamine and tryptophan. If the *amber* triplet is (UAC), then one of these must be (CAC) and the other (UGC); if it is (UAG), then the corresponding codons are (CAG) and (UGG). Nirenberg *et al.*[16] have shown that poly AC does not code for tryptophan, but does for glutamine. However, they find that the triplet for glutamine clearly has the composition (CAA) and is definitely not (CAC). Since this latter triplet corresponds neither to glutamine nor to tryptophan we can eliminate the first alternative. We note with satisfaction that (UGG) is the composition of a codon assigned to tryptophan[16,17], and this assignment of (UAG) to the *amber* triplet suggests that glutamine is (CAG).

We can also make a reasonable assignment of the order of the bases in the triplet. Our original argument was based on deductions from a few known triplets and from amino-acid replacement data; it will not be given here. The order of the bases follows directly from a recent demonstration by Nirenberg *et al.*[18] that the triplet CAG does, in fact, correspond to glutamine. The *amber* triplet is therefore UAG and the *ochre* triplet UAA. This assignment is supported by the following additional evidence. We have found a tyrosine replacement in 21 independent spontaneous mutants of the head *amber* mutant, *H*36 (ref. 19). This change must be due to a transversion because we have already accounted for all the transitions of the *amber* triplet. In support of this, we find that the change is not induced by 2-aminopurine. There are six possible transversions of the *amber* triplet, namely, AAG, GAG, UUG, UCG, UAU and UAC. It has recently been shown that both UAU and UAC correspond to tyrosine[20] which confirms the order. The spontaneous revertants of the *amber* mutants to leucine, serine and glutamic acid found by Weigert and Garen[21] are further evidence for the assignment. UUG, a transversion of the *amber* triplet, does in fact code for leucine[22], and reasonable allocations for serine and glutamic acid are UCG and GAG, respectively. Weigert and Garen[21] also find revertants of an *amber* mutant to either lysine or arginine. This may be the final transversion expected since AAG is a codeword for lysine[20].

It should be noted that in the foregoing discussion it has been tacitly assumed that the *amber* and *ochre* signals are triplets. Examination of revertants of *amber* mutants has supported this assumption, since in 41 independent revertants of *H*36 the amino-acid replaced is always at the site of mutation, and never in adjacent positions. The 21 revertants that Weigert and Garen[21] isolated reinforce this conclusion.

Function of amber and ochre triplets and the mechanism of suppression. According to present-day ideas of protein synthesis, it is expected that the termination of the growth of the polypeptide chain should involve a special mechanism. Since the terminal carboxyl group of the growing peptide chain is esterified to an *s*RNA (ref. 23), chain termination must involve not only the cessation of growth, but also the cleavage of this bond. Since the *amber* mutants have been shown to result in efficient termination of polypeptide chain synthesis, it is reasonable to suppose that this special mechanism may be provided by the *amber* and *ochre* triplets.

We postulate that the chain-terminating triplets UAA and UAG are recognized by specific *s*RNAs, just like other codons. These *s*RNAs do not carry amino-acids but a special compound which results in termination of the growing polypeptide chain. There are many possible ways of formulating the mechanism in detail, but all are speculative and will not be considered here. The essential feature of this hypothesis is to make the process of chain termination exactly congruent with that of chain extension.

In suppressing strains, a mechanism is provided for competing with chain termination; it is easy to visualize this process as being due to two ways of recognizing the nonsense codon—one by the chain-terminating *s*RNA, and the other by an *s*RNA carrying an amino-acid. Mechanisms of suppression can be classified according to which *s*RNA carries the amino-acid to the nonsense codon.

Alteration in the recognition of the chain-terminating *s*RNA might allow the attachment of an amino-acid to this *s*RNA. This could be brought about either by modifying normal activating enzymes so as to widen their specificity, or by changing the chain terminating *s*RNAs to allow them to be recognized by activating enzymes.

Another possibility is that the region of an amino-acyl *s*RNA used for triplet recognition is modified so that it can recognize the nonsense triplet. Clearly, this alteration must not affect the normal recognition of its own codon by the amino acyl *s*RNA because such a change would be lethal. Either there must be more than one gene for the given *s*RNA, or else the change must produce an ambiguity in the recognition site so that it can read both its own codon and the nonsense codon. Such ambiguity could result not only from mutation in the *s*RNA gene but also by enzymatic modification of one of the bases in the recognition site. The ambiguity, however, must be narrowly restricted to prevent the suppression from affecting codons other than the *amber* and *ochre* triplets. Moreover, the amino-acids which are inserted by the *amber* suppressors must be those the codons of which are connected to UAG. It should be noted that this condition is fulfilled by su^{+}I which inserts serine, since serine has been found as a reversion of an *amber* mutant. This theory does not easily explain the *ochre* suppressors. Since these recognize both *amber* and *ochre* mutants the *s*RNA must possess this ambiguity as well.

Another quite different possibility for suppression that has been considered is that the suppressors alter a component of the ribosomes to permit errors to occur in the reading of the messenger RNA (ref. 24). This is probably the explanation of streptomycin suppression[25], but suppression of *amber* and *ochre* mutants cannot be readily explained by this theory. It is scarcely likely that such a mechanism could be specific for only one or two triplets, and for this reason it might be expected to give us suppression of mutants which are not nonsense, but missense, and this has not been found[1,8]. Moreover, the efficiency of *amber* suppression argues strongly against such a mechanism. It is unlikely that a generalized error in reading nucleotides could produce the 60 per cent efficiency of suppression found for su^{+}I without seriously affecting the viability of the cell.

It is a consequence of our theory that normal chain termination could also be suppressed in these strains. Since the *amber* suppressors are efficient we have to introduce the *ad hoc* hypothesis that the UAG codon is rarely used for chain termination in *Escherichia coli* and bacteriophage *T*4 and that UAA is the common codon. This is supported by the fact that all *ochre* suppressors thus far isolated are weak[7,8]. Another possibility is that neither is the common chain terminating triplet. We cannot exclude the existence of other chain terminating triplets which are not suppressible.

To summarize: we show that the triplets of the *amber* and *ochre* mutants are UAG and UAA, respectively. We suggest that the 'nonsense' codons should be more properly considered to be the codons for chain termination. In essence, this means that the number of elements to be coded for is not 20 but more likely 21. We propose that the recognition of the chain-terminating codons is carried out by two special *s*RNAs.

We thank our colleagues for their advice, and Dr. M. Nirenberg for allowing us to quote his unpublished results.

1 Benzer, S., and Champe, S. P., *Proc. U.S. Nat. Acad. Sci.*, **47**, 1025 (1961); **48**, 1114 (1962).

2 Garen, A., and Siddiqi, O., *Proc. U.S. Nat. Acad. Sci.*, **48**, 1121 (1962).

3 Campbell, A., *Virology*, **14**, 22 (1961).

4 Epstein, R. H., Bolle, A., Steinberg, C. M., Kellenberger, E., Boy de la Tour, E., Chevalley, R., Edgar, R. S., Susman, M., Denhardt, G. H., and Lielausis, A., *Cold Spring Harbour Symp. Quant. Biol.*, **28**, 375 (1963).

5 Sarabhai, A., Stretton, A. O. W., Brenner, S., and Bolle, A., *Nature*, **201**, 13 (1964).

6 Stretton, A. O. W., and Brenner, S., *J. Mol. Biol.* (in the press).

7 Kaplan, S., Stretton A. O. W., and Brenner, S. (in preparation).

8 Brenner, S., and Beckwith, J. R. (in preparation).

9 Brenner, S., and Stretton, A. O. W. (in preparation).

10 Brown, D. M., and Schell, P., *J. Mol. Biol.*, **3**, 709 (1961). Freese, E., Bautz-Freese, E., and Bautz, E., *J. Mol. Biol.*, **3**, 133 (1961). Schuster, H., *J. Mol. Biol.*, **3**, 447 (1961). Freese, E., Bautz, E., and Freese, E. B., *Proc. U.S. Nat. Acad. Sci.*, **47**, 845 (1961).

11 Tocchini-Valentini, G. P., Stodolsky, M., Aurisicchio, A., Sarnat, M., Graziosi, F., Weiss, S. B., and Geiduschek, E. P., *Proc. U.S. Nat. Acad. Sci.*, **50**, 935 (1963). Hayashi, M., Hayashi, M. N., and Spiegelman, S., *Proc. U.S. Nat. Acad. Sci.*, **50**, 664 (1963). Marmur, J., Greenspan, C. M., Palacek, E., Kahan, F. M., Levene, J., and Mandel, M., *Cold Spring Harbor Symp. Quant. Biol.*, **28**, 191 (1963). Bautz, E. K. F., *Cold Spring Harbor Symp. Quant. Biol.*, **28**, 205 (1963). Hall, B. D., Green, M., Nygaard, A. P., and Boezi, J., *Cold Spring Harbor Symp. Quant. Biol.*, **28**, 201 (1963).

12 Champe, S. P., and Benzer, S., *Proc. U.S. Nat. Acad. Sci.*, **48**, 532 (1962). Tessman, I., Poddar, R. K., and Kumar, S., *J. Mol. Biol.*, **9**, 352 (1964).

13 Freese, E., *J. Mol. Biol.*, **1**, 87 (1959). Freese, E., *Proc. U.S. Nat. Acad. Sci.*, **45**, 622 (1959). Howard, B. D., and Tessman, I., *J. Mol. Biol.*, **9**, 372 (1964).

14 Garen, A., and Weigert, M. G., *J. Mol. Biol.* (in the press).

15 Notani, G. W., Engelhardt, D. L., Konigsberg, W., and Zinder, N., *J. Mol. Biol.* (in the press).

16 Nirenberg, M., Jones, O. W., Leder, P., Clark, B. F. C., Sly, W. S., and Pestka, S., *Cold Spring Harbor Symp. Quant., Biol.*, **28**, 549 (1963).

17 Speyer, J. F., Lengyel, P., Basilio, C., Wahba, A. J., Gardner, R. S., and Ochoa, S., *Cold Spring Harbor Symp. Quant. Biol.*, **28**, 559 (1963).

18 Nirenberg, M., Leder, P., Bernfield, M., Brimacombe, R., Trupin, J., and Rottman, F., *Proc. U.S. Nat. Acad. Sci.* (in the press).

19 Stretton, A. O. W., and Brenner, S. (in preparation).

20 Trupin, J., Rottman, F., Brimacombe, R., Leder, P., Bernfield, M., and Nirenberg, M., *Proc. U.S. Nat. Acad. Sci.* (in the press). Clark, B. F. C., presented before the French Biochemical Society, February, 1965.

21 Weigert, M. G., and Garen, A., *Nature* (preceding communication).

22 Leder, P., and Nirenberg, M. W., *Proc. U.S. Nat. Acad. Sci.*, **52**, 1521 (1964).

23 Gilbert, W., *J. Mol. Biol.*, **6**, 389 (1963). Bretscher, M. S., *J. Mol. Biol.*, **7**, 446 (1963).

24 Davies, J., Gilbert, W., and Gorini, L., *Proc. U.S. Nat. Acad. Sci.*, **51** 883 (1964).

25 Gorini, L., and Kataja, E., *Proc. U.S. Nat. Acad. Sci.*, **51**, 487 (1964).

Reprinted from PROCEEDINGS OF THE NATIONAL ACADEMY OF SCIENCES, Vol. 57, 1967

TRANSLATION OF THE GENETIC MESSAGE, IV. UAA AS A CHAIN TERMINATION CODON*

BY JEROLD A. LAST,† WENDELL M. STANLEY, JR.,‡ MARGARITA SALAS,§ MERRILL B. HILLE,** ALBERT J. WAHBA, AND SEVERO OCHOA

DEPARTMENT OF BIOCHEMISTRY, NEW YORK UNIVERSITY SCHOOL OF MEDICINE, NEW YORK, NEW YORK

Communicated February 27, 1967

The discovery that certain mutations result in the production of unfinished, amino-terminal peptide fragments of the corresponding wild-type proteins suggested that "nonsense" codons may normally signal termination of polypeptide chain synthesis.[1-3] Moreover, the finding that these "amber" mutations are suppressible, whereby the resulting polypeptide contains serine (codon, UCG) in place of glutamine (CAG) or tryptophan (UGG),[4-6] led to the prediction[2, 3] that UAG is the chain terminator of "amber" mutants. The study of other nonsense ("ochre") mutations suggested further that UAA also functions as a chain termination codon.[3] These suggestions were in accord with the fact that no amino acids are coded by UAG or UAA.[7, 8] There are indications that UAA rather than UAG may be the codon normally concerned with chain termination.[9]

In vitro experiments with synthetic polynucleotides of random base sequence lent some support to the view that UAA is involved in chain termination. The polypeptides synthesized under the direction of U- and A-containing copolymers or natural messengers consist largely of free, i.e., released peptide chains, whereas the peptide chains formed with polynucleotides which do not contain U and A residues are attached to tRNA.[10-12]

In order to determine unambiguously whether UAA is a chain-termination codon, we have studied the *in vitro* translation of a series of oligonucleotides having, along with the chain-initiating formylmethionine codon (AUG) and lysine codons (AAA), leucine (UUA), phenylalanine (UUU), or "ochre" (UAA) codons in specified sequences as follows: (*a*) AUGAAAAAAAAA...AAA ($AUGA_n$); (*b*) AUGUAAAAAAAA...AAA ($AUGUA_n$); (*c*) AUGUUAAAAAAA...AAA ($AUGU_2A_n$); (*d*) AUGUUUAAAAAA...AAA ($AUGU_3A_n$); (*e*) AUGUUUUAAAAA...AAA ($AUGU_4A_n$).

The (*a*) polymers promoted the incorporation into acid-insoluble products of methionine and lysine; the (*c*) polymers, that of methionine, leucine, and lysine; and the (*d*) polymers, that of methionine, phenylalanine, and lysine. On the other hand, the (*b*) and (*e*) polymers promoted the incorporation of only small amounts of lysine, with little or no incorporation of methionine or phenylalanine. The inference that the (*e*) oligonucleotides directed the synthesis and release of *acid-soluble* formylmethionyl-phenylalanine was proved correct by isolation of the dipeptide. This work provides conclusive proof that UAA is a chain-termination codon. It also provides a suitable system for study of the mechanism of chain termination.

Materials and Methods.—These were as in previous work[13, 14] unless otherwise noted.

Amino acid incorporation: The low-nuclease system consisting of purified *Escherichia coli* Q13 ribosomes and *Lactobacillus arabinosus* supernatant,[15] supplemented with initiation factors,[13, 16] was used throughout. In experiment 1 of Table 1 the samples contained the following components in a final volume of 0.125 ml: Tris-HCl buffer, pH 7.8, 60 mM; ammonium chloride, 50 mM;

2-mercaptoethanol, 16 mM; magnesium acetate, 14 mM; ATP, 1.2 mM; GTP, 0.3 mM; phosphocreatine, 18 mM; creatine kinase, 9 μg; *E. coli* W tRNA, 0.60 mg; purified *E. coli* Q13 ribosomes, 7 A_{260} units; *L. arabinosus* supernatant, 0.2 mg of protein; N^{10}-formyltetrahydrofolic acid (FTHF), 8 mM; initiation factors (F_1 and F_2), 6 and 21 μg of protein, respectively; 20 amino acids (one of them with C^{14} label), each 0.2 mM; without or with oligonucleotide messenger, 15–20 μg. The specific radioactivities (μc/μmole) of the labeled amino acids were: lysine, 2; methionine, 20; leucine, 22; and phenylalanine, 40. The composition of the samples in experiment 2 of Table 1 was similar except that only three amino acids, namely, lysine (0.2 mM), methionine (0.02 mM), and phenylalanine (0.02 mM) were present, one of them having C^{14} label. Their specific radioactivities (μc/μmole) were: lysine, 10; methionine, 186; and phenylalanine, 176. After incubation at 37° the reaction was terminated by addition of KOH (final concentration, 0.5 *M*). The mixtures were kept at 37° for 3 hr, followed by neutralization, and the acid-insoluble radioactivity was determined as previously described.[16]

Polynucleotides: The methods used in this laboratory for the preparation of various kinds of oligonucleotides of specified base sequence have been described.[17] The preparation of the polymers used in the present work is described below.

$AUGA_n$ polynucleotides were prepared[17] with ADP and polynucleotide phosphorylase, using ApUpG[18] (AUG) as primer. For the other polymers, the series of primers ApUpGpU (AUGU), ApUpGpUpU ($AUGU_2$), ApUpGpUpUpU ($AUGU_3$), and ApUpGpUpUpU ($AUGU_4$) was prepared by limited addition of uridylic acid residues from UDP to ApUpG primer with polynucleotide phosphorylase and chromatographic isolation of the individual oligonucleotides. Adenylic acid residues were then added to each of these primers, by incubation with ADP and polynucleotide phosphorylase. For characterization purposes the polymers were prepared with H^3-labeled guanylic acid residues. In the following description this label will be indicated by an asterisk.

ApUpG* was prepared from ApU (lot no. B 3503, Gallard-Schlesinger) as described previously.[17] ApUpG*pU, ApUpG*pUpU, ApUpG*pUpUpU, and ApUpG*pUpUpUpU were synthesized by the addition of uridylic acid residues to ApUpG* under the following conditions. The reaction mixture (pH 8.2) contained the following components (in μmoles/ml unless otherwise indicated) in a final volume of 4.0 ml. ApUpG*, 7.3; UDP (lot no. 1116, Pabst), 40; Tris-HCl, 150; magnesium acetate, 10; EDTA, 1; sodium chloride, 500; polynucleotide phosphorylase, 300 μg of protein/ml. After incubation for 20 min at 30°, the solution was held at 100° for 5 min, cooled, diluted 50-fold, and applied to a column (1.8 × 39 cm) of DEAE-cellulose (lot no. 1675, 1 mEq/gm, Brown). Oligonucleotides of the form ApUpG* with 1, 2, 3, etc. residues of uridylic acid added to the 3′-end were separated at 25° by gradient elution with ammonium bicarbonate (0.01 *M* ammonium bicarbonate, pH 8.6, in a 2-liter mixer; 0.5 *M* ammonium bicarbonate, pH 8.6, in a reservoir) at 1 ml/min. In order to remove water and salt, the oligonucleotides were recovered by flash evaporation and lyophilization. The yields were, in moles of oligonucleotide/mole input ApUpG*: ApUpG*, 0.142; ApUpG*pU, 0.295; ApUpG*pUpU, 0.166; ApUpG*pUpUpU, 0.119; and ApUpG*pUpUpUpU, 0.072. The structure and purity of each oligonucleotide was confirmed by paper chromatography in two solvents (solvent 1, 1.0 *M* ammonium acetate, pH 7.5/95% ethanol, 6/4, v/v; solvent 2, 40 gm ammonium sulfate plus 100 ml of 0.1 *M* potassium phosphate, pH 7.0), by degradation with RNase T_1 followed by identification of ApUpG*p plus the expected length of uridylic acid residues (i.e., ApUpG*p plus UpUpU were recovered from ApUpG*pUpUpU), and by the determination of the ratio of uridine to uridylic acid following complete alkaline hydrolysis.

Adenylic acid residues were added to the 3′-ends of each of the purified oligonucleotide primers, and the resulting polymers were fractionated according to chain length and recovered as previously described.[17, 19] The structure was confirmed by the identification of the low-molecular-weight products arising from the degradation of each polymer with RNase T_1 and by pancreatic RNase. For example, ApUpG*pUpU$(pA)_n$ gave rise to ApUpG*p following RNase T_1 digestion, and to ApUp, G*pUp, and Up, in the molar ratio 1/1/1 following digestion with pancreatic RNase.

The polynucleotide phosphorylase was a preparation from *Micrococcus lysodeikticus* with an absolute requirement for primer oligonucleotides. The enzyme catalyzed the polymerization of 6.5 μmoles of ADP/min/mg protein at 37° under the following conditions (reagents in μmoles/ml):

ADP, 40; ApApA, 1; sodium chloride, 500; magnesium acetate, 10; EDTA, 1; and Tris-HCl buffer, pH 8.2, 150.

Isolation and characterization of (H^3)*formyl-*(C^{14})*methionyl-*(C^{12})*phenylalanine and* (H^3)*formyl-*(C^{12})*methionyl-*(C^{14})*phenylalanine:* The incubation mixtures were as described under *Amino Acid Incorporation* (expt. 1 of Table 1) with the following exceptions: H^3-FTFH (formate-labeled, 2000 $\mu c/\mu$mole), 0.022 mM; either C^{14}-methionine (186 $\mu c/\mu$mole) or C^{14}-phenylalanine (352 $\mu c/\mu$mole), 0.02 mM; nonlabeled amino acids, either 19 amino acids, each 0.2 mM, or lysine (0.2 mM) and phenylalanine or methionine (0.02 mM); without or with $AUGU_4A_{\overline{10}}$, 15–20 μg. Pancreatic DNase (electrophoretically purified, Worthington), 25 μg, was also added to the samples to prevent any synthesis of endogenous messenger RNA.

Prior to adding the ribosomes, the samples were incubated for 5 min to allow time for transfer of H^3-formate from H^3-FTFH to met $\sim$ $tRNA_F$.[20] The ribosomes were then added and the incubation was continued for 15 min. The short incubation was chosen to minimize enzymatic removal of labeled formyl residues from the peptides synthesized.[21] After incubation, the samples were chilled in ice and treated with 0.127 ml of 6% perchloric acid. The precipitate was removed by centrifugation and washed three times with 30-μl portions of ice-cold 3% perchloric acid. The combined supernatant and washings were neutralized with 0.168 ml of 1.0 *N* KOH. The potassium perchlorate precipitate was removed by centrifugation and washed repeatedly with 50-μl portions of cold water until the washings were free of radioactivity. The supernatant and the washings were combined and the pH of the solution was adjusted to 7.8, with Tris-HCl buffer, prior to chromatography on a Dowex-1 (Cl^-) column (0.6 × 5 cm). The column was washed with 10 ml of water, containing 1–2 mg of nonlabeled phenylalanine or methionine, and the formylated amino acids and peptides were then eluted with 8.0 ml of 0.5 *N* acetic acid adjusted to pH 2.7 with pyridine. This solution was passed through a Dowex-50 (H^+) column (0.6 × 20 cm). The formylated compounds were washed through with 8.0 ml of 0.001 *N* HCl. For separation of formylmet and formylmet-phe, the eluate was lyophilized, the residue taken up in 0.5 ml of water, and aliquots of the solution were subjected to high-voltage paper electrophoresis, along with markers of formylmet and formylmet-phe, as previously described,[14] except that the electrophoresis was run for 7 hr. For measurement of radioactivity after electrophoresis, the paper was cut into strips (1 cm long, 2.5 cm wide) which were placed in the bottom of scintillation vials. The strips were covered with solvent (4 gm of 2,5-diphenyloxazole (PPO) and 50 mg of p-bis [2-(5-phenyloxazolyl)]-benzene (POPOP) per liter of toluene) and the radioactivity was measured in a Packard Tri-Carb scintillation spectrometer. Under these conditions, the counting efficiency of the two isotopes on Whatman 3MM paper was about 29% for C^{14} and 0.9% for H^3, with 12% of the C^{14} counts appearing in the H^3 channel, and less than 0.1% of the H^3 counts in the C^{14} channel. The total amount of formylmet-phe calculated from the electrophoretic data was corrected for small aliquots of material withdrawn, during the processing of the reaction mixtures, for determination of the acid-soluble radioactivity. This correction was never greater than 6%.

Formylmet-phe was prepared by formylation of L-methionyl-L-phenylalanine.[22] L-methionyl-L-phenylalanine was obtained from the Mann Research Laboratories, Inc., New York.

Results and Discussion.—Incorporation of amino acids into acid-insoluble products: As shown in Table 1, oligonucleotides of the series $AUGA_n$, $AUGU_2A_n$, and $AUGU_3A_n$ promote the incorporation of lysine, along with smaller amounts of methionine ($AUGU_{\overline{18}}$), methionine and leucine ($AUGU_2A_{\overline{10}}$), and methionine and phenylalanine ($AUGU_3A_{\overline{10}}$). It may further be noted that the incorporation of methionine and leucine, or methionine and phenylalanine, promoted by poly $AUGU_2A_{\overline{10}}$ or $AUGU_3A_{\overline{10}}$, respectively, is roughly equimolar. Moreover, there is negligible or no incorporation of phenylalanine with poly $AUGU_2A_{\overline{10}}$ and the same is true of leucine in the case of poly $AUGU_3A_{\overline{10}}$. This shows that these polymers are read in phase from the 5′-end.

In contrast to the above results, the incorporation of the various amino acids into acid-insoluble material is markedly diminished when oligonucleotides of the series

TABLE 1

POLYPEPTIDE SYNTHESIS WITH OLIGONUCLEOTIDE MESSENGERS OF SPECIFIED BASE SEQUENCE*

Expt. no.	Incubation time (min)	Messenger	Amino Acid Incorporation† Lysine	Methionine	Leucine	Phenylalanine
1	40	None (blank)	81	43	24	18
		AUGA$_{\overline{18}}$	1261 (1521)	163 (152)	0	0
		AUGUA$_{\overline{11}}$	162 (142)	10 (10)	0	0
		AUGU$_2$A$_{\overline{10}}$	644 (563)	65 (58)	56 (51)	0
		AUGU$_3$A$_{\overline{10}}$	543 (425)	56 (47)	2	62 (50)
		AUGU$_4$A$_{\overline{10}}$	282 (153)	10 (10)	2	4 (9)
2	15	None (blank)	43	3	—	5
		AUGU$_3$A$_{\overline{10}}$	371	29	—	21

* Conditions as described under *Materials and Methods*. In expt. 2 the samples were preincubated for 5 min prior to starting the reaction by addition of the ribosomes.

† Values expressed in $\mu\mu$moles/sample. The first line for each expt. gives the blank values in the absence of added oligonucleotide. All others are net values, the blanks without polymer having been subtracted from the values with polymer. The net values given in parentheses are the combined averages of expt. 1 and three other experiments carried out at different times between August 1966 and February 1967. Duplicate samples were run in each experiment and the results averaged.

AUGU$_4$A$_n$ (AUGU$_4$A$_{\overline{10}}$), having the UAA triplet in the third position from the 5′-terminus, are used as messengers. Although there is some lysine incorporation, that of methionine and phenylalanine is very low. This incorporation is referable to slight contamination of poly AUGU$_4$A$_n$ with AUGU$_3$A$_n$. Table 1 also lists an experiment with poly AUGUA$_{\overline{11}}$, having the UAA triplet in the second position from the 5′-end. Some lysine but little if any methionine is incorporated.

Release of formylmethionyl-phenylalanine upon translation of AUGU$_4$A$_n$ polymers: In order to determine whether acid-soluble formylmet-phe is synthesized and released upon translation of the AUGU$_4$A$_n$ messengers, incubations with and without polymer were conducted with either H^3-FTHF and C^{14}-methionine or H^3-FTHF and C^{14}-phenylalanine. The nonlabeled amino acids were either 19 amino acids (no methionine or no phenylalanine) or lysine with either phenylalanine or methionine. After incubation, any labeled acid-insoluble products, including formylmet ~

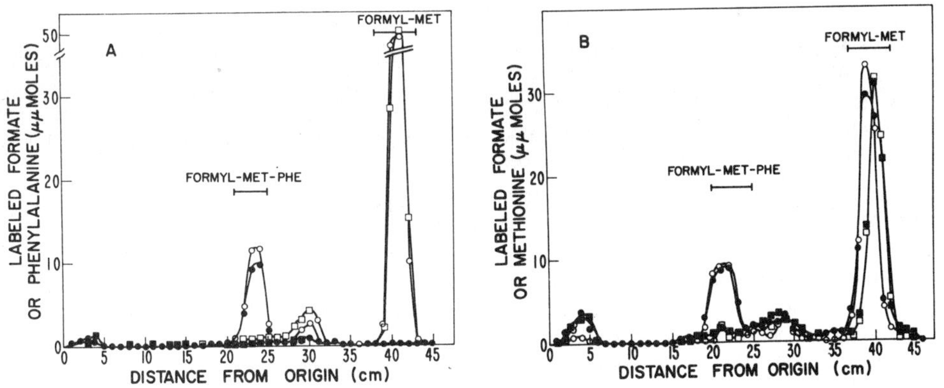

FIG. 1.—Electrophoretic patterns of labeled peptides synthesized with AUGU$_4$A$_{\overline{10}}$ oligonucleotide messenger. Aliquots corresponding to 10% of the sample were used for electrophoresis. The anode is to the right. The position of formylmet-phe and formylmet markers is indicated. (*A*) Incubation with H^3-FTHF and C^{14}-phenylalanine. (*B*) Incubation with H^3-FTHF and C^{14}-methionine. In this experiment the samples contained the remaining 19 C^{14}-amino acids. Two other experiments were carried out with similar results. *Squares*, blank samples (no oligonucleotide added); *circles*, samples with oligonucleotide added. *Open symbols* s(□, O), H^3 label; *solid symbols* (■, ●), C^{14} label.

TABLE 2

H^3 AND C^{14} RADIOACTIVITY IN THE FORMYLMETHIONYL-PHENYLALANINE AREA OF THE ELECTROPHEROGRAMS OF FIGURE 1*

Expt.	Radioactive label	Blank sample† (*a*)	Sample with $AUGU_4A_{\overline{1}}$ (*b*)	Polymer-promoted (*b* − *a*)
Fig. 1*A*	H^3 (formate)	3.4	32.1	28.7
	C^{14} (phenylalanine)	2.5	26.2	23.7
Fig. 1*B*	H^3 (formate)	4.8	32.6	27.8
	C^{14} (methionine)	7.3	34.2	26.9

* Values expressed as $\mu\mu$moles of formate, phenylalanine, or methionine/sample.
† Sample with no oligonucleotide added.

$tRNA_F$, met ~ $tRNA_M$, and phe ~ tRNA, were removed by precipitation with perchloric acid and the supernatants worked up as described under *Materials and Methods.*

The electrophoretic patterns of the H^3- and C^{14}-labeled products (Fig. 1) show two main peaks. One of them is in the position of formylmet-phe, the other in the position of formylmet. The presence of formylmet would appear to reflect the occurrence of some deacylation of formylmet ~ $tRNA_F$ during the incubation. It may be noted that the blank incubations, with no added oligonucleotide, show roughly the same amount of radioactivity in the position of formylmet but little or no radioactivity in the region of the dipeptide.

The H^3 and C^{14} radioactivity under the formylmet-phe peak, expressed as $\mu\mu$-moles of product per incubation sample, is given in Table 2. Reference to experiment 2 of Table 1 shows that the formation of formylmet-phe with poly $AUGU_4A_{\overline{10}}$ is of the same order of magnitude as the incorporation of methionine and phenylalanine into acid-insoluble material with $AUGU_3A_{\overline{10}}$ messenger.

The above results indicate that the $AUGU_4A_n$ polymers are translated with the same efficiency as the others except for termination of chain growth and release once formylmet-phe is synthesized. It may be emphasized in this connection that formylmet-phe is not released as peptidyl ~ tRNA (a compound that would be acid-insoluble) but as the free dipeptide. Thus, complete chain termination is achieved. The isolation of formylmet-phe in good yields, upon translation of the 5′-end sextet AUGUUU, also provides further proof for the triplet nature of the genetic code.[23]

We are indebted to Mr. Norton A. Elson for his help in the preparation and characterization of the oligonucleotides used here. Our thanks are also due to Miss Martha Miller and Mr. Horace Lozina for help in various phases of this work.

* Aided by grants AM-01845, AM-08953, and FR-05399 from the National Institutes of Health, U.S. Public Health Service, the Jane Coffin Childs Fund for Medical Research, and E. I. du Pont de Nemours and Company, Inc.

† Postdoctoral fellow of the American Cancer Society.

‡ Present address: Division of Biological Sciences, Department of Molecular and Cell Biology, University of California, Irvine, California.

§ Postdoctoral fellow of the Jane Coffin Childs Fund for Medical Research. Permanent address: Instituto Marañón, Centro de Investigaciones Biológicas, C.S.I.C., Madrid, Spain.

** Postdoctoral fellow of the National Institutes of Health, U.S. Public Health Service.

[1] Sarabhai, A. S., A. O. W. Stretton, S. Brenner, and A. Bolle, *Nature*, **201**, 13 (1964).

[2] Weigert, M. G., and A. Garen, *Nature*, **206**, 992 (1965).

[3] Brenner, S., A. O. W. Stretton, and S. Kaplan, *Nature*, **206**, 994 (1965).

[4] Weigert, M. G., and A. Garen, *J. Mol. Biol.*, **12**, 448 (1965).

[5] Stretton, A. O. W., and S. Brenner, *J. Mol. Biol.*, **12**, 456 (1965).

[6] Webster, R. E., D. L. Engelhardt, and N. D. Zinder, these PROCEEDINGS, **55**, 155 (1966).

[7] Nirenberg, M. W., P. Leder, M. Bernfield, R. Brimacombe, J. Trupin, F. Rottman, and C. O'Neal, these PROCEEDINGS, **53**, 1161 (1965).

[8] Söll, D., E. Ohtsuka, D. S. Jones, R. Lohrman, H. Hayatsu, S. Nishimura, and H. G. Khorana, these PROCEEDINGS, **54**, 1368 (1965).

[9] Garen, A., S. Garen, and R. C. Wilhelmi, *J. Mol. Biol.*, **14**, 167 (1965).

[10] Takanami, M., and Y. Yan, these PROCEEDINGS, **54**, 1450 (1965).

[11] Bretscher, M. S., H. M. Goodman, J. R. Menninger, and J. D. Smith, *J. Mol. Biol.*, **14**, 634 (1965).

[12] Ganoza, M. C., and T. Nanamoto, these PROCEEDINGS, **55**, 162 (1966).

[13] Salas, M., M. B. Hille, J. A. Last, A. J. Wahba, and S. Ochoa, these PROCEEDINGS, **57**, 387 (1967).

[14] Viñuela, E., M. Salas, and S. Ochoa, these PROCEEDINGS, **57**, 729 (1967).

[15] Salas, M., M. A. Smith, W. M. Stanley, Jr., A. J. Wahba, and S. Ochoa, *J. Biol. Chem.*, **240**, 3988 (1965).

[16] Stanley, W. M., Jr., M. Salas, A. J. Wahba, and S. Ochoa, these PROCEEDINGS, **56**, 290 (1966).

[17] Stanley, W. M., Jr., M. A. Smith, M. B. Hille, and J. A. Last, in *Cold Spring Harbor Symposia on Quantitative Biology*, vol. 31 (1967), in press.

[18] Shorthand writing of polynucleotides and most of the abbreviations used are standard. Other abbreviations are as follows: FTHF, formyltetrahydrofolic acid; tRNA, transfer RNA; met $\sim$ $tRNA_F$, the species of methionyl tRNA whose methionine residue can be formylated; met $\sim$ $tRNA_M$, the nonformylatable species; phe $\sim$ tRNA, phenylalanyl tRNA; formylmet, formylmethionine; formylmet-phe, formylmethionyl-phenylalanine.

[19] Stanley, W. M., Jr., in *Nucleic Acids*, a volume of *Methods in Enzymology*, ed. K. Moldave and L. Grossman (New York: Academic Press, 1967), in press.

[20] Clark, B. F. C., and K. A. Marcker, *Nature*, **211**, 378 (1966).

[21] Capecchi, M. R., these PROCEEDINGS, **55**, 1517 (1966).

[22] Sheehan, J. C., and D. M. Yang, *J. Am. Chem. Soc.*, **80**, 1154 (1958).

[23] Jones, D S., S. Nishimura, and H. G. Khorana, *J. Mol. Biol.*, **16**, 454 (1966).

Reprinted from JOURNAL OF MOLECULAR BIOLOGY, Vol. 17, 1966

The Role of N-Formyl-methionyl-sRNA in Protein Biosynthesis

B. F. C. CLARK AND K. A. MARCKER

Medical Research Council Laboratory of Molecular Biology
Hills Road, Cambridge, England

(*Received 23 December 1965*)

Bacterial sRNA has been fractionated into two methionine-accepting species, only one of which could be formylated. The binding of each charged species to ribosomes under the direction of a variety of trinucleoside diphosphates has been studied. The codeword AUG could be assigned to both species, whereas the related trinucleoside diphosphates GUG and UUG caused binding only of the species which could be formylated. The presence of the formyl group on this methionyl-sRNA did not alter its binding characteristics. The results of the binding studies were supported by investigations using a cell-free system directed by poly UAG or poly UG. Methionine was incorporated from the charged species, which could be formylated into the N-terminal position of polypeptides by both polymers whether this species was formylated or not. Poly UAG, but not poly UG, stimulated the incorporation of methionine into polypeptide from the methionyl-sRNA species, which could not be formylated. The analysis of this polypeptide product identified methionine in internal positions.

1. Introduction

The discovery that N-formyl-methionyl-sRNA was formed by the formylation of methionyl-sRNA in a bacterial cell-free system (Marcker & Sanger, 1964) initiated a study of the possible role of F-met-sRNA† in protein biosynthesis. At least two species of met-sRNA were shown to exist in bacterial cell extracts and at least one of these species could be formylated (Marcker, 1965). A random heteropolyribonucleotide, poly UG, was found to stimulate the incorporation of methionine mostly into the N-terminal position of a polypetide product in a cell-free system from *Escherichia coli* (Clark & Marcker, 1965). When the coding characteristics for the binding of met-sRNA to ribosomes were studied, a striking ambiguity of codewords was observed (Clark & Marcker, 1965). This paper reports on further investigations of the coding characteristics and the positions of incorporation of methionine using fractionated met-sRNA. A counter-current distribution of *E. coli* sRNA separated two methionine-accepting species from each other. The faster moving species which was charged to give met-sRNA$_1$ could not be formylated, whereas the slower moving

† Abbreviations used: F-met-sRNA, N-formyl-methionyl-sRNA; met-sRNA, methionyl-sRNA. Poly UG and poly UAG refer to random heteropolyribonucleotides containing nucleosides as indicated by initial letters. Trinucleoside diphosphates and dinucleoside monophosphates are designated by initial letters of nucleosides in sequence with phosphates omitted; e.g. AUG represents adenylyl-(3′,5′)-uridylyl-(3′,5′)-guanosine. The oligonucleotides are measured in terms of A_{260} units; 1 A_{260} unit in 1 ml. of aqueous solution gives an optical density of 1 when measured in a silica cell of 1 cm light path at 260 mμ in a spectrophotometer. The standard three-letter abbreviations are used for amino acids. PEP is phosphoenolpyruvate.

met-$sRNA_2$ could be formylated completely. If met-$sRNA_1$ was used as a source in a cell-free system stimulated by poly UAG, methionine was incorporated into polypeptide mostly internally. It was found that met-$sRNA_1$ corresponded to the codeword AUG by the binding assay of Nirenberg & Leder (1964). However, the codeword corresponding to met-$sRNA_2$ could not be assigned specifically to one codeword by the same assay. If met-$sRNA_2$ was used as a source in a cell-free system stimulated by poly UAG or by poly UG, the methionine was incorporated into the N-terminal positions of the polypeptide products.

2. Materials and Methods

(a) *Preparation of cell extracts*

Cell extracts were prepared by the method of Nirenberg & Matthaei (1961). The S-30 fraction was treated with DNase but pre-incubated without any addition at 37°C for 45 min. The ribosomes from 50 g of frozen cells (*E. coli* MRE600 or CA265) were resuspended in 40 ml. of the standard buffer made 0·5 M in tris at pH 8·1. To this suspension 1 ml. of a 5% w/v aqueous solution of sodium deoxycholate was added. After being collected by centrifugation, the ribosomes were washed twice in the buffer solution, 0·01 M-tris–HCl (pH 7·8), 0·5 M-NH_4Cl, 0·05 M-KCl, 0·01 M-magnesium acetate and 0·006 M-2-mercaptoethanol. The resulting ribosomes were suspended in 9 ml. of a solution containing 0·01 M-tris–HCl (pH 7·8), 0·06 M-KCl, 0·01 M-magnesium acetate and 0·006 M-2-mercaptoethanol. After a low-speed centrifugation to remove ribosomal aggregates, the ribosomes were ready to use. A protein fraction containing activating enzymes was isolated from the S-100 fraction according to Carbon (1964). The activating enzyme fraction contained little transformylase activity. Transfer factors for protein synthesis were purified as described by Nathans & Lipmann (1961), except that the active protein was eluted from a DEAE-cellulose column by a stepwise procedure. The protein eluted at pH 7·4 between potassium phosphate concentrations of 0·1 M and 0·25 M was precipitated with ammonium sulphate at 80% saturation. The precipitate was dissolved in 0·01 M-tris–HCl (pH 7·4) containing 0·005 M-2-mercaptoethanol and the solution dialysed against the same buffer. Samples of the protein solution were stored at −20°C; they were found to be free of transformylase activity. The preparation of protein fractions associated with transformylase activity has already been reported (Marcker, 1965).

(b) *Preparation of sRNA*

To obtain unfractionated sRNA almost free of contaminants such as rRNA, DNA and polysaccharides, the sRNA was isolated by a method essentially as described by Zubay (1962), but instead of purification by repeated precipitation with isopropanol, the crude sRNA was chromatographed on a DEAE-cellulose column as described by Holley *et al.* (1961). A further departure from Zubay's procedure has recently been used with the stripping of the sRNA in 1·8 M-tris (pH 8) (Sarin & Zamecnik, 1964).

(c) *Fractionation of sRNA*

The sRNA was fractionated to separate two methionine-accepting activities (met-$sRNA_1$ and met-$sRNA_2$) by counter-current distribution in a 1·7 M-phosphate solvent system, as reported by Goldstein, Bennett & Craig (1964). A steady-state distribution machine (Quickfit & Quartz Ltd.) of 10-ml. phase volume was loaded with 10 tubes with 6250 A_{260} units of sRNA from *E. coli* CA265. After 330 phase transfers (264 upper and 66 lower), the distribution pattern of significant amounts, as measured by ultraviolet absorption, in the collected upper phase fractions is shown in Fig. 1. The sRNA recovered from these fractions amounted to about 5200 A_{260} units. The amino acid-accepting capacity of the sRNA fractions was surveyed by an assay similar to that used by Gold-

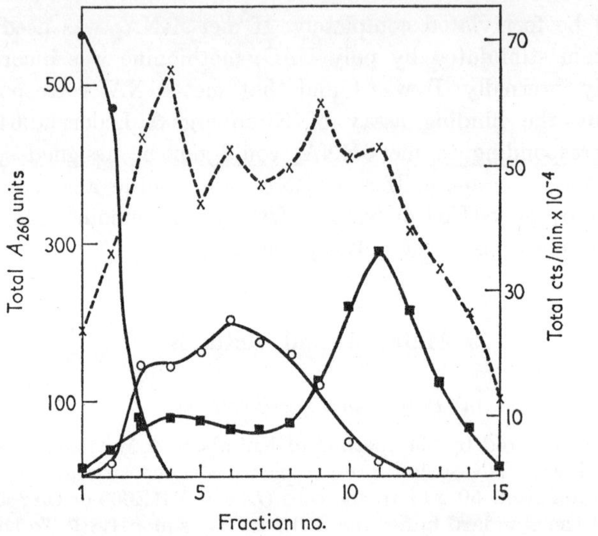

FIG. 1. Fractionation of *E. coli* CA265 sRNA in a steady-state distribution machine as described under Materials and Methods.

The fractions are numbered in order of collection in the upper-phase fraction collector. Thus the lower numbers correspond to the faster-moving sRNA species. The dotted line curve (--×--×--) depicts the pattern of total A_{260} units. The total charging ability of each fraction is shown by the solid line curves for Met (—■—■—), and as examples of other amino acids, for Tyr (—●—●—) and Lys (—○—○—). Preparations of met-sRNA (see Materials and Methods) as used for binding or incorporation experiments had a specific activity usually of at least 4 times that of unfractionated met-sRNA.

stein *et al.* (1964). A reaction mixture (0·70 ml.) which was incubated at 37°C for 15 min, contained 0·1 M-tris–HCl (pH 7·4), 0·016 M-$MgCl_2$, 0·004 M-ATP, 0·0001 M-CTP, 0·016 M-2-mercaptoethanol, 0·2 μC of radioactively labelled amino acid of specific activity about 20 μC/μmole, 0·01 μmole each of 19 other unlabelled amino acids, 0·3 to 1·0 A_{260} units of sRNA and ~ 500 μg of activating enzyme protein. The assay involving precipitation with perchloric acid was employed, and the precipitate was collected on glass paper filters. A sample of *E. coli* B sRNA (General Biochemical Co.) has also been separated by counter-current distribution to give two methionine-accepting species similar to those found for the separation of *E. coli* CA265 sRNA.

To charge sRNA, the reaction mixture for incubation (again at 37°C for 15 min) contained 2 mg of unfractionated sRNA or 0·5 to 1·0 mg of fractionated sRNA, 0·1 M-tris–HCl (pH 7·4), 0·016 M-magnesium acetate, 0·08 M-KCl, 0·0008 M-PEP, 2 μl. pyruvate kinase, 10 mg/ml. (Boehringer Corporation Ltd.) 0·002 M-ATP, 0·0002 M-CTP, 0·0035 M-2-mercaptoethanol, 5 μC [^{14}C]- or [^{35}S]methionine, 19 other [^{12}C]amino acids (25 mμmoles each), 1250 μg of activating enzyme protein or 0·1 ml. of S–100 cell fraction. After incubation, the mixture was chilled in ice and then shaken twice with an equal volume of cold phenol saturated with 0·2 M-potassium acetate (pH 5). The aqueous phase was separated and made 0·2 M in potassium acetate. The phenol layers were extracted in turn with 0·5 ml. of cold 0·2 M-potassium acetate. The resulting aqueous phase was combined with the aqueous phase from the first extraction. Addition of 2 ml. of cold absolute ethanol to the combined aqueous phases precipitated the charged sRNA, which was collected and then dissolved in 0·5 ml. of 0·2 M-potassium acetate. The sRNA was again precipitated by ethanol (3 vol.), and after collection it was dissolved in 0·4 ml. of 0·002 M-potassium acetate (pH 5·0). The solution was dialysed against the same buffer and was characterized by counting its radioactivity and by measuring its absorption at a wavelength of 260 mμ. To prepare samples of met-sRNA$_1$ and met-sRNA$_2$, fractions 5 and 12, respectively (see Fig. 1), were charged as above.

(d) *Preparation of oligomers*

The trinucleoside diphosphates AUG, UAG, UUG, AAG and CUG, were isolated from the products of RNase-T_1 digestion of yeast RNA (Seravac Laboratories, Ltd.), poly UG or poly AUG. The conditions of hydrolysis were found to give complete splitting of the RNA chain after G-3′-p residues (Dr J. D. Smith, personal communication). The isolation procedure involved paper chromatographic and electrophoretic techniques. A reaction mixture for the digestion contained 125 mg of polynucleotide and 1 mg RNase-T_1 (Sankyo Company Ltd.) in 3 ml. of buffer solution, which was 0·1 M in tris–HCl (pH 7·5) and 0·001 M in disodium EDTA. When the reaction mixture had been incubated at 37°C for 2 hr, its pH was adjusted to 8 by the addition of a 2 M-tris solution to achieve the optimal conditions for a further incubation for 2 hr in the presence of 25 units of chromatographically purified bacterial alkaline phosphatase (Worthington Biochemical Corp.). The resulting solution was streaked at the origins of 2 sheets of Whatman 17 paper for descending chromatography (10 hr) in *n*-propanol–water–concn NH_4OH = (55:35:10 by vol.) using AGU (from a pancreatic RNase digestion of RNA) as a marker. The band corresponding to the trinucleoside diphosphate position was excised and its ultraviolet-absorbing material eluted with water adjusted to pH 8 with dilute NH_4OH. If the chromatographic separation was not definitive, the concentrated eluate was rechromatographed on Whatman 3 MM paper in the same solvent. The trinucleoside diphosphate mixture was isolated as before and applied to the origin of a sheet of Whatman 3 MM paper for flat-plate electrophoresis in 0·05 M-ammonium formate (pH 3·5) at 3 kv for 3 hr. At least 3 major bands separated from the origin towards the anode. These bands, in order from the anode to the origin, contained (1) UUG; (2) AUG, UAG, CUG, UCG; (3) AAG, ACG, CAG. Sometimes band 2 was subdivided into bands containing (i) AUG, UAG and (ii) CUG, UCG. The trinucleoside diphosphates were isolated and separated on DEAE-paper (90 cm long) by electrophoresis at pH 1·9 according to the method of Sanger, Brownlee & Barrell (1965). This involved running for about 14 hr at 2 kv. The ultraviolet-absorbing material was eluted from the electrophoretogram with 1 M-triethyl-ammonium bicarbonate (pH 8·1). The salt was removed from the concentrated solution by repeated evaporation and rechromatography on Whatman 31ET in the same solvent as before. The nucleoside sequences of the trinucleoside diphosphates were confirmed by digesting them with venom phosphodiesterase (Worthington Biochemical Corp.) (Bernfield & Nirenberg, 1965). An average yield of a trinucleoside diphosphate from RNA was at least 0·5%. An example of the mobilities of selected trinucleoside diphosphates on electrophoresis at pH 1·9 is shown in Table 1.

TABLE 1

Electrophoretic mobilities on DEAE-paper at pH 1·9

Trinucleoside diphosphate	$R_B = \frac{\text{mobility of compound}}{\text{mobility of blue marker}}$
UUG	0·87
AUG	1·42
UAG	1·62
CUG	1·78
UCG	1·88

The compounds GUG and AUA were enzymically synthesized by the method of Thach & Doty (1965). The dinucleoside phosphate AU and samples of highly purified polynucleotide phosphorylase (Singer & O'Brien, 1963) were generous gifts from Dr Beverly Griffin. We purchased GU, GDP and ADP from the Sigma Chemical Company. Our grateful thanks are expressed to Dr Marianne Grunberg-Manago for numerous gifts of polynucleotides and for experienced advice on the preparation of polynucleotide phosphorylase. The dinucleoside phosphate UG was obtained from a RNase-T_1 digestion of poly UG. A controlled degradation of poly U (Miles Chemical Company) with pig liver nuclease (Dr L. Heppel, personal communication) to yield short-chain oligonucleotides enabled us to isolate the trinucleotide which on dephosphorylation gave UUU. It is a pleasure to record our appreciation of the expert help provided by Drs Jeremy Hayward and John S. Anderson in the preparation of the trinucleoside diphosphates. In addition, we thank Drs D. Söll and M. W. Nirenberg for samples of the trinucleoside diphosphates GUG and UUG, respectively.

(e) *N-Terminal end-group determination of polypeptides*

End-group determinations were performed using a modified Edman procedure (Drs R. N. Perham & J. I. Harris, personal communication). The polypeptide precipitate was dissolved in 3 ml. of 8 M-urea solution which was 0·1 M in N-ethyl morpholinium acetate (pH 9). Phenylisothiocyanate was added to the solution and the reaction was allowed to proceed at 40°C for 4 hr. The resulting solution was dialysed for 20 hr against deionized water and then lyophilized. A solution of the residual product in 2 ml. of trifluoroacetic acid was kept at 22°C for 2 hr. The acid was then evaporated off and the residue suspended in 1 ml. of 2 N-aqueous HCl at 37°C for 2 hr. An equal volume of water was added and the resulting aqueous solution extracted 3 times with 2-ml. portions of ethyl acetate. The combination of the ethyl acetate phases was extracted with 2 ml. of water. The solvents were evaporated from both organic and aqueous phases, before the thiohydantoin derivative and the polypeptide were dissolved in small portions of acetone and formic acid, respectively. Suitable duplicate samples were dried for counting on planchets in a gas-flow counter. The total recovery of radioactive material corresponded to about 70% of that present at the start of the determination.

3. Results

Binding of fractionated met-sRNA to ribosomes

E. coli sRNA was fractionated by the method described by Goldstein *et al.* (1964). To characterize the distribution pattern the accepting activities for Met, Lys and Tyr were determined with the results shown in Fig. 1. Two distinct methionine-accepting species were separated from each other and were charged to give preparations of met-$sRNA_1$ and met-$sRNA_2$ (see Materials and Methods). The binding of these fractionated species to ribosomes was investigated to test the specificity of binding under optimal conditions. Table 2 shows the results of binding in the presence of AUG, UUG or CUG under varying magnesium concentrations. The best binding of met-$sRNA_1$ due to AUG occurred in 0·03 M-Mg^{2+}. In addition, larger amounts than the usual 0·15 to 0·2 A_{260} units (Brimacombe *et al.*, 1965; Söll *et al.*, 1965) of UUG and CUG did not cause significant met-$sRNA_1$ binding. The best binding of met-$sRNA_2$ to ribosomes in the presence of AUG occurred at a different Mg^{2+} concentration (namely 0·01 M) from that of met-$sRNA_1$. At an incubation temperature of 24°C, UUG caused significant met-$sRNA_2$ binding, with 0·02 M being the most favourable Mg^{2+} concentration. It is noteworthy that AUG caused more than 30% of the added met-$sRNA_2$ to be bound to the ribosomes, indicating that met-$sRNA_2$ binds especially well when compared with other amino-acyl sRNA's (cf. Brimacombe *et al.*, 1965; Söll *et al.*, 1965). The different optimal magnesium concentrations found for the two fractions strongly suggest different sRNA structures.

The interesting binding ambiguity found for met-$sRNA_2$ was studied at various temperatures of incubation. Binding of met-$sRNA_2$ by UUG was observed at Mg^{2+} concentrations as low as 0·01 M-Mg^{2+} and at elevated temperatures such as 37°C.

TABLE 2

Effect of magnesium concentration on the binding of fractionated met-sRNA to ribosomes

$[Mg^{2+}]$ in molarity	Additions (A_{260} units) for binding in $\mu\mu$moles of $[^{35}S]$methionine			
	None	AUG (0·26)	UUG (0·32)	CUG (0·31)
(a) Met-$sRNA_1$				
0·01	0·14	0·85		
0·02	0·21	3·00	0·45	0·27
0·03	0·24	3·86	0·48	0·25
0·05	0·56	2·00		
0·07	0·61	1·58		
(b) Met-$sRNA_2$				
0·01	0·55	11·60	2·15	
0·02	1·74	11·60	5·10	
0·03	1·90	6·60	3·34	

The Mg^{2+} concentration was varied as indicated, but the other reaction components and the method of assay were as described by Nirenberg & Leder (1964). The reaction mixture contained 2·5 A_{260} units of *E. coli* MRE600 ribosomes and either (a) 0·17 A_{260} units of met-$sRNA_1$ charged with 19·6 $\mu\mu$moles of $[^{35}S]$methionine (specific activity 1030 μc/μmole; 1 $\mu\mu$mole = 570 cts/min in a Nuclear-Chicago thin-window low-background gas-flow counter); or (b) 0·14 A_{260} unit of met-$sRNA_2$ charged with 33 $\mu\mu$moles of $[^{35}S]$methionine. Incubation at 24°C was for 25 min.

TABLE 3

Specificity of binding of met-sRNA₁ to ribosomes

Addition	Experiment 1		Experiment 2	
	A_{260} units	$\mu\mu$moles methionine	A_{260} units	$\mu\mu$moles methionine
None		0·24		0·13
Poly UG	0·22	1·03	0·22	0·43
Poly UG †	0·22	0·24		
Poly UAG	0·24	1·67	0·24	0·60
Poly UAG †	0·24	0·24		
AUG	0·26	3·86	0·13	0·79
GUG	0·10	0·56	0·10	0·27
UUG	0·32	0·48	0·13	0·17
CUG	0·31	0·25	0·12	0·15
UAG	0·16	0·30		
AUA	0·24	0·33		
AAG	0·30	0·24		

The reaction conditions and the method of assay were essentially the same as in Table 2, with a Mg^{2+} concentration of 0·03 M, but the reaction components varied in the different experiments. In experiment 1, the ribosomes, sRNA and incubation were as described in Table 2. In experiment 2, met-$sRNA_1$ (0·37 A_{260} unit) charged with 13·3 $\mu\mu$moles of $[^{14}C]$methionine with a specific activity of 200 μc/μmole (New England Nuclear Corporation) was added to each reaction mixture.

† Control incubations which omitted ribosomes.

However, the best conditions for binding by UUG involved incubation at 10°C in 0·01 M-Mg^{2+}. Incubation was necessary for binding, which was complete in less than 15 minutes. Incubation periods for longer than 30 minutes gradually reduced the binding. This type of ambiguity was much more pronounced than any observed in previous binding studies (e.g. Brimacombe *et al.*, 1965; Söll *et al.*, 1965). The conclusion was reached that UUG could be a codeword for methionine under certain conditions.

The specificity of codeword recognition by the met-sRNA fractions was investigated in detail under optimum or near optimum binding conditions with the results shown in Tables 3 and 4. Only AUG out of the related trinucleoside diphosphates tested caused binding of met-$sRNA_1$ to ribosomes (Table 3). The unassigned codewords UAG, AUA did not cause either fraction of met-sRNA to bind to ribosomes. However, the polymer poly UG unexpectedly caused a small amount of binding of met-$sRNA_1$. This was unexplained in terms of codeword binding, but perhaps was due to met-$sRNA_1$ containing a small amount of met-$sRNA_2$ (~5%).

The specificity of directed met-$sRNA_2$ binding was not as well defined as for met-$sRNA_1$. The results using the related triplets AUG, GUG, UUG and CUG are shown

TABLE 4

Specificity of binding of met-sRNA₂ to ribosomes

Addition	Experiment 1		Experiment 2		Experiment 3		Experiment 4	
	A_{260} units	μμmoles methionine	A_{260} units	μμmoles methionine	A_{260} units	μμmoles methionine	A_{260} units	μμmoles F-met
None		0·21		1·30		0·15		0·09
Poly UG	0·22	3·94	0·22	3·92			0·22	1·23
Poly UG †			0·22	0·61			0·22	0·10
Poly UAG			0·24	3·80			0·24	1·52
Poly UAG †			0·24	0·47				
AUG	0·21	7·45	0·13	6·35	0·10	5·27	0·10	6·00
AUG †			0·13	0·29				
GUG			0·10	4·30	0·15	4·30	0·10	4·40
GUG †			0·10	0·53				
UUG	0·24	1·77	0·13	2·90	0·30	2·00	0·24	2·20
UUG †			0·13	0·40	0·30	0·06		
CUG	0·12	0·55			0·30	0·46	0·12	0·48
CUG †					0·30	0·08		
UAG	0·16	0·30					0·18	0·25
AUA	0·24	0·27						
UUU					0·20	0·15		
AAG	0·30	0·22					0·30	0·11
UG			0·20	1·14				

The method of assay was similar to that described in Table 2. However, in experiment 1 met-$sRNA_2$ (0·18 A_{260} unit) from *E. coli* B charged with 31 μμmoles of [^{35}S]methionine was added to the reaction mixture which had a Mg^{2+} concentration of 0·01 M. The met-$sRNA_2$ in experiment 2 was the same as described in Table 2, but the Mg^{2+} concentration was 0·02 M. For experiment 3, at a Mg^{2+} concentration of 0·01 M in the presence of 2·6 A_{260} units of *E. coli* CA265 ribosomes, met-$sRNA_2$ (0·27 A_{260} unit) charged with 20 μμmoles of [^{14}C]methionine was added to the reaction mixture for incubation at 12°C for 30 min. Experiment 4 had similar conditions to those of experiment 3, except for incubation at 20°C and the addition of met-$sRNA_2$ (0·09 A_{260} unit) from *E. coli* B charged with 21·6 μμmoles [^{35}S]methionine under conditions which ensured formylation.

† Control incubation which omitted ribosomes.

in Table 4. Rather surprisingly, GUG was found to cause binding as efficiently as AUG did at relatively low concentrations (compared with $0{\cdot}2\,A_{260}$ unit). Higher concentrations of UUG were needed to give effects comparable with those achieved with AUG and GUG. A general ambiguity NUG (where N could be any of the four usual nucleosides) was not observed, because CUG caused only small binding of met-sRNA$_2$ at high concentrations. However, other workers (Brimacombe *et al.*, 1965; Söll *et al.*, 1965) have observed that trinucleoside diphosphates with the common formula CUN, in general, directed the binding of sRNA to ribosomes poorly when compared with that directed by other trinucleoside diphosphates. Perhaps the same effect of poor binding with CUG is found using met-sRNA$_2$. Controls tested for binding were UUU and the dinucleoside phosphate UG, but with negative results. Other triplets which also did not cause binding of met-sRNA$_2$ when tested were AAC, UUA, UGA and GAU. However, controls omitting ribosomes from the incubation mixtures were found to give a slight binding of met-sRNA to the Millipore filters (see Tables 3 and 4).

The presence of the formyl group on met-sRNA$_2$ did not alter the pattern of binding already found using non-formylated met-sRNA$_2$ (cf. experiment 4 with 1, 2 and 3 of Table 4). We checked on any variation in binding patterns due to the [^{35}S]methionine by using [^{14}C]methionine in similar experiments (cf. experiment 2 with 1 of Table 3, and experiment 3 with 1 and 2 of Table 4).

Formylation of fractionated met-sRNA

Fractionated and unfractionated methionine-accepting sRNA's were charged under strongly formylating conditions. The charged sRNA fractions were digested with RNase-T$_1$ to give the results depicted in Plate I. The fraction designated met-sRNA$_2$ was capable of complete formylation and did not contain met-sRNA$_1$. However, met-sRNA$_1$ could not be formylated under these conditions, but did appear to be contaminated with a small amount of overlapping met-sRNA$_2$ ($\sim$5%). The amino-acyl terminal oligonucleotide from the two fractions met-sRNA$_1$ and met-sRNA$_2$ could be identified with the two species already found in unfractionated met-sRNA (Marcker, 1965).

Incorporation of methionine into polypeptide in a cell-free system

Similar to the results already reported by us (Clark & Marcker, 1965), poly UG stimulated the incorporation of F-met into polypeptides in a cell-free system in which formylation of met-sRNA could take place. A cell extract of *E. coli* MRE600 which permitted formylation was used to make polypeptides from added charged unfractionated sRNA from *E. coli* B. The polypeptide products were enzymically digested with pronase, a non-specific degradative enzyme, or chymotrypsin, with the results shown in Plate II. A series of F-met-containing peptides were obtained from each digestion. These were easily identifiable because the formylated peptides were expected to travel towards the anode. This assumption was confirmed by a treatment of the anionic peptides with carboxypeptidase liberating F-met but not free methionine. In the case of the pronase digestion, F-met itself was released, proving its N-terminal position in the polypeptide. Clearly, almost all the ^{35}S-label incorporated had been present in the polypeptide capable of digestion as N-terminal F-met.

Further knowledge about the position of the [^{35}S]methionine incorporated into polypeptide in an S-30 cell extract was obtained by determination of the N-terminal group with the modified Edman technique (described under Materials and Methods). The formyl group was removed by an acidic treatment for this determination. The results of several determinations are indicated in Table 5.

An averaged result gave 52% of the incorporated [^{35}S]methionine as N-terminal F-met. The disparity between this result and the evidence of Plate II can be excused by only partial reaction of the N-terminal groups, because of only partial solubilization of the polypeptide.

TABLE 5

Determination of the N-terminal amino acid of the polypeptide product radioactively labelled from unfractionated [^{35}S]met-sRNA

Experiment	% of [^{35}S]methionine incorporation as N-terminal
1	49
2	54
3	53
4	51

Incorporation of [^{35}S]methionine into polypeptide by poly UG was measured under the same condition as those which gave the results shown in Plate II. After the incubation, 10% trichloroacetic acid was added to a concentration of 5% in the reaction mixture, which was then heated at 90°C for 15 min. The precipitate was washed as described for Plate II, and when dry it was suspended in 1·5 N-anhydrous HCl gas in methanol. The mixture was incubated at 40°C for 1½ hr. This acidic treatment has been found to remove more than 90% of formyl groups from F-met. The solvent was removed from the polypeptide by evaporation, so that N-terminal end-group determination could be carried out as described under Materials and Methods.

Clearly, under conditions where formylation of the met-sRNA could occur, and especially where no normal AUG codeword was present in the polypeptide, most of the incorporated [^{35}S]methionine appeared in the N-terminal position as F-met. Therefore it was decided to investigate the incorporation of [^{35}S]methionine by polymers in cell extracts where formylation of met-sRNA was avoided. To attain these conditions, the resolved system described in Fig. 2 was used. The effect of using ^{35}S-label in either met-sRNA$_1$ or met-sRNA$_2$ in the presence of other [^{12}C]amino-acyl-sRNA's was studied using poly UG and poly UAG. When poly UG was used as mRNA, only met-sRNA$_2$ allowed the incorporation of [^{35}S]methionine into protein. This result is clear in Fig. 2 (a) from the difference between curves 1 and 4. A comparison of curves 2 and 3 in the same Figure indicates the negligible amount of incorporation when met-sRNA$_1$ is used as a source of [^{35}S]methionine. This small amount can be allocated to the minor contamination of met-sRNA$_1$ by met-sRNA$_2$ (see Plate I). When poly UAG was used as mRNA (see Fig. 2(b)) both met-sRNA$_1$ and met-sRNA$_2$ could serve as [^{35}S]methionine sources (curves 1 and 2, respectively). However, with this polymer met-sRNA$_1$ was more efficient as a source than was met-sRNA$_2$. To confirm that the resolved system used did not cause formylation of met-sRNA$_2$ the polypeptide product was digested with pronase. This treatment, as described in Plate II, liberated free methionine and methionine-containing peptides but no F-met

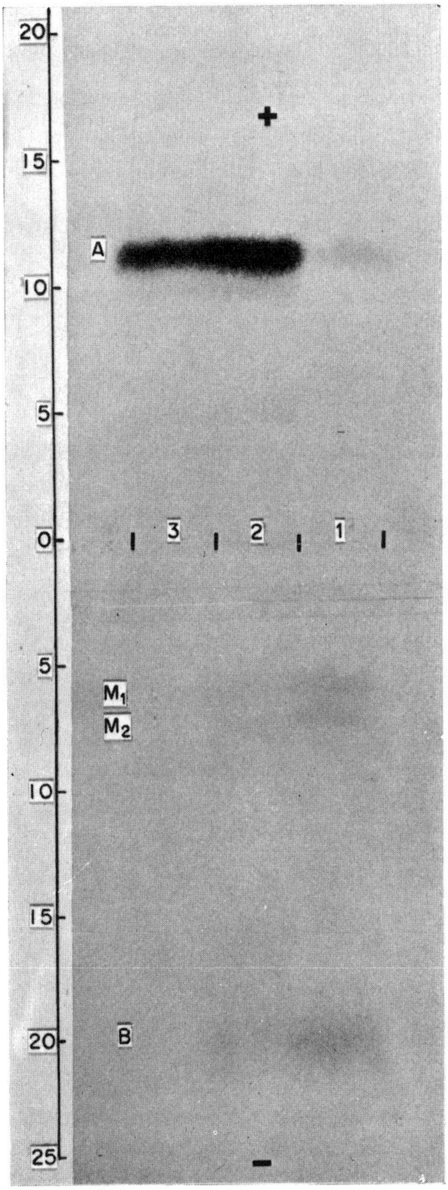

PLATE I. Radioautograph of an electrophoretogram (pyridinium acetate pH 3·5,3 kv, 2·5 hr) (Sanger *et al.*, 1965) of RNase-T_1 digests of 1, [^{35}S]met-$sRNA_1$ charged under strongly formylating conditions; 2, unfractionated [^{35}S]met-sRNA charged as for 1; 3, [^{35}S]met-$sRNA_2$ charged as for 1. The digestions were performed at pH 5·4 as described previously (Marcker, 1965). A indicates the F-met-containing oligonucleotide; B, the methionine-containing oligonucleotide; M_1, methionine sulphoxide; and M_2, methionine. The scale is in cm.

[*facing p. 402*

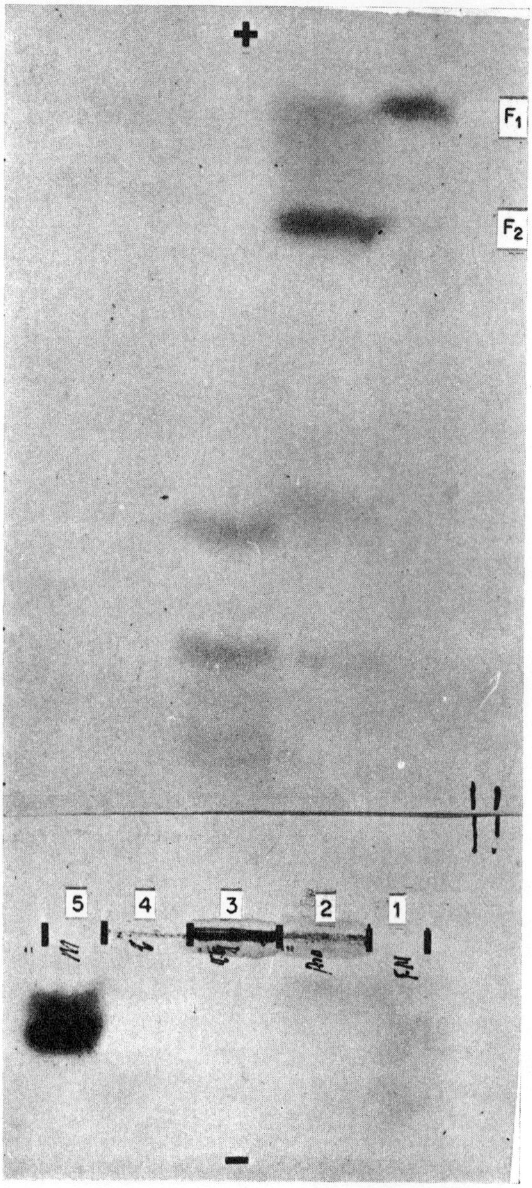

Plate II. Pronase and chymotryptic digestions of the polypeptide product formed during incubation of an *E. coli* cell-free system in the presence of unfractionated [^{35}S]met-sRNA and poly UG. The reaction mixture (0·5 ml.) contained 0·05 M-tris–HCl (pH 7·4), 0·1 M-ammonium acetate, 0·01 M-$MgCl_2$, 0·001 M-ATP, 0·0002 M-GTP, 0·004 M-PEP, 10 μg pyruvate kinase, 0·01 M-2-mercaptoethanol, 40 μg of poly UG (of S_{20} at pH 7·2 = 3·2 and base ratio U/G = 1·6/1) unfractionated met-sRNA (5 A_{260} units charged with 280 $\mu\mu$moles [^{35}S]methionine in the presence of 19 other [^{12}C]amino acids) and 0·2 ml. of a pre-incubated S-30 cell fraction from *E. coli* MRE600. After the reaction mixture had been incubated at 37°C for 20 min, EDTA at pH 7·0 was added to give a concentration of 0·05 M to attain the conditions for a further incubation of 10 min in the presence of 100 μg of pancreatic RNase (5 times crystallized, Calbiochem). The protein was then precipitated with 10% trichloroacetic acid and the precipitate was washed 5 times with 5% trichloroacetic acid containing 3% Casamino acids, twice with ethanol–ether, 1:1, (v/v) and finally dried with ether.

Digestions with pronase (Calbiochem) or α-chymotrypsin (3 times crystallized, Worthington Biochemical Corp.) were carried out in 1% ammonium bicarbonate (enzyme–protein, 1:25) at 37°C for 15 hr. After the digestion, the solutions of solubilized material were concentrated to permit their application to Whatman 3 MM paper for electrophoresis under conditions described in Plate I but for 1 hr. The numbers refer to the following applications: 1, a mixture of F-met (not visible) and F-met sulphoxide; 2, the pronase digest; 3, the chymotryptic digest; 4, an incubation mixture without the digestive enzymes; 5, a mixture of methionine and methionine sulphoxide (slower moving band); F_1 locates F-met sulphoxide and F_2, F-met.

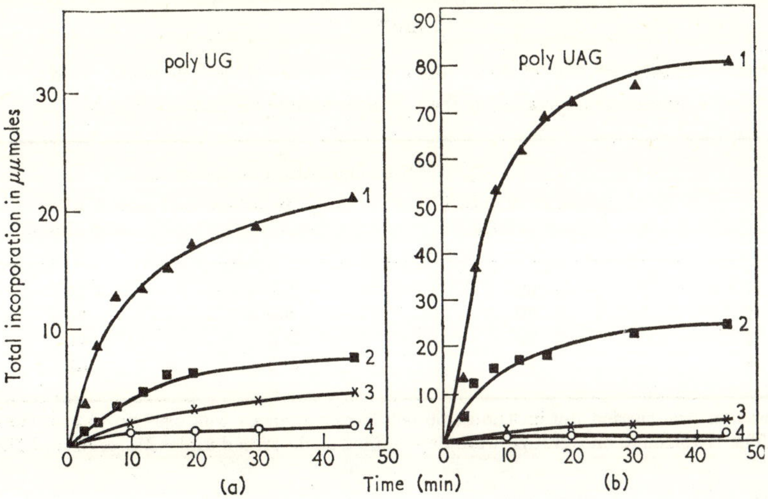

FIG. 2. Incorporation of [^{35}S]methionine from [^{35}S]met-sRNA$_1$ and [^{35}S]met-sRNA$_2$ into polypeptide in a cell-free system containing no transformylase activity.

The buffer and energy-providing components of the system (0·5 ml.) were the same as those described in Plate II. In addition, the incubation mixture contained 0·005 M-[^{12}C]methionine and 0·0002 M 19 other [^{12}C]amino acids, 32·4 A_{260} units of ribosomes from *E. coli* MRE600 (described under Materials and Methods), 250 μg of transfer factor protein, 8 A_{260} units of unfractionated *E. coli* B sRNA charged with the 20 [^{12}C]amino acids, 40 μg of the polymer and either 2·75 A_{260} units of [^{35}S]met-sRNA$_1$ (235 μμmoles [^{35}S]methionine) or 1·35 A_{260} units of [^{35}S]met-sRNA$_2$ (245 μμmoles [^{35}S]methionine). At the indicated times during incubation at 30°C, 50-μl. samples were removed and added to tubes containing 100 μl. each of 0·5% solutions of bovine serum albumin. The polypeptide product was precipitated in 5% trichloroacetic acid and the suspended precipitate heated at 90°C for 15 min before being collected on a cellulose acetate filter for counting in a Nuclear-Chicago thin-window gas-flow counter.

(a) Addition of poly UG. Curve 1, with [^{35}S]met-sRNA$_2$; curve 2, with [^{35}S]met-sRNA$_1$; curve 3, with [^{35}S]met-sRNA$_1$ but no polymer; curve 4, with [^{35}S]met-sRNA$_2$ but no polymer. (b) Addition of poly UAG (S_{20} at pH 7·2 = 2·6 and a base ratio of U/A/G = 2·8/1·0/1·5). Curve 1, with [^{35}S]met-sRNA$_1$; curve 2, with [^{35}S]met-sRNA$_2$; curve 3, with [^{35}S]met-sRNA$_1$ but no polymer; curve 4, with [^{35}S]met-sRNA$_2$ but no polymer.

or anionic [^{35}S]labelled peptides. The results of the methionine incorporation experiments can be explained in terms of the data obtained with the binding studies of fractionated met-sRNA. The species [^{35}S]met-sRNA$_1$, which responded to the codeword AUG alone, was available as a source of [^{35}S]methionine in the system directed by poly UAG and was not available where poly UG was used. Since [^{35}S]met-sRNA$_2$ responded to the codewords AUG, GUG and UUG, it could act as a [^{35}S]methionine source in systems directed by poly UAG or by poly UG. The results in Fig. 2 indicated that the absence of the formyl group did not alter the pattern of incorporation due to expected messenger codeword–sRNA interactions. The fact that met-sRNA$_2$ was less efficient than was met-sRNA$_1$ as a source for polypeptide synthesis under the conditions for Fig. 2(b) suggested that met-sRNA$_2$ had the special function of incorporating methionine into only the N-terminal position of the polypeptide. This would preclude the possibility that the process of formylation had a locating effect on the position of incorporation of methionine from met-sRNA$_2$. This evidence, indicating that the sRNA carries the information for the position of the amino acid, was confirmed by N-terminal end-group determinations with the results shown in Table 6. At least 70% of the incorporated [^{35}S]methionine from [^{35}S]met-sRNA$_2$, under

TABLE 6

Determination of the N-terminal amino acid of polypeptide products radioactively labelled from fractionated [^{35}S]met-sRNA

Experiment	% of ^{35}S-incorporation as N-terminal		
	in presence of poly UG	in presence of poly UAG	
	met-sRNA$_2$	met-sRNA$_1$	met-sRNA$_2$
1	65	2·0	72
2	70	3·0	75
3	70	3·5	69
4	68	3·2	73

Incorporation was carried out in the same cell-free system as was used to obtain the results of Fig. 2. The N-terminal end groups were determined as described under Materials and Methods.

conditions excluding formylation, appeared in N-terminal positions of the polypeptide products directed by poly UAG and by poly UG. This figure is in agreement with all or nearly all incorporation into N-terminal positions. No significant amounts of [^{35}S]methionine were detected in internal positions of the polypeptide under these conditions. A small proportion (about 3%) of the [^{35}S]methionine incorporated from met-sRNA$_1$ appeared in an N-terminal position. Most of this could be explained by the small contamination (~5%) of met-sRNA$_1$ by met-sRNA$_2$ (see Plate I).

We have extended our observations on the incorporation of F-met into protein to studies of protein synthesis *in vitro* using a natural mRNA. When [^{35}S]met-sRNA$_2$ was used as a source in a non-pre-incubated S-30 cell extract from *E. coli* MRE600, F-met was again incorporated into the N-terminal ends of polypeptide products, whereas methionine was incorporated from [^{35}S]met-sRNA$_1$ into internal positions.

4. Discussion

In an attempt to understand the role which F-met-sRNA plays in protein synthesis, we decided to fractionate met-sRNA species and to study their relation *in vitro* with synthetic polynucleotides and oligonucleotides carrying coding information. The met-sRNA was separated into two species, met-sRNA$_1$ (faster moving) and met-sRNA$_2$, and these on formylation and degradation were shown to be the two species previously suggested by similar experiments on unfractionated met-sRNA (Marcker, 1965). Only met-sRNA$_2$ could be formylated and contained no met-sRNA$_1$. However, met-sRNA$_1$ which could not be formylated was contaminated with met-sRNA$_2$ to about 5%. This contamination accounted for some minor ambiguities of binding and incorporation when met-sRNA$_1$ was used. The same formylation patterns of fractionated met-sRNA were obtained with sRNA from the two *E. coli* strains B and CA 265.

A study of the recognition of codewords by the separated met-sRNA species did not give completely unambiguous results. The binding efficiency obtained in our experiments appeared to be significantly greater than that observed in recent surveys of codeword assignments (Brimacombe *et al.*, 1965; Söll *et al.*, 1965). From our binding studies we were able to assign the codeword AUG to met-sRNA$_1$. This was in agreement with the results of the experiments in which polypeptides were synthesized *in vitro*, since poly UAG caused incorporation of methionine from met-sRNA$_1$ but poly UG did not. As expected, using [^{14}C]met-sRNA or [^{35}S]met-sRNA in the binding studies did not give different results, nor did the use of ribosomes from different *E. coli* strains.

The species met-sRNA$_2$ which could be formylated was also bound strongly by AUG and in addition showed an anomalously high level of ambiguity for binding by the GUG and UUG codewords. The conditions for its optimal binding to ribosomes were also different from those for met-sRNA$_1$. Both poly UG and poly UAG stimulated the incorporation of methionine mostly into the N-terminal position of polypeptides from met-sRNA$_2$, whether this was formylated or not. In addition, the binding studies showed no significant differences in the binding of met-sRNA$_2$ in the formylated or unformylated state. The conclusion is reached that the location of the methionine in the N-terminal position of the polypeptide is determined by the nature of the met-sRNA and presumably by the usual mRNA–sRNA interaction. The part which formylation plays is still obscure. Perhaps a rate effect is involved, as already noted (Clark & Marcker, 1965). The met-sRNA$_2$ clearly carries the information for a special mechanism involving chain initiation. The signal for the chain initiation is not clear, since it should be specific if it is a sequence of nucleotides. In order to investigate this further, we intend to design oligonucleotides of known sequences as better models for the beginning of mRNA. In addition, evidence is being collected about the structure of met-sRNA$_2$ with a view to the elucidation of the codeword–sRNA reading mechanism.

Our view that F-met-sRNA is a chain initiator has recently found strong support in other *in vitro* systems. In RNA-phage infected bacterial systems which synthesize phage coat protein, F-met has been found by Adams & Capecchi (1966) in an N-terminal position of the protein product. Webster, Engelhardt & Zinder (1966) have identified F-met also in the N-terminal position of the protein fragment formed in a cell-free system directed by coat protein mRNA which contains an amber codeword (Weigert & Garen, 1965; Brenner, Stretton & Kaplan, 1965). These discoveries were surprising, in view of the finding that the N-terminal amino acid of the coat protein from intact virus particles was alanine (see Adams & Capecchi, 1966). Significantly, both groups working with the *in vitro* systems found alanine next to F-met. Some mechanism as yet unknown must therefore remove the formyl or F-met from the nascent protein *in vivo*.

We are grateful for the skilled technical assistance of Messrs W. A. Whybrow, I. S. Palmer and A. I. Matus and for the encouragement given to us by Dr F. Sanger.

REFERENCES

Adams, J. M. & Capecchi, M. R. (1966). *Proc. Nat. Acad. Sci., Wash.* **55**, 147.
Bernfield, M. R. & Nirenberg, M. W. (1965). *Science*, **147**, 479.
Brenner, S., Stretton, A. O. W. & Kaplan, S. (1965). *Nature*, **206**, 994.
Brimacombe, R., Trupin, J., Nirenberg, M., Leder, P., Bernfield, M., & Jaouni, T. (1965). *Proc. Nat. Acad. Sci., Wash.* **54**, 954.
Carbon, J. A. (1964). *Biochem. Biophys. Res. Comm.* **15**, 1.
Clark, B. F. C. & Marcker, K. A. (1965). *Nature*, **207**, 1038.
Goldstein, J., Bennett, T. P. & Craig, L. C. (1964). *Proc. Nat. Acad. Sci., Wash.* **51**, 119.
Holley R. W., Apgar, J., Doctor, B. P., Farrow, J., Marini, M. A. & Merrill, S. H. (1961). *J. Biol. Chem.* **236**, 200.

Marcker, K. A. (1965). *J. Mol. Biol.* **14**, 63.
Marcker, K. A. & Sanger, F. (1964). *J. Mol. Biol.* **8**, 835.
Nathans, D. & Lipmann, F. (1961). *Proc. Nat. Acad. Sci., Wash.* **47**, 497.
Nirenberg, M. W. & Leder, P. (1964). *Science,* **145**, 1399.
Nirenberg, M. W. & Matthaei, J. H. (1961). *Proc. Nat. Acad. Sci., Wash.* **47**, 1588.
Sanger, F., Brownlee, G. G. & Barrell, B. G. (1965). *J. Mol. Biol.* **13**, 373.
Sarin, P. S. & Zamecnik, P. C. (1964). *Biochim. biophys. Acta,* **91**, 653.
Singer, M. F. & O'Brien, B. M. (1963). *J. Biol. Chem.* **238**, 328.
Söll, D., Ohtsuka, E., Jones, D. S., Lohrmann, R., Hayatsu, H., Nishimura, S. & Khorana, H. G. (1965). *Proc. Nat. Acad. Sci., Wash.* **54**, 1378.
Thach, R. E. & Doty, P. (1965). *Science,* **148**, 632.
Webster, R. E., Engelhardt, D. L. & Zinder, N. D. (1966). *Proc. Nat. Acad. Sci., Wash.* **55**, 155.
Weigert, M. G. & Garen, A. (1965). *Nature,* **206**, 992.
Zubay, G. (1962). *J. Mol. Biol.* **4**, 347.

Reprinted from THE JOURNAL OF BIOLOGICAL CHEMISTRY, Vol. 240, 1965

Direction of Reading of the Genetic Message*

MARGARITA SALAS,† MARVIN A. SMITH,‡ WENDELL M. STANLEY, JR.,‡ ALBERT J. WAHBA, AND SEVERO OCHOA

From the Department of Biochemistry, New York University School of Medicine, New York, New York 10016

(Received for publication, June 9, 1965)

The assembly of polypeptide chains during protein biosynthesis is believed to proceed from the NH_2-terminal through the COOH-terminal amino acid (1–4). Hence, the most direct method for ascertaining the direction in which the genetic message is read is to determine the end location of a given amino acid in polypeptide chains synthesized in a cell-free system under the direction of synthetic polynucleotides having a codon of specified base sequence at one end of the chain. Previous experiments (5) were inconclusive because of (*a*) presence of nucleases in the system, (*b*) insufficient characterization of the polynucleotide messenger, and (*c*) difficulty of performing end group assays because of the insolubility of the phenylalanine peptides formed. All of these obstacles have now been removed through (*a*) the use of a system low in nuclease activity consisting of purified *Escherichia coli* ribosomes and *Lactobacillus arabinosus* supernatant and (*b*) the preparation and unequivocal characterization of short polyadenylic acid messengers with 1 cytidine residue (and therefore an AAC codon) at the 3′-end. Lysine polypeptides are soluble in water and can be readily characterized.

Experiments to be reported in this paper have shown that polynucleotides of the structure ApApAp . . . pApApC (Ap_nC),[1] with 21 to 23 nucleotide residues, directed the synthesis of a family of asparagine-containing peptides of increasing chain length. The most abundant of these had 1 asparagine per 3 to 5 lysine residues, with NH_2-terminal lysine and COOH-terminal asparagine. Hence, if the polypeptide is indeed assembled from the NH_2- to the COOH-terminal end, the polynucleotide message must be read in the direction from the 5′- to the 3′-end or from left to right in the currently accepted fashion of writing polynucleotide chains. Removal of the terminal cytidine residue, through periodate oxidation followed by aminolysis, suppressed the incorporation of asparagine leaving that of lysine essentially unchanged. These results also establish pApApC (AAC) directly and unequivocally as the nucleotide sequence for one of the asparagine coding triplets.

EXPERIMENTAL PROCEDURE

Preparations

Ribosomes and Supernatant

The conventional cell-free system of *Escherichia coli* for protein synthesis contains very active nucleases and could not be used as such for this work. *E. coli* has a ribonuclease (endonuclease) which, until recently, was believed to be associated with the 30*S* ribosomal subunits (6). More recently it has been shown (7) that the bulk of the *E. coli* ribonuclease is located between the cell membrane and the cell wall and is released to the medium when the cells are converted to spheroplasts. However, when the cells are disrupted, most of the ribonuclease is bound by the 30*S* ribosomes. *E. coli* cells also contain an exonuclease (8) that is believed to act like snake venom phosphodiesterase, cleaving polynucleotides from the 3′-end with release of nucleoside 5′-monophosphates. About 75% of the exonuclease in *E. coli* extracts is present in the supernatant fluid after high speed centrifugation, the remaining 25% being carried down with the ribosomes. The ribosomes can be obtained essentially free of nuclease activity by washing with ammonium chloride followed by chromatography on DEAE-cellulose. Attempts to remove nucleases from the *E. coli* supernatant were unsuccessful, but it was found that the high speed supernatant fluid of *Lactobacillus arabinosus* is largely devoid of nuclease activity and can be used together with the purified *E. coli* ribosomes for polynucleotide-directed polypeptide synthesis.[2] This system will henceforth be referred to as the *E. coli-L. arabinosus* system. In Table I the nuclease activity of *E. coli* ribosomes, just washed with ammonium chloride, and *E. coli* supernatant is compared with that of *E. coli* ribosomes, that were both washed with ammonium chloride and chromatographed on DEAE-cellulose and *L. arabinosus* supernatant. It may be seen in Table II that, under the conditions of amino acid incorporation, the nuclease activity of the system of purified *E. coli* ribosomes and *L. arabinosus* supernatant fluid was negligible. This was tested both with poly A-^{14}C in which all of the adenine residues were labeled with ^{14}C, and with a polynucleotide containing a sequence

* Aided by Grants AM-01845, AM-08953, and 1-SO1-FR-05099 from the National Institutes of Health, United States Public Health Service, and E. I. Du Pont de Nemours and Company, Inc. A preliminary report of this work was presented at the Second Meeting of the Federation of European Biochemical Societies (symposium on "Ribonucleic Acid—Structure and Function"), Vienna, April 21 to 24, 1965 (*Abstracts*, p. 275).

† International Postdoctoral Fellow of the National Institutes of Health, United States Public Health Service. Permanent address, Instituto Marañon, Centro de Investigaciones Biologicas, C.S.I.C., Madrid, Spain.

‡ Postdoctoral Fellow of the National Institutes of Health, United States Public Health Service.

[1] The abbreviations used are: Ap_nC and Ap_nU, oligoribonucleotides containing *n* adenylic acid residues and either 1 cytidine or 1 uridine residue at the 3′-end (no 5′-phosphate at opposite end of chain); Lys_7-^{14}C, hepta-L-lysine uniformly labeled with ^{14}C; tRNA, transfer RNA; poly A, polyadenylic acid; poly U, polyuridylic acid; poly AC, random copolymer of adenylic acid and cytidylic acid; poly AU, random copolymer of adenylic and uridylic acid; DNP-, 2,4-dinitrophenyl-; an asterisk indicates labeling with ^3H.

[2] *E. coli* ribosomes rather than *L. arabinosus* ribosomes were used because the latter were of low activity in amino acid incorporation.

of 29 to 31 nonlabeled adenylic acid residues with 1 ^{3}H-labeled cytidine residue at the 3′-end of the chain.

E. coli Ribosomes—Ribosome pellets were suspended at 4° in Buffer A (0.5 M NH_4Cl, 2 mM magnesium acetate, 0.01 M Tris-HCl, pH 7.8) for 8 to 12 hours. The suspension was then centrifuged at low speed, and the sediment was discarded. The supernatant was centrifuged at 110,000 × *g* and 4° for 4 hours. A dark brown upper layer was carefully scraped off with a spatula and discarded. The pellets were resuspended in Buffer A for 8 to 12 hours and recentrifuged at 110,000 × *g* for 4 hours. They were then suspended in Buffer B (0.3 M NH_4Cl, 5 mM magnesium acetate, 0.01 M Tris-HCl, pH 7.8) at a nucleoprotein concentration of 10 mg per ml. About 50 ml of this ribosomal suspension were applied at 4° to a DEAE-cellulose column (1 meq per g, 2.5 × 40 cm) which had previously been equilibrated with Buffer B. The column was washed with about 1 liter of the same buffer, and the ribosomes were subsequently eluted (flow rate, 120 ml per hour) with 1.0 M NH_4Cl, 0.01 M magnesium acetate, 0.01 M Tris-HCl, pH 7.8. The turbid fractions containing the ribosomes, recovered in 40 to 50 ml, were dialyzed overnight at 4° against Buffer A. The ribosomes were then sedimented by high speed centrifugation and the pellets resuspended in Buffer A at a nucleoprotein concentration of 40 mg per ml. These ribosomes were active for amino acid incorporation after storage at 4° for 2 to 3 months.

L. arabinosus Supernatant—*Lactobacillus arabinosus* 17-5 (ATCC 8014) was grown at 37° in Micro Inoculum broth (Difco) for 16 to 20 hours. The cells were harvested by centrifugation, washed once with 0.2 M NH_4Cl, pH 7.0, and twice with a buffer containing 0.02 M Tris-HCl, pH 7.8, 0.01 M magnesium acetate, and 0.01 M mercaptoethanol. They were then ground with twice their weight of Alcoa alumina A-301 and extracted with 2 volumes of the same buffer. Cell fragments and alumina were removed by centrifugation at 30,000 × *g* for 30 min. Supernatants used for incorporation experiments, obtained by centrifuging the extract for 4 hours at 110,000 × *g*, were dialyzed for 16 hours against the above buffer. All operations were carried out at 4–5°.

TABLE I

Nuclease activity of E. coli and L. arabinosus cell fractions

Values are expressed in millimicromoles per min per mg of protein at 37°. Samples contained the following components, in micromoles, in a final volume of 0.25 ml: Tris-HCl buffer, pH 7.8, 3.5; KCl, 25; magnesium acetate, 0.5; ^{14}C-labeled polymer (2.5 × 10^4 cpm per μmole of nucleotide), 0.034 and 0.043 of nucleotide in poly A and poly U, respectively. After incubation the samples were chilled in ice, treated with 2 volumes of ice-cold ethanol and 0.5 mg of yeast RNA in NaCl to make a final concentration of NaCl of 0.1 M, and centrifuged after standing 15 min at 0°. Aliquots of the supernatant were taken for measurement of ethanol-soluble radioactivity in a gas flow window counter. Blanks without ribosomes or supernatant were run to correct for any ethanol-soluble radioactivity not due to nuclease action.

Cell fraction	Polynucleotide	Nuclease activity
E. coli supernatant	Poly A-^{14}C	30
	Poly U-^{14}C	23
L. arabinosus supernatant	Poly A-^{14}C	0.06
	Poly U-^{14}C	0.07
E. coli ribosomes washed with NH_4Cl	Poly A-^{14}C	21
	Poly U-^{14}C	18
E. coli ribosomes washed with NH_4Cl and chromatographed on DEAE-cellulose	Poly A-^{14}C	0.2
	Poly U-^{14}C	0.2

TABLE II

Nuclease activity under conditions of amino acid incorporation

Values are expressed in millimicromoles of radioactive nucleotide and (in parentheses) as percentage of radioactivity released as ethanol-soluble material upon incubation for 40 min at 37°. They are corrected for blanks without ribosomes or supernatant. The incubation samples were as described under "Amino Acid Incorporation System," except for no addition of amino acids and for the addition of either poly A-^{14}C (1.4 × 10^5 cpm per μmole of nucleotide), 10 μg (34 mμmoles of labeled nucleotide) or oligo $Ap_{29-31}C$ (cytidine residue (C) labeled with ^{3}H, 7.2 × 10^6 cpm per μmole of cytidine), 20 μg (1 mμmole of labeled nucleoside). Samples were worked up for measurement of ethanol-soluble radioactivity as in Table I. Radioactivity was measured in a Packard Tri-Carb liquid scintillation spectrometer.

Ribosomes	Supernatant	Nuclease activity on	
		Poly A-^{14}C	$Ap_{29-31}C$-^{3}H
E. coli (washed with NH_4Cl)	*E. coli*	13.5 (39)	0.34 (34)
E. coli (washed with NH_4Cl)	*L. arabinosus*	4.0 (12)	0.08 (8)
E. coli (washed and chromatographed on DEAE-cellulose) ...	*E. coli*	14.6 (42)	0.32 (32)
E. coli (washed and chromatographed on DEAE-cellulose)....	*L. arabinosus*	0.1 (0.3)	0.006 (0.6)

The *L. arabinosus* supernatant is deficient in of some the aminoacyl ~ tRNA synthetases. We found it to be specifically low in glutaminyl ~ tRNA synthetase so that only traces of glutamine are incorporated into acid-insoluble material by the *E. coli-L. arabinosus* system with poly AC, a polymer that contains a glutamine codon.

Oligoadenylic Acids

Poly A and poly U were prepared with *Azotobacter* polynucleotide phosphorylase (specific activity, 30) as in previous work (9). For preparation of poly A-^{14}C, ADP-8-^{14}C was used as substrate. For poly U-^{14}C, the substrate was UDP-2-^{14}C. All nucleoside diphosphates were obtained from Schwarz BioResearch. Oligoadenylic acids were prepared by partial alkaline hydrolysis of poly A followed by treatment with HCl at pH 2.0, to hydrolyze terminal 2′,3′-cyclic phosphodiester bonds, and with phosphomonoesterase to remove terminal 2′(3′)-phosphate residues. They were subsequently chromatographed in 8.0 M urea on DEAE-cellulose (10) to yield a homologous series of oligoadenylic acids of increasing chain length.

^{14}C-Labeled L-Lysine Oligopeptides

These were prepared with the *E. coli-L. arabinosus* system with either poly A or oligoadenylic acid messengers and lysine-^{14}C as substrate. The incubation conditions are described under

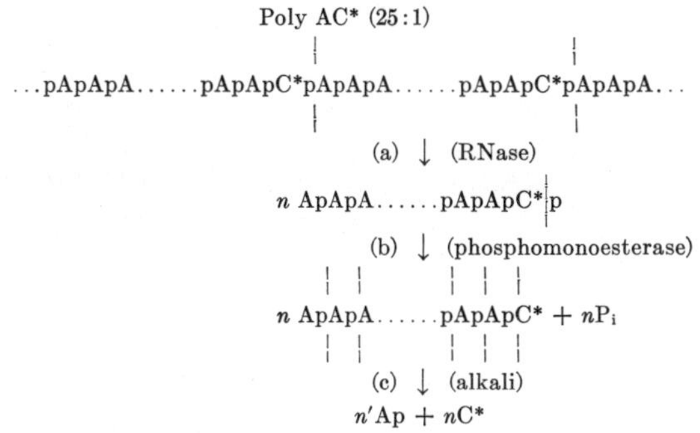

FIG. 1. Diagram illustrating the preparation of ApApAp . . . pApApC* oligonucleotides. The *asterisk* denotes ^{3}H label in the cytidine residue.

"Methods." The peptides were fractionated according to chain length by chromatography on carboxymethyl cellulose as described by Stewart and Stahmann (11). The *E. coli-L. arabinosus* system appears to be free from peptidases acting on lysine oligopeptides. This is evidenced by the fact that Lys$_7$-^{14}C when incubated under the conditions of amino acid incorporation for 40 min at 37°, except for the absence of added polynucleotide or amino acids, was recovered intact upon re-isolation by carboxymethyl cellulose chromatography.

Ap$_n$C Oligonucleotides*

These are short polynucleotides consisting of a stretch of adenylic acid residues with a cytidine residue at the 3′-end; their 3′-terminal codon is therefore AAC. For convenience in characterization, and to minimize interference with the assay for incorporation of ^{14}C-labeled amino acids, the cytidine residue was labeled with ^{3}H. These polynucleotides were obtained from poly AC* (^{3}H-labeled cytidylic acid residues) which was prepared with *Azotobacter* polynucleotide phosphorylase (9) from a mixture of nonlabeled ADP and ^{3}H-labeled CDP in a molar ratio of 15:1 or 25:1. The preparation and characterization of these oligonucleotides is outlined in Fig. 1. The copolymer (60 mg) was digested with 150 μg of pancreatic RNase A (Worthington) for 12 hours at 37° in 60 ml of a solution containing 0.05 M Tris-HCl, pH 7.5, 0.1 M KCl, and 0.002 M EDTA, to give a mixture of oligonucleotides of the type ApApAp . . . pApApC*p (Fig. 1*a*). After extraction with 80% aqueous phenol and dialysis, first against 0.15 M sodium chloride, 0.015 M sodium citrate, pH 7.5, and then against distilled water, the oligonucleotides were treated with yeast phosphomonoesterase, in the presence of 0.02 M potassium acetate, pH 4, to remove the terminal 3′-phosphate (Fig. 1*b*). Re-extraction with phenol and dialysis against water yielded approximately 25 mg of Ap$_n$C*. These oligonucleotides were chromatographed on a column (1.4 × 70 cm) of DEAE-cellulose (0.9 meq per g) equilibrated with 0.01 M Tris-HCl-8 M urea, pH 7.8, with an exponential gradient of NaCl, 0 to 0.6 M, obtained from a 1-liter constant volume mixer. The concentration of the Tris-urea buffer was held constant during the elution, which was carried out at a flow rate of 18 ml per hour, and monitored continuously at 260 mμ. A family of peaks representing a homologous series of oligonucleotides of increasing chain length was obtained. The fractions to be used for amino acid incorporation were dialyzed exhaustively against water. These oligonucleotides were characterized (Fig. 1*c*) for chain length and composition by alkaline hydrolysis (0.5 N KOH, 18 hours at 37°). Ascending paper chromatography, in ethanol:1 M ammonium acetate, pH 7.5 (7:3 v/v) (12), of the neutralized hydrolysate yielded all the adenylic acid residues as 2′(3′)-AMP and 94% of the tritium-labeled cytidylic acid residues as cytidine. Only 6% of the ^{3}H label was recovered as 2′(3′)-CMP.

Ap$_n$U* (uridine residues labeled with ^{3}H) was similarly prepared starting with poly AU, synthesized from a mixture of ADP and ^{3}H-labeled UDP in a molar ratio of 25:1. Alkaline hydrolysis and chromatography indicated that 98% of the radioactivity in Ap$_n$U* was in the 3′-terminal position.

For control experiments the terminal cytidine residue was removed from Ap$_{26-28}$C* and Ap$_{24-25}$C* by periodate oxidation and aminolysis with lysine, according to the procedure of Neu and Heppel (13). The oligonucleotides were then treated with yeast phosphomonoesterase to remove the 3′-phosphate, extracted with phenol, and dialyzed exhaustively, first against 0.15 M sodium chloride-0.015 M sodium citrate, pH 7.5, and then against distilled water; 85 to 90% of the terminal cytidine residue was lost by this treatment as determined by residual radioactivity.

Other Preparations

^{14}C-Labeled L-asparagine and L-glutamine (specific radioactivity, 29.8 and 32.2 μC per μmole, respectively) were purchased from Nuclear-Chicago. All other ^{14}C-labeled amino acids were from the New England Nuclear Corporation. ^{3}H-Labeled lysine (specific radioactivity, 1000 μC per μmole) was obtained from Schwarz BioResearch. Dinitrophenyl derivatives of the amino acids were purchased from the Mann Research Laboratories. Poly-L-lysine hydrobromide was from Pilot Chemicals, Inc. Carboxypeptidase B (treated with diisopropyl fluorophosphate to inactivate contaminating trypsin) was purchased from Worthington. We are greatly indebted to Dr. Gerhard Schmidt, Tufts University, and to Professor Hans Neurath, University of Washington, for generous gifts of phosphomonoesterase (highly purified yeast acid phosphatase (14)) and crystalline carboxypeptidase A (free of carboxypeptidase B), respectively. *N*-Carbobenzoxyglycyl-L-phenylalanine and *N*-carbobenzoxyglycyl-L-aspartic acid were obtained from the Mann Research Laboratories. Other preparations, such as nucleoside triphosphates, phosphocreatine, creatine kinase, and *E. coli* tRNA, were as in previous work (5).

METHODS

Amino Acid Incorporation System

Samples contained the following components (in micromoles unless otherwise specified) in a final volume of 0.25 ml: Tris-HCl buffer, pH 7.9, 15; NH_4Cl, 20; 2-mercaptoethanol, 4; magnesium acetate, 4.5; ATP, 0.3; GTP, 0.075; phosphocreatine, 4.5; creatine kinase, 16 μg; *E. coli* tRNA, 2.5 mg; purified *E. coli* ribosomes with 0.17 mg of protein; *L. arabinosus* supernatant with 0.35 mg of protein; lysine, asparagine, threonine, and glutamine (one of which was labeled with ^{14}C) each 0.05; without or with polynucleotide (either poly AC or oligo Ap$_n$C*), 50 to 60 μg. The specific radioactivity of the labeled amino acids used was (in micro-Curies per μmole), lysine, 2; asparagine, 29.8; threonine, 33. After incubation for 40 min at 37°, 0.5 ml of

3% NH_3 and 5 μmoles of nonlabeled amino acid (the one that was labeled in the incubation mixture) were added, and the mixture was kept at 37° for 1 hour. This hydrolyzes aminoacyl (or peptidyl)~tRNA linkages and dilutes the labeled amino acid in the sample 100-fold. The pH was then lowered to about 7.0 and the polypeptides were precipitated (15) by addition of 6 ml of a solution containing 0.25% sodium tungstate and 5% trichloroacetic acid (pH 2.0). The precipitate, collected by centrifugation (8000 × *g*, 10 min), was washed twice with the above tungstate reagent and dissolved in 0.5 ml of 3% NH_3, plated, and its radioactivity was measured in a window gas flow counter. Duplicate samples were run in each experiment and the values averaged.

The same procedure was applied in amino acid incorporation experiments with poly AU (25:1) or with oligo Ap_nU^* except that the amino acids, lysine, asparagine, and isoleucine (one of which was labeled with ^{14}C), were present in the incubation mixtures.

Isolation of Peptides

For end group assays of the peptides synthesized with oligo Ap_nC messengers, the size of the incubation mixtures described in the preceding section was increased 10- to 15-fold, and two samples were run simultaneously, one without and one with polynucleotide. Both samples contained ^{14}C-labeled asparagine and nonlabeled lysine as the only added amino acids. After incubation (40 min, 37°), KOH was added to a final concentration of 0.3 M, the mixtures were kept for 1 hour at 37°, and the pH was then adjusted to 7.0 with $HClO_4$. The precipitates were discarded and the supernatants were diluted with water to 20-fold the volume of the original incubation mixtures. A partial acid hydrolysate (400 μg) of polylysine (11) was added as carrier, and the two samples were loaded onto identical carboxymethyl cellulose columns (0.58 meq per g, 0.6 × 19 cm) at a flow rate of 36 ml per hour. After washing overnight with water, elution of the peptides was carried out simultaneously with an exponential gradient from 0 to 1.0 M NH_4HCO_3. A common mixer with 500 ml of water was used. Fractions corresponding to oligopeptides with about 3 to 8 lysine residues were pooled, evaporated to dryness, and dissolved in a small volume of water. This procedure yielded salt-free aqueous solutions of the peptides. The ^{14}C-asparagine radioactivity of the peptides from the samples without added polynucleotide, *i.e.* the blank value, was one-sixth to one-fifteenth that of the peptides from the samples with added polynucleotides. For the preparation of doubly labeled peptides, ^{14}C-asparagine and ^3H-lysine (specific radioactivities, 29.8 and 250 μC per μmole, respectively) were used. The above described isolation procedure was followed.

End Group Assays

Chemical—The fluorodinitrobenzene method was used for assay of NH_2-terminal amino acids. ^{12}C-Lysine-^{14}C-asparagine peptide (0.074 to 0.1 mμmole; 41,900 cpm per mμmole) from an incubation with $Ap_{17-19}C$ was treated with 1-fluoro-2,4-dinitrobenzene for 2 hours at room temperature. Samples of the peptide material formed in the absence of added polynucleotide, in addition to controls of Lys_7-^{14}C (18 mμmoles; 150,200 cpm per μmole) and ^{14}C-asparagine (0.16 mμmole; 17,500 cpm per mμmole) were similarly treated. After dinitrophenylation, the excess 1-fluoro-2,4-dinitrobenzene was removed with alcohol from the DNP-peptides and with ether from the DNP-asparagine. The DNP-derivatives were then hydrolyzed in 5.7 N HCl in sealed tubes for 10 hours at 105°. The hydrolysates were diluted with water to a concentration of 1.0 N HCl. The DNP-amino acids were extracted three times with ether. Both the ether and residual water phases, after concentration, were chromatographed (16) ascendingly on Whatman No. 1 for 15 to 18 hours in the upper layer from 1-butanol-acetic acid-water (4:1:5, by volume). Authentic samples of amino acids and their DNP-derivatives were used as markers. Marker amino acid spots were located with ninhydrin spray, and radioactivity was scanned with a strip counter (Nuclear-Chicago Actigraph). The radioactive spots were eluted with water from the paper and counted in a Packard Tri-Carb liquid scintillation spectrometer.

TABLE III

Effect of carboxypeptidases on ^{14}C*-labeled heptalysine*

Conditions as described under "End Group Assays; Enzymatic" with 13 μg of carboxypeptidase A, 1.5 μg of carboxypeptidase B; final volume, 0.1 ml. Incubation for 30 min at 37°.

Enzyme	Initial (Lys_7-^{14}C)	Final (lysine-^{14}C + dilysine-^{14}C)	
	cpm	*cpm*	%
None	5337	72	2
Carboxypeptidase A	5329	415	8
Carboxypeptidase B	5334	5124	95
Carboxypeptidase A + B	5323	4923	93

Enzymatic—Carboxypeptidase A was used for assay of COOH-terminal asparagine in the ^{12}C-lysine-^{14}C-asparagine peptides. Control experiments showed that carboxypeptidase A rapidly released asparagine from L-phenylalanyl-L-glutaminyl-L-asparagine. On the other hand the enzyme had little activity on Lys_7-^{14}C, a compound which was rapidly cleaved (to a mixture of lysine and dilysine) by carboxypeptidase B (Table III), and no activity on *N*-carbobenzoxy-L-aspartic acid. Carboxypeptidase B did not act on peptides with COOH-terminal asparagine. Carboxypeptidase A was routinely assayed with *N*-carbobenzoxyglycyl-L-phenylalanine. Phenylalanine was quantitatively released from 3 μmoles of substrate per ml by 0.88 μg per ml of enzyme in 90 min at 37°. The samples for carboxypeptidase assay of the peptides contained (in a final volume of 0.1 ml), Tris-HCl buffer, pH 7.8, 2.5 μmoles; NaCl, 5 μmoles; peptide, either Lys_7-^{14}C, 18 mμmoles, or ^{12}C-lysine-^{14}C-asparagine, 18 to 30 μμmoles of incorporated asparagine, the latter peptide containing, in addition, 15 to 20 mμmoles of carrier lysine peptides; and either carboxypeptidase A, 13 μg, or carboxypeptidase B, 1.5 μg, or both. After incubation at 37° the samples were subjected to paper chromatography in the Waley and Watson (17) system (descending, Whatman No. 1 paper, 1-butanol-acetic acid-water-pyridine (30:6:24:20, by volume, for 65 to 72 hours). This system effectively resolves lysine peptides, from the monomer that moves farthest from the origin, to the heptamer that moves but a short distance from the origin. Asparagine moved slightly ahead of a lysine marker. Amino acid markers were located with ninhydrin spray. Radioactivity was routinely scanned with the strip counter. Quantitation of the results was obtained by cutting out sections of the paper strips (3 × 4 cm) and measuring their radioactivity in the Packard Tri-Carb scintillation spectrometer.

TABLE IV

Amino acid incorporation with poly AC and Ap_nC oligonucleotides

Conditions as described under "Amino Acid Incorporation System." Values (blanks without added polynucleotide subtracted from values with polynucleotide) expressed in micromicromoles of amino acid incorporated per sample. The cytidine (C) residue in Ap_nC oligonucleotides was labeled with ^3H. Samples with labeled oligonucleotide, but without labeled amino acid, were run simultaneously with the experimental samples to correct for the slight radioactivity of the polymer. All the experiments have been repeated at least twice. The oligonucleotides $Ap_{25-27}A$ (Experiment 1) and $Ap_{23-24}A$ (Experiment 2) were prepared from $Ap_{26-28}C$ and $AP_{24-25}C$, respectively, by periodate oxidation, aminolysis, and phosphomonoesterase treatment as described under "Ap_nC* Oligonucleotides."

Polynucleotide	Amino acid incorporation			Ratio of lysine to asparagine
	Lysine	Asparagine	Threonine	
Experiment 1				
None (blank)	(115)	(14)	(15)	
Poly AC (25:1)	2467	122	129	
$Ap_{26-28}C$	815	20	0	41
$Ap_{25-27}A$	701	1		
Experiment 2				
None (blank)	(51)	(18)		
$Ap_{24-25}C$	1960	40		49
$Ap_{23-24}A$	1692	2		
Experiment 3				
None (blank)	(66)	(21)	(18)	
Poly AC (15:1)	1934	126	108	
$Ap_{19-20}C$	205	11	0	18
Experiment 4				
None (blank)	(152)	(13)		
Poly AC (25:1)	3367	135		
$Ap_{16-17}C$	166	23		7

RESULTS AND DISCUSSION

Amino Acid Incorporation—Typical results of amino acid incorporation experiments in the *E. coli-L. arabinosus* system, with poly AC or with Ap_nC oligonucleotides of various chain lengths, are shown in Table IV. Poly AC promoted the incorporation of lysine (AAA codon), asparagine (2A1C codon), and threonine (2A1C codon). Glutamine was not used because, as previously mentioned, the *L. arabinosus* supernatant is deficient in glutaminyl ~ tRNA synthetase. As expected asparagine and threonine were incorporated not only to the same extent but to an extent relative to that of lysine (4 to 5% with poly AC (25:1), 6% with poly AC (15:1)) corresponding to the calculated frequency of 2A1C triplets relative to AAA triplets in the AC copolymers used.

The Ap_nC oligonucleotides promoted the incorporation of lysine and asparagine, but not threonine (Table IV). Although the incorporation of asparagine in the presence of oligonucleotide was on the average only about 1.5 times higher than the blank, *i.e.* the asparagine incorporation with no added polymer, the results were highly reproducible. Moreover, removal of the 3′-terminal cytidine residue wiped out the incorporation of asparagine leaving that of lysine virtually unchanged (Table IV, Experiments 1 and 2). These experiments thus provide direct and unequivocal proof that the 2A1C asparagine codon (18) has the nucleotide sequence pApApC (AAC). Freedom from exonuclease was essential for asparagine incorporation mediated by Ap_nC oligonucleotides. Because of the presence of this enzyme these polymers failed to promote asparagine incorporation in the conventional *E. coli* system of protein synthesis.

It may be seen from the last column of Table IV that with polynucleotides of increasing chain length the ratio of lysine incorporation to that of asparagine increased faster than the ratio of AAA content to AAC content within this series of oligonucleotides. This is due to the fact that homopeptides of lysine are produced along with those containing both asparagine and lysine (*cf.* Fig. 3). The lysine homopeptides may arise from ribosome-bound unfinished polypeptide chains or because the Ap_nC messengers are not always read throughout their entire length, or both. These effects would be more pronounced with messengers of increasing chain length. Formation of lysine peptides of various sizes has been observed with individual oligoadenylic acids (19, 20).

Table V shows the results of an experiment with poly AU (25:1) and Ap_nU oligonucleotide. Because of the small amount of oligonucleotide prepared, several peaks from the DEAE-cellulose chromatography in urea were pooled to give the heterogeneous fraction $Ap_{16-21}U$ used. Whereas poly AU stimulated the incorporation of lysine, asparagine (2A1U codon), and isoleucine (2A1U codon) in the *E. coli-L. arabinosus* system, oligo $Ap_{16-21}U$ promoted the incorporation of lysine and asparagine but not that (or only traces) of isoleucine. Hence, the nucleotide sequence of the 2A1U asparagine codon (18) is pApApU (AAU).

End Group Assays of ^{12}C-Lysine-^{14}C-Asparagine Peptides—For this purpose the peptides were isolated as described under "Isolation of Peptides." It may be recalled that the isolation procedure greatly reduced the blank, *i.e.* the asparagine radioactivity in the peptide fractions from samples incubated without added oligonucleotide. This blank may represent asparagine introduced into polypeptides (not retained by carboxymethyl cellulose) formed under the direction of endogenous messenger.

Dinitrophenylation and hydrolysis of the peptide fraction derived from an incubation with oligo $Ap_{17-19}C$ yielded 40% of the radioactivity as aspartic acid but essentially none as DNP-aspartic acid (Table VI). While the recovery of label was low, the same procedure applied to asparagine-^{14}C yielded only 30% of the counts as DNP-aspartic acid and virtually none as free aspartic acid. Similar low recoveries of bis-DNP-lysine and ε-DNP-lysine were obtained from Lys_7-^{14}C. From these results it may be concluded that the ^{14}C-asparagine-containing peptide had no NH_2-terminal asparagine.

Digestion with carboxypeptidase A rapidly released up to 70% of the radioactivity of the ^{12}C-lysine-^{14}C-asparagine peptides as asparagine whereas Lys_7-^{14}C (*cf.* Table III) was very poorly hydrolyzed. The time course of carboxypeptidase A

TABLE V

Amino acid incorporation with poly AU and Ap_nU oligonucleotides

Conditions as in Table IV.

Polynucleotide	Amino acid incorporation		
	Lysine	Asparagine	Isoleucine
None (blank)	(42)	(10)	(22)
Poly AU (25:1)	6098	237	273
$Ap_{16-21}U$	738	15	2

digestion is shown in Fig. 2. In a separate experiment, peptides derived from an incubation with $Ap_{23}C$ and labeled in either asparagine or lysine residues with ^{14}C were exposed to carboxypeptidase B and reisolated by carboxymethyl cellulose chromatography (see Fig. 3 and below for the effect of carboxypeptidase B on these peptides). Subsequent treatment of this material

TABLE VI

Dinitrophenylation and hydrolysis of ^{12}C-lysine-^{14}C-asparagine peptide and controls

Conditions for dinitrophenylation and hydrolysis as described under "End Group Assays; Chemical."

Sample	DNP-aspartic acid	Aspartic acid	Bis-DNP-lysine	ε-DNP-lysine
	moles/mole		*moles/mole*	
^{12}C-Lysine-^{14}C-asparagine peptide				
Found	0.03*	0.4*		
Calculated	0†	1.0†		
Asparagine-^{14}C control‡				
Found	0.3	0.03		
Calculated	1	0		
Lys_7-^{14}C control				
Found			0.08	0.4
Calculated			0.14	0.86

* These values have been corrected for the contribution of peptide material formed in the absence of added oligonucleotides. These corrections in no case amounted to more than 10% of the values given above.

† Assuming that asparagine is COOH-terminal.

‡ Average of two experiments.

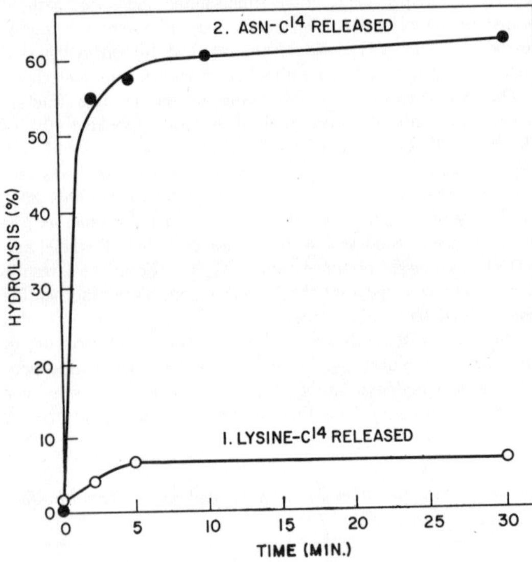

FIG. 2. Action of carboxypeptidase A on Lys_7-^{14}C (*Curve 1*) and ^{12}C-lysine-^{14}C-asparagine peptide (*Curve 2*) as a function of time. Incubation conditions and working up procedure as described under "End Group Assays; Enzymatic" with 13 μg of enzyme in a final volume of 0.1 ml. Temperature, 37°. The ^{14}C-asparagine-containing peptide was derived from an incubation with $Ap_{20-22}C$ oligonucleotide messenger.

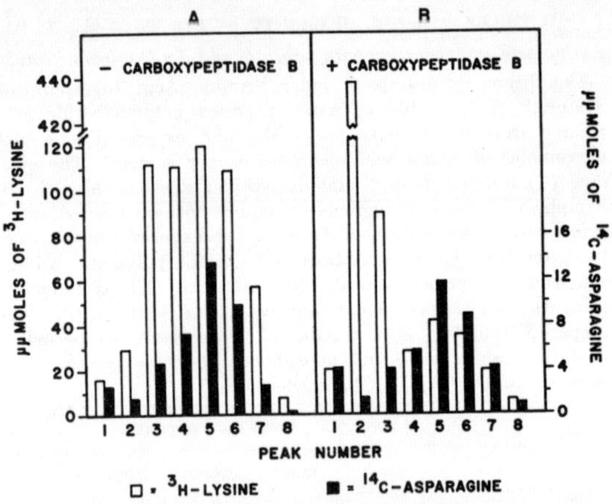

FIG. 3. Effect of carboxypeptidase B on ^{3}H-lysine-^{14}C-asparagine peptides. Peptides containing 63 μμmoles of ^{14}C-asparagine and 656 μμmoles of ^{3}H-lysine were incubated with carboxypeptidase B as described under "End Group Assays; Enzymatic" for 30 min at 37°. A control incubated without enzyme was also prepared. After boiling for 2 min, 200 μg of a polylysine partial hydrolysate was added as carrier, the samples were diluted with water 10-fold the volume of the original incubation mixtures and separated on columns (0.25 × 20 cm) of carboxymethyl cellulose (0.58 meq per g) with an exponential NaCl gradient (0 to 0.8 M) at 3.4 ml per hour. A mixer containing 45 ml of water was used. The eluate was monitored continuously at 220 mμ. Fractions of 0.85 ml eluted from the column were directly collected in vials and counted in the presence of 10 ml of Bray's scintillation mixture (21). ^{3}H and ^{14}C radioactivities were measured simultaneously in a Packard Tri-Carb scintillation spectrometer. Free asparagine was recovered in *Peak 1* which was not retained by the carboxymethyl cellulose. Free lysine was found in *Peak 2*. Successive peaks represent oligopeptides in order of their increasing chain length.

TABLE VII

Effect of carboxypeptidases on ^{12}C-lysine-^{14}C-asparagine peptide

Conditions as in Table III. ^{14}C-Asparagine-containing peptide derived from an incubation with $Ap_{17-19}C$ oligonucleotide messenger.

Enzyme	Initial* (^{14}C-peptide)	Final* (^{14}C-asparagine)	
	cpm	*cpm*	*%*
None	606	0	0
Carboxypeptidase A	534	333	62
Carboxypeptidase B	606	11	2
Carboxypeptidase A + B	564	377	66

* These values have been corrected for the contribution of peptide material formed in the absence of added oligonucleotide. These corrections in no case amounted to more than 10% of the above values.

with carboxypeptidase A caused a rapid release of free asparagine but no concomitant detectable formation of lysine.

Since the release of asparagine was never greater than 70%, the possibility was considered that the radioactivity not released by carboxypeptidase A could be due to asparagine located internally in the oligopeptide chains. Therefore, use was made of the fact (Table III) that carboxypeptidase B, which does not cleave asparagine residues, rapidly releases COOH-terminal lysine. If the biosynthetic peptides contained any internal asparagine-^{14}C besides that in COOH-terminal position, all of the radioactivity in these peptides should be released through simultaneous digestion with carboxypeptidases A and B. As shown in Table VII this was not the case, for carboxypeptidases A and B together released no more asparagine-^{14}C than carboxypeptidase A alone. Although this result makes the presence of internal asparagine rather doubtful, the unlikely possibility that carboxypeptidase B fails to hydrolyze off COOH-terminal lysine adjacent to asparagine, whereby the action of both enzymes would come to a standstill without release of internal asparagine, cannot be ruled out. However, we think that some alteration of asparagine residues in peptide linkage (racemization ?) during incubation or isolation of the peptides is more likely to be responsible for the incomplete recovery.

An experiment was done with a doubly labeled ^{3}H-lysine-^{14}C-asparagine peptide derived from an incubation with $Ap_{21}C$ and isolated as described under "Isolation of Peptides." The distribution of ^{3}H-lysine and ^{14}C-asparagine in the different peptides before and after treatment with carboxypeptidase B, is shown in Fig. 3. It can be seen that upon treatment with the enzyme, although the ^{14}C-asparagine distribution in the different peptides remained essentially unchanged, homopeptides of ^{3}H-lysine were degraded with accumulation of free lysine which appeared under Peak 2 of Fig. 3. A plausible explanation for the nature of the distribution of lysine and asparagine in the peptides before and after treatment with carboxypeptidase B is that initiation and termination of translation has occurred in a nonspecific manner. Since the presence of NH_2-terminal or internal asparagine would have produced a modification in the distribution of the asparagine-containing peptides, and such was not found, the conclusion that asparagine was present in COOH-terminal position is strongly supported. Consistent with this conclusion is the observation that following treatment with carboxypeptidases A and B of the peptide material, the bulk of asparagine radioactivity appeared under Peak 1, the position of free asparagine.

Peptide fractions from 4 through 6 amino acid residues (3 to 5 lysine residues), isolated by chromatography on carboxymethyl cellulose after treatment with carboxypeptidase B as described in Fig. 3, were diluted 10-fold with water, and freed of salt by adsorption to and elution from carboxymethyl cellulose with 1 M NH_4HCO_3. The eluates were repeatedly taken to dryness by rotary evaporation and dissolved in a small volume of water. An aliquot containing 148 $\mu\mu$moles of ^{3}H-lysine and 46 $\mu\mu$moles of ^{14}C-asparagine yielded 14 $\mu\mu$moles as bis-DNP-^{3}H-lysine (from NH_2-terminal lysine), 48 $\mu\mu$moles as ϵ-DNP-^{3}H-lysine (from non-NH_2-terminal lysine), 23 $\mu\mu$moles as ^{14}C-aspartic acid (from COOH-terminal asparagine), and 1 $\mu\mu$mole as DNP-^{14}C-aspartic acid (most probably derived from free asparagine and aspartic acid generated during the dinitrophenylation reaction), with a total recovery in ^{3}H of 41% and in ^{14}C of 52%. If a correction is made for the difference in yield, a value not far from 1 mole of NH_2-terminal lysine per mole of COOH-terminal asparagine is obtained. Thus, it may be concluded that peptides of the structure Lys–Lys–Lys . . . Lys–Asn are synthesized by the cell-free bacterial system under the direction of Ap_nC oligonucleotide messengers.

The results presented in this paper provide strong evidence that an AAC codon at the 3′-end of an oligoadenylic acid chain directs the incorporation of asparagine at the COOH terminus of a lysine oligopeptide. From this fact, and from the accepted NH_2-terminal → COOH-terminal polarity of protein biosynthesis, it may be concluded that the genetic message is read from the 5′- to the 3′-end of the messenger. This conclusion is at variance with the views expressed in recent papers by Williamson and Schweet (22) and Eikenberry and Rich (23). Utilizing the fact that reticulocyte ribosomes contain hemoglobin messenger, these investigators treated the ribosomes either with spleen phosphodiesterase (an exonuclease that cleaves RNA from the 5′-end) or snake venom phosphodiesterase (an exonuclease that cleaves RNA from the 3′-end) and observed the effect of these treatments on the synthesis of hemoglobin in the cell-free reticulocyte system. The results indicated that the snake venom interfered predominantly with the initiation of hemoglobin chains suggesting that translation began at the 3′-end of the messenger. These results are contrary to ours if the biological assembly of polypeptides proceeds from the NH_2- to the COOH-end. The reason for the discrepancy is at present not clear, but it is not inconceivable that the results of the experiments on ribosomes treated with exonucleases (which were present during the subsequent incubation period with radioactive amino acids) may have been affected by contaminating proteases in the nuclease preparations.

Since the code transcription step, *i.e.* the synthesis of messenger RNA on the DNA template, proceeds in a 5′ → 3′ direction (24, 25), transcription and translation would have the same polarity. This would permit translation to begin as soon as a few nucleotide sequences of nascent messenger were available.

SUMMARY

1. A cell-free bacterial system with very low nuclease activity was obtained by combining purified *Escherichia coli* ribosomes with dialyzed supernatant fluid obtained by high speed centrifugation of extracts of *Lactobacillus arabinosus*. The nuclease activity of the ribosomes was lowered by washing in ammonium chloride solutions followed by chromatography on DEAE-cellulose. The *L. arabinosus* supernatant is naturally almost free of nucleases.

2. Oligonucleotides of the type ApApAp . . . pApApC, with an AAC codon at the 3′-end of the chain, directed in the above system the synthesis of oligopeptides of the structure Lys–Lys–Lys . . . Lys–Asn with NH_2-terminal lysine and COOH-terminal asparagine.

3. The results show that if the biological assembly of the polypeptide chains of proteins proceeds, as currently believed, from the NH_2-terminal through the COOH-terminal amino acid, the genetic code is translated by reading the messenger in the direction from the 5′- to the 3′-end of the polynucleotide chain.

Acknowledgments—We wish to thank Mr. M. C. Schneider and Mr. H. Lozina for their invaluable help with the nuclease assays and with the preparation of tRNA and of ribosomal and supernatant fractions.

REFERENCES

1. Bishop, J. A., Leahey, J., and Schweet, R., *Proc. Natl. Acad. Sci. U. S.*, **46**, 1030 (1960).
2. Naughton, M. A., and Dintzis, H. M., *Proc. Natl. Acad. Sci. U. S.*, **48**, 1822 (1962).
3. Goldstein, A., and Brown, A. G., *Biochim. et Biophys. Acta*, **53**, 438 (1961).

4. Nathans, D., *Proc. Natl. Acad. Sci. U. S.*, **51**, 585 (1964).
5. Wahba, A. J., Basilio, C., Speyer, J. F., Lengyel, P., Miller, R. S., and Ochoa, S., *Proc. Natl. Acad. Sci. U. S.*, **48**, 1683 (1962).
6. Elson, D., and Tal, M., *Biochim. et Biophys. Acta*, **36**, 281 (1959).
7. Neu, H. C., and Heppel, L. A., *J. Biol. Chem.*, **239**, 3893 (1964).
8. Spahr, P. F., *J. Biol. Chem.*, **239**, 3716 (1964).
9. Grunberg-Manago, M., Ortiz, P., and Ochoa, S., *Biochim. et Biophys. Acta*, **20**, 269 (1956).
10. Tomlinson, R. V., and Tener, G. M., *J. Am. Chem. Soc.*, **84**, 2644 (1962).
11. Stewart, J., and Stahmann, M. A., *J. Chromatog.*, **9**, 233 (1962).
12. Pabst Laboratories, Circular OR-10, 1956.
13. Neu, H. C., and Heppel, L. A., *J. Biol. Chem.*, **239**, 2927 (1964).
14. Schmidt, G., Bartsch, G., Laumont, M. C., Herman, T., and Liss, M., *Biochemistry*, **2**, 126 (1963).
15. Gardner, R. S., Wahba, A. J., Basilio, C., Miller, R. S., Lengyel, P., and Speyer, J. F., *Proc. Natl. Acad. Sci. U. S.*, **48**, 2087 (1962).
16. Consden, R., in C. Long (Editor), *Biochemists' handbook*, Van Nostrand, London, 1961, p. 151.
17. Waley, S. J., and Watson, J., *Biochem. J.*, **55**, 328 (1953).
18. Wahba, A. J., Gardner, R. S., Basilio, C., Miller, R. S., Speyer, J. F., and Lengyel, P., *Proc. Natl. Acad. Sci. U. S.*, **49**, 116 (1963).
19. Smith, J. D., *J. Mol. Biol.*, **8**, 772 (1964).
20. Smith, M. A., Salas, M., Stanley, W. M., Jr., and Wahba, A. J., *Federation Proc.*, **24**, 409 (1965).
21. Bray, G. A., *Anal. Biochem.*, **1**, 279 (1960).
22. Williamson, A. R., and Schweet, R., *Nature*, **206**, 29 (1965).
23. Eikenberry, E. F., and Rich, A., *Proc. Natl. Acad. Sci. U. S.*, **53**, 668 (1965).
24. Bremer, H., Konrad, M. W., Gaines, K., and Stent, G. S., *J. Mol. Biol.*, in press.
25. Maitra, U., and Hurwitz, J., *Proc. Natl. Acad. Sci. U. S.*, in press.